Melanism

Evolution in Action

Michael E.N. Majerus

Department of Genetics, University of Cambridge

OXFORD NEW YORK TOKYO
OXFORD UNIVERSITY PRESS
1998

This book has been printed digitally in order to ensure its continuing availability

OXFORD
UNIVERSITY PRESS

Great Clarendon Street, Oxford OX2 6DP

Oxford University Press is a department of the University of Oxford.
It furthers the University's objective of excellence in research, scholarship,
and education by publishing worldwide in

Oxford New York

Auckland Bangkok Buenos Aires Cape Town Chennai
Dar es Salaam Delhi Hong Kong Istanbul Karachi Kolkata
Kuala Lumpur Madrid Melbourne Mexico City Mumbai Nairobi
São Paulo Shanghai Singapore Taipei Tokyo Toronto
with an associated company in Berlin

Oxford is a registered trade mark of Oxford University Press
in the UK and in certain other countries

Published in the United States
by Oxford University Press Inc., New York

A catalogue record for this book is available from the British Library

Library of Congress Cataloging in Publication Data
Majerus, M. E. N.
Melanism: evolution in action/Michael E. N. Majerus.
Includes bibliographical references and index.
1. Melanism. I. Title.
QL767.M25   1997   576.5'4—dc21   97-28434

ISBN 0-19-854983-0 (Hbk)
ISBN 0-19-854982-2 (Pbk)

For Tamsin, Kara, Nicolas, and Kai,
and in memory of my parents,
Fernand and Muriel Majerus

Preface

I have no idea why I have been fascinated by moths and other insects for as long as I can remember. I only know that I caught my first butterfly when I was four, because my mother told me so. But I do remember starting to make a natural history collection when I was six, and making my first moth trap when I was nine (I think dad made most of it). The following year, I found E.B. Ford's two wonderful books, *Butterflies* and *Moths* (Ford 1945a, 1954), and I began to collect and score moth varieties. I also began to breed moths, trying to apply the principles of genetics explained so clearly by Ford. The idea that the features of animals were not haphazard, but were the product of a sort of competition to survive and reproduce, was not new to me. My mother had read Kipling's *Just-so Stories* to me and my brothers at an early age, and I clearly understood why the Leopard got his spots (although Kipling seems to have been more Lamarckian than Mendelian, see p. 101). However, that the development of adaptations, those characteristics of an organism that suit it to the habitats in which it lives, and to its way of life, could be studied by looking at the moths in my own back garden, had not really occurred to me.

It is on record (Berry 1990) that I bought Ford's *Moths* in May 1964. Four weeks later I caught my first black peppered moth, *Biston betularia* f. *carbonaria*, in our garden in Northwood, Middlesex. Later that summer I found another black peppered moth on the trunk of a hornbean in the local woods, and following Ford's advice, I wrote down precise details of the find. So began the collection of one of the slowest accumulating data sets I have compiled: the resting sites of peppered moths in the wild, a data set that comprises notes on 47 moths found over 34 years (see Howlett and Majerus 1987 and p. 122).

In *Moths*, Ford gives an account of an explanation of the evolution of melanism (chapter 13), and stresses the need to both record the frequencies of melanic and non-melanic forms of moths, and to breed from the different forms to discover how the forms are inherited. This I tried to do, but it was not until 1973, when Bernard Kettlewell's *The Evolution of Melanism*, a compendious review of the subject, was published, that I gained some insight into which of my studies were novel and which had been done before. By that time I was an undergraduate at Royal Holloway College, and I was fortunate enough to hear lectures from several of the scientists whose work I will describe. Among these were Ford and

Kettlewell themselves, Sir Cyril Clarke, Professor Philip Sheppard, Dr Lawrence Cook, The Rt Hon. Miriam Rothschild, and Dr Robert Creed. The latter examined my Ph.D. thesis on the causes and consequences of melanic polymorphism in larvae of the angleshades moth, *Phlogophora meticulosa*, a study inspired by Ford's *Moths* (pp. 58–9).

Although for much of the time since 1980, my work has centred on a beetle, the 2 spot ladybird, *Adalia 2-punctata*, which, as I shall describe, shows melanic polymorphism, my fascination for Lepidoptera, and moths in particular, has never waned. This book, although not exclusively on moths, gives me the opportunity to express some of that fascination.

I believe that this book is timely for three interconnected reasons. A significant section of the book concentrates on the evolution of melanism in environments changed by industrialisation. The realisation that industrial pollutants are harmful, and the recent increase in environmental conscientiousness, has led to a reduction in some pollutants, so that the melanic forms of many species are in rapid decline. Most of this decline has occurred since *The Evolution of Melanism* (Kettlewell 1973) was published. Over the last decade, several authors have predicted that the black form of the peppered moth will all but disappear, within the next two decades, in Britain. Thus, we may have only a relatively short period now when we may study this most famous example of evolution in action, and it is essential that we do take this opportunity.

Many advances have been made in the last 23 years, and part of the reason for this book, when first conceived, was to 'update Kettlewell'. But I also hope to demonstrate that much still needs to be done. The techniques for studying evolution have advanced hugely over the last two decades. In particular, innovations in molecular genetics have provided a powerful set of tools to look at mutation rates, evolutionary divergence, reproductive behaviour, paternity, and migration. However, few of these advances have been much used to expand our knowledge of melanism in moths. It is almost as though scientists, and the bodies that fund science, regard melanism as a closed chapter, because we know the answer for the peppered moth. In my opinion, this is a misguided view for two reasons. First, as I hope to demonstrate in Chapters 5 and 6, we do not have all the answers in respect of the peppered moth yet. Second, and more crucially, in many ways melanism in the peppered moth is a special case, at least in part, because it is so recent, and so is not representative of species with melanic forms, for most have had melanic forms since long before widespread industrialisation. Consequently, in this book I will try to place as much emphasis on gaps in our knowledge as on what we think we do know.

It is because of the many who seem to feel that 'as we know about melanism in the peppered moth, we know about melanism' that I have resisted the advice of some, in the planning stage of this book, to focus exclusively on moths. Melanism is a common phenomenon in many different taxa, for many different reasons. Each, from *Daphnia* to panthers, has its own evolutionary story to tell.

The book is divided into 10 chapters. Chapter 1 defines melanism and places the history of investigations of melanism into context, while Chapter 2 considers the diversity of melanism, both in respect of the taxa that exhibit melanism and the reasons for being melanic. Chapters 3 and 4 explain the principles of genetics and evolution, respectively, using melanism as illustrative material. The 'textbook' story of industrial melanism in the peppered moth is related in Chapter 5, and expanded upon in Chapter 6, to show the complexity of the system. The diversity of reasons for melanism, specifically in cryptic Lepidoptera, is discussed in Chapter 7, as part of a revised classification of melanism in the Lepidoptera. One category, that concerning melanism in conspicuous species of Lepidoptera, is the focus of Chapter 8. Chapter 9 extends this theme by considering melanism in ladybirds, where the complexity of interacting factors affecting the evolution and maintenance of melanic polymorphism is revisited. In the final chapter I try to pull some of the foregoing threads together, to show how the study of melanism has contributed to the field of biology and in particular the understanding of evolution, and I suggest where the study of melanism should go in the future.

Throughout the book I have tried to keep technical terms to a minimum, but some have been essential to avoid verbosity and explanatory repetition. I have, therefore, produced a glossary of these terms. As my intention has been to produce a readable text, that will be accessible to amateur naturalists, as well as professional scientists, I have used the common names of organisms (where such are known to me) throughout, giving the scientific name only on first usage in the main text. However, I have included all the scientific names in the index, cross-referenced to their common names.

Finally, although one of the reasons for this book was initially to 'update Kettlewell', I know I have not achieved this. Had I done so, the book would have been three times the length it is. I have made no effort to include all known, or probable examples of a particular type of melanism, nor have I attempted to cover more than some of that small proportion of species showing melanism that have received scientific scrutiny. Rather, I have used only those examples known to me which best illustrate the points I have tried to make. The real aim of this book now, as it is ready for publication, is to engender renewed interest in the subject, so that the many unanswered questions, thrown up by this ubiquitous phenomenon, are addressed before we lose, possibly forever, some of the best material we have ever had available for the study of evolution in action.

Cambridge   M.E.N.M.
February 1997

Acknowledgements

There are many people who I should thank for their help in the production of this book. I am indebted to Dr Cathy Kennedy of Oxford University Press for several reasons. First, for listening to me when the subject of this book was originally mentioned. Second, for thinking that it was a good idea. Third, for being available whenever I needed her advice on style, format, approach, and what was possible, as well as what was not.

Over the years, many people have help me in my work on melanism either by collecting data, running experiments, giving data, or just discussing aspects of the subject. They include Dr Peter O'Donald, Dr Rory Howlett, Dr Carys Jones, Tamsin Majerus, Professor Bruce Grant, Sir Cyril Clarke, Dr Robert Creed, Dr David Lees, Dr John Pontin, Professor Sam Berry, Mary Louisa Hartley, Dr John Barrett, Hinrich graf von de Schulenburg, Drs Greg and Laurence Hurst, Linda Walker, Irene Geoghegan, Jim Stalker, and Dr Clair Brunton.

I also thank Dr William Foster for giving me access to the insect collections in the Department of Zoology Museum, University of Cambridge, and allowing me to photograph specimens. Dr David Carter, also allowed me, and some of my research students access to the collections of British Lepidoptera at the British Natural History Museum in South Kensington.

Most of my work on melanism has been conducted without formal funding support. Consequently, I am inordinately grateful to the Department of Genetics, University of Cambridge, for providing laboratory and other facilities when necessary and allowing me to run moth traps at various sites around Cambridge, particularly the University's Ecological Reserve at Madingly Wood. I am also grateful to the Royal Holloway College, University of London, and the University of Keele for providing both funding and facilities for my work on melanism in the 1970s. My thanks are also due to the Science and Engineering Research Council for supporting some of my work with Dr Peter O'Donald on sexual selection in the polymorphic 2 spot ladybird during the 1980s. In addition, I thank the Natural Environment Research Council for funding more recent work on differential habitat selection in the Lepidoptera, and the Trustees of the Cockayne Research Fund for supporting my work on the genetics of melanism in the mottled beauty moth. Finally in this vein, I owe my deepest gratitude to my late parents for encouraging and indeed funding my early interest in moths and butterflies. My mother warrants particular mention for chauffeuring me to many places to collect, often staying in bed and breakfast accommodation while I spent the night moth-trapping in some isolated spot, and getting up soon after dawn to collect me.

A number of people have helped in the actual preparation of the book. Naomi Davis and my wife Tamsin did many of the literature searches, turning up obscure entomological references which I would certainly have missed. Sarah Briant, Jane Snaith, Joanne Purser, and Tamsin typed many of the tables, and Tamsin constructed many of the genetical line figures for Chapter 3. Dominique Bertrand and Linda Walker produced the ladybird diagrams for some of the figures in Chapter 9. Brian Curtis advised on many aspects of the preparation of photographic illustrations, and developed many of the black and white photographs.

The majority of the photographs are my own, but I am indebted to Sir Cyril Clarke, and Professor Bruce Grant for both some of the colour and black and white prints, and for permission to use Derek Whiteley's excellent illustrations of the forms of the peppered moth from Britain and North America. I am also grateful to many of those who have researched into the subject of melanism, for permission to reproduce or adapt data and illustrations from their learned publications.

Finally, Tamsin has read the whole book, made constructively critical comments throughout, and took great pains to assess the readability of the text. For an author, this is frequently not an easy thing to do as one becomes too intimately involved in one's subject matter. If, as I hope, the readers of this book do find its style clear, interesting, and stimulating, much of the thanks for that should go to Tamsin rather than me.

Contents

1 What is melanism?

Introduction

The 'colour' black is a strong motif in life. It conjures dark, mysterious, and frequently frightening images. It is associated with death through 'The Black Angel of Death', 'The Black Rabbit' of Richard Adams' (1972) *Watership Down*, and J.R.R. Tolkien's (1954) Black Riders of *The Lord of the Rings*, who were themselves wraiths. Indeed, Shakespeare, in sonnet 147, describes one who has driven their lover to madness and death as being 'as black as hell'. The association with death extends further to the clothes worn in many cultures when mourning the passing of a loved one. The blackening of skin in bruising is a short lived reminder of painful injury, while a black mark results from bad behaviour. Black in several Oriental nations is a badge of servitude, slavery, and low birth. In the City of London, Black Monday was so-named to commemorate a stock exchange crash, although Black Monday also refers to 14 April 1360 when many men and horses of Edward III's army, lying outside Paris, died of exposure. We refer to a relative who behaves antisocially as the black sheep of the family. Our word blackguard seems to point to this meaning; the Latin *niger* sometimes being used to mean bad or unpropitious. To be black-balled generally means to be excluded from some club or society. The term originates from the practice of voting on whether or not a prospective candidate should be admitted by placing coloured balls into a box, a white or red ball being usually used to indicate acceptance, and a black ball to indicate rejection. Judges used to wear a black cap when they passed the sentence of death on a prisoner. Black arts were and are practised by conjurors, wizards, witches, and warlocks, who professed to have dealings with the devil; black here meaning diabolical or wicked, possibly derived from nigromancy, a corruption of necromancy.

In the animal kingdom, black is commonly one of the colours used by toxic or distasteful species to warn off potential predators. Some of the most feared organisms are black, such as the black mamba and the black widow spider. The black widow seems to have a particular place in human fears, and has been featured widely in literature and myth. For example, Lovell in *Mr Edwards and the Spider* writes:

I saw the spiders marching through the air,
Swimming from tree to tree in that mildewed day
In later August when the hay
Came creeking into the barn.

Let there pass
A minute, ten, ten trillion; but the blaze
Is infinite, eternal, this is death,
To die and know it. This is the Black Widow, death.

But black also has some positive associations. In blazonry, black represents constancy, wisdom, and prudence. Black may be an attractive colour in some circumstances. The berries of many plants turn black when they ripen so that they are obvious to birds who eat them and help to disperse the seeds once they have passed through the bird's digestive tract. A number of film actresses, and for a time, many other women, found a black spot or 'beauty spot' to be an attractive feature. Indeed, the colour black has considerable sexual power. For example, in Western cultures at least, black satin sheets and black underwear carry strong connotations. Again there are parallels in other animals, although these are somewhat tenuous. Some female 2 spot ladybirds, *Adalia 2-punctata* (Linnaeus), prefer to mate with black rather than red males (Majerus *et al.* 1982a). The extravagant tail of the male long-tailed widow bird, *Euplectes progne*, which is used as a courtship signal, is jet black (Andersson 1982).

In these instances, whether positive or negative, black is a strong and memorable colour. However, black is also a very common colour in the environment, and one that, though we may be frightened of the dark because of what it may hide, we are not frightened of in itself. From an early age, we get used to the dark. Indeed many young organisms spend much of their early life in the dark because there is safety in the darkness. As most environments contain many dark or black patches of varying sizes, many animals, particularly small ones, have evolved dark colours themselves to blend in with these dark elements of the environment.

There is one other negative attitude towards the colour black which is particularly pertinent to this book: that is, it is associated with being dirty. On an environmental scale, the association is with certain types of pollution, particularly that from the burning of fossil fuels. The particulate air pollution that has resulted from industrialisation (in Britain since the eighteenth century), has caused profound and rapid changes to the environments of many of our population centres. Although the situation has improved over the last four decades, many industrial centres in Britain could have been fairly described as grimy for some two hundred years, between the middle of the eighteenth and twentieth centuries. One of several manifestations of the pollution was in the darkening, by soot fallout, of many natural and man-made surfaces.

The rapidity of the changes to surfaces in the environment meant that species of organisms that have become adapted over the aeons to live on these surfaces, were no longer well adapted. This imposed a pressure for change. One mechanism that could result in species changing in appropriate ways, i.e. in ways that restored their adaptedness, was that proposed by Darwin (1859): natural selection. Those individuals of a species that had traits that caused their bearers to be more well suited to the new characteristics of their environment would survive and as long as the traits were heritable, they would pass these advantageous characteristics on to future generations. Less well adapted individuals would be less likely to survive to reproduce.

The response of many organisms to the rapid and visible changes wrought on the environment by the industrial revolution was to become darker, or melanic, through the action of natural selection on heritable variation in their populations. In some instances the changes in organisms have been monitored scientifically to a greater or lesser extent, and some of the causes and consequences of biological change, which we may call evolution, have been elucidated. It is these studies of the darkening of populations of organisms in smoke-polluted regions that have provided some of the most dramatic examples of evolution in action.

What is melanism?

Before expanding on these examples it is necessary to define what is meant by the term melanism. The word melanic is adjectival, meaning simply black, from the Greek *melos*, and so, strictly speaking, melanism is the phenomenon whereby a species has abnormally large amounts of black colouring at the expense of other colours. Were I to adhere to this rigorous definition, I would have rather few cases of melanism to discuss and this would be a very short book. However, since the nineteenth century, it has been common practice to apply the term melanism to any situation in which there is, on average, a general darkening of the ground colour or patterning of an organism. Some efforts were made in the late nineteenth century to split truly black melanics from those showing just an overall darkening. For example, White (1877) suggested that the latter be referred to as melanochroism. However, although this term is still occasionally used, most authorities include both types of darkening under the term melanism. Indeed, the great Victorian lepidopterist, J.W. Tutt (1891) notes that the use of melanism to cover cases that were in fact examples of melanochroism was already widespread by the latter part of the nineteenth century. Furthermore, Bernard Kettlewell (1973), in *The Evolution of Melanism*, specifically justifies the use of melanism as a blanket term to cover all cases:

> So-called melanic, melanistic, and melanochroic forms refer to heterogeneous and genetically quite indeterminate groups. In all species where an increase in dark pigmentation occurs, though not necessarily producing an all-black insect, such instances must be recorded as coming under the heading of melanism.

I will follow Kettlewell's example and not differentiate between examples that would rigorously be defined as either melanism or melanochroism. Kettlewell's definition of melanism is:

> The occurrence in a population of a species, of some individuals which are darker than the typical form due to an increase in the epidermis of certain polymerised products of tyrosine substances which produce the complex of pigments collectively known as the melanins.

This appears to be a reasonable definition, although in some ways it is rather restrictive. Undoubtedly, Kettlewell (1973) discovered this himself, for he does not adhere strictly to this definition, citing numerous examples of melanism that do not fall within its limits. In particular, Kettlewell discusses many cases of monomorphic melanism in which no paler typical form exists. Furthermore, the use of the comparison to the 'typical' form is problematic, for it suggests that the inclusion or exclusion of a species from being described as showing melanism depends on the frequency of the melanic relative to the commonest other form.

My inclination is to follow Kettlewell's practice rather than his specific definition of melanism. Therefore, in this book, I take melanism to mean simply: the occurrence in a species of dark or black forms. I exclude specific mention of the mechanism by which darkness is achieved, for in the majority of species with such forms, the pigments or other mechanisms responsible for the darkness have not been analysed. While I use as wide a definition of melanism as seems sensible, in most of this book I will be discussing examples which would fall within Kettlewell's definition. The reason for this is that monomorphic melanism, as seen in species such as the chimney sweeper moth, *Odezia atrata* (Linnaeus), or the kidney-spot ladybird, *Chilocorus renipustulatus* (Scriba), are not of great value in unravelling the advantages and disadvantages of being melanic. More relevant are the many species that show changes in melanism over geographic distance in a whole host of situations. Most relevant of all are those species which have shown change either in the degree of melanisation, or in the frequency of melanic forms, or both, in response to environmental changes resulting from the industrial revolution, for these species have been seen to have evolved. Here evolution has actually been witnessed in action.

The moths as material for the study of evolution in action

Foremost among the examples of industrial melanism are those involving moths. There are good reasons why this should be the case. First, most moths rest by day, actively flying mainly at night. When at rest during the day, the majority use some means of concealment, either hiding low down in dense vegetation or some other out of the way place where they will not easily be seen; or they will be camouflaged having a shape and colour pattern that makes them difficult to pick out from their background or surroundings. To effect successful camouflage, moths have to 'hide' their wings. The name Lepidoptera comes from the Greek words *lepidos* and *pteron* meaning scaled wing. This order of insects includes both butterflies and moths and boasts some of the brightest and most beautiful of animals. The colours are produced either chemically through pigments in the individual scales, or physically as a result of the defraction of light in the microscopic gaps between the rows of scales. Thousands upon thousands of scales on each wing give the potential for an infinite variety of colours and patterns on this natural parchment. The surface structure of the wings of a species of moth lends itself to the production of countless variations on a theme, whether it be by dint of changes in the positioning of scales, or the pigments in the scales. It is not then surprising that those species which use camouflage as their primary defence against predators should evolve darker wings when the backgrounds that they rest against become darkened by pollution.

There are also good reasons why melanism in moths, and particularly industrial melanism, has received so much attention in Britain. These are both cultural and biological. The industrial revolution began in Britain, the general consensus being that it started in the early to mid-eighteenth century, although the impact on the environment was not immediate. The origins of this revolution were based on the rise, partly due to advances in technology, of a narrow range of manufacturing industries, such as textiles, metal goods, glass making, and steam engine construction, concentrated in a few localities. In textiles, the invention of Kay's flying shuttle, in 1733, was crucial, leading to concerted efforts to mechanise spinning. Improvements in iron manufacturing introduced by the Darbys at Coalbrookedale from 1713, and the invention of the steam engine by James Watt, were also critical. The rise of the manufacturing industries was paralleled by expansion in transport, overseas trade and banking, and by more efficient agriculture. At the heart of the industrial revolution lay abundant supplies of accessible coal. The north-eastern collieries and later those in the Midlands, produced huge amounts of cheap coal for foundries and to power the new machine engines. In the penultimate decade of the seventeenth century, on average, less than 3 million tonnes of coal were mined in Britain per year. A century later, the figure had risen to over 10 million tonnes, and during the first decade this century, over 240000000 tonnes were mined, although about one-third of this was exported.

The increase in pollution in the late eighteenth century and particularly in the nineteenth century was largely the result of burning this fossil fuel for heat and power. Thus Britain was the first country to experience widespread industrial pollution, and so was probably the first country in which an organism evolved in a readily discernible way in response to industrial pollution.

That an evolutionary change should have been first observed in environments that had recently changed dramatically should not surprise us. As John Maynard Smith (1975) observed:

We should expect to find the most rapid evolutionary changes in populations suddenly exposed to new conditions. It is therefore natural that one of the most striking changes which has been observed in a wild population has occurred in industrial regions of Britain and western Europe during the past hundred years. This is the phenomenon of 'industrial melanism', the appearance and spread of dark forms of a number of species of moths.

In Britain, industrialisation, together with the profits made from interests in the Empire, gave some people great wealth which in turn gave them, or their descendants, leisure time for hobbies. The British have the strongest tradition of collecting insects of any nationality in the world. The earliest known collection still surviving in its original form is that of James Petiver, now housed in the British Natural History Museum, in London; it dates from the late seventeenth and early eighteenth centuries, just prior to the birth of the industrial age. And of course, for the study of variation over time, the Lepidoptera have the advantage that they are easy to preserve, and that most specimens, if kept in the right conditions, retain their colours with little deterioration over time.

However, it was not just that industrial melanism first became apparent in Britain that led to particular interest in melanic moths in this country. Already, by the mid-nineteenth century, before the observed rise of black peppered moths, *Biston betularia* (Linnaeus), it was realised that melanism was generally more prevalent here than elsewhere, and particularly towards western coasts and in the Highlands and Islands of Scotland. For example, Dobrée (1887), observing this fact, attempted to analyse the causes of melanism by comparing the melanic forms of British species with continental melanics and their distribution. Unfortunately, Dobrée's conclusions are of little value for they were based on the erroneous belief that melanism scarcely ever occurred at high latitude. This error is illustrated in a letter written by Professor Christiana Schöyen to Tutt in March 1890 (see Tutt 1891). In response to a specific request from the latter about the occurrence of melanics in Scandinavia, Schöyen wrote:

> The tendency to melanism in our lepidoptera is certainly, as a rule, more apparent on the western shores than in the eastern districts of the country, even at a less height above sea level; in some cases this is marked as in mountainous tracts, e.g. *Cidaria flavicinctata* occurs at the town Molde in Romsdale Amt, close to the sea-shore, much darker than at Dovrefjeld, about 2000–3000 feet above the sea: *Agrotis candelarum* occurs at Bergen as dark as on the Jotünfjelds (var. *jotunensis, mihi*). But in the majority of species, the darker colour is generally found, however, either at a greater height above the sea, viz., on the mountains, or in a more northern latitude. Yet this may also be chiefly caused by the more humid condition of the atmosphere, for the mountains are generally covered with masses of cloud, making the atmosphere much more humid than in the valleys and lowlands, and in northern latitudes, large quantities of rainfall and fog and mist are prevalent. Under these circumstances the grey colour of many moths will generally be more or less darkened.

The combination of distinctive features of the British lepidopteran fauna and the Victorian obsession with making collections, particularly of unusual items, led many entomologists, both professional and amateur, to visit regions where different and particularly melanic forms were common. Thus many of the collections in our major natural history museums have series of specimens from the Highlands and Islands of Scotland, from Ireland, and from the south-western counties of England and Wales, dating from the second half of the nineteenth century. Not only that, but the entomological literature of the last three decades of the nineteenth century is liberally scattered with articles discussing the causes of melanism.

It is not then surprising that Kettlewell's (1973) comprehensive treatise on melanism was not the first book devoted to the subject. In 1891, Tutt, then editor of the *Entomologist's Record and Journal of Variation*, published a book of 66 pages entitled *Melanism and Melanochroism in British Lepidoptera* (which was first published in serialised form in the *Entomologist's Record and Journal of Variation*, 1890, 1891). In this, Tutt discusses the prevalence of melanic forms among the British Lepidoptera and points to at least some of the reasons for them. In particular, Tutt, following Cockerell (1887), considers that humidity was highly influential in the presence of melanic forms on western coastlines, at high altitude, and at high latitude. Most of the book was devoted to arguing that humidity plays a central role in the distribution of melanic forms in the Lepidoptera. However, although he did not fully recognise the association between melanic forms of some species and industrialisation until 5 years later (Tutt 1896), he came agonisingly close. In considering the various factors put forward to account for the existence of melanic forms in various parts of Britain, Tutt discussed the high incidence of melanism in Yorkshire and Lancashire. These counties presented Tutt with a problem, for their rainfall was considerably lower than that of Cumberland where melanic forms were less prevalent. To resolve this difficulty, Tutt combines observations of Chapman (1888) that wetting surfaces darkens them, and the obvious fact that rainwater is rarely pure. He writes in a most illuminating passage:

It is well known to the most elementary student of chemistry that rainwater rarely occurs pure, and that in large towns the quantity of impurities is very great. The vast quantity of smoke, gases, fumes, etc., in manufacturing towns, brought down by rain, is scarcely credible, and it is from these impurities I consider the permanent darkening comes. When the water evaporates, the solid matter is left behind, and as a result the impurities are left to darken the surfaces to which they have been carried by the rainwater. The theories of 'natural selection' and 'protection' now apply in their fullest sense, the insects become darkened, 'hereditary tendency' perpetuating and intensifying the melanism. I believe from this (and it appears a fair deduction), that Lancashire and Yorkshire melanism produced by humidity and smoke, is intensified by 'natural selection' and 'hereditary tendency'. As example I would cite the species *Amphidasys betularia, Tephrosia biundularia, Boarmia rhomboidaria, B. repandata, Diurnœa fagella* and *Hibernia marginaria* (*progemmaria*), which, occurring on trees, fences and similar objects would thus be affected. Taking this view, I consider the melanism of these counties 'extraordinary'; I also consider it due to 'humidity'. The melanism of

Staffordshire, Derbyshire, and other localities of a similar character, I consider,
is produced much in the same way, and the protective melanism of London Lepidoptera and of other large towns I would refer to the same cause. Probably no
better instance of this kind occurs among London leipodoptera than *Eupithecia
rectangulata* var. *nigrosericeata* (another species which in the imago state fre-
quents tree-trunks).

Here then Tutt already identified one of the causes of melanism in industrial
areas as the result of moths adapting to the blackened surfaces in the environment for protective purposes. Thereafter, in the same volume, he places emphasis on humidity for almost all of the cases of melanism he cites, which appears to
result from a desire to find a single unifying factor at the root of all cases of
melanism in the Lepidoptera. As we shall see, such search must be in vain, for a
single factor does not exist. A great many influences, both biotic and abiotic, are
involved in the evolution of melanism in the Lepidoptera.

Tutt's (1891) book shows that a considerable number of lepidopterists were
interested in the phenomenon of melanism in the latter part of the nineteenth
century. There is not space here to give a full history of the study of melanism in
animals, or even in the Lepidoptera. Suffice it to say, a wide range of explanations was offered to explain the existence of melanic forms. These included: pollutants ingested by larvae with their food (N. Cooke 1877a); that the soils in which
larval food plants grew affected the form of the adult insect (Prest 1877; E. Robinson 1877; N. Cooke 1877b); that lack of light in dense woodlands acting on larvae
causes resulting adults to be dark (Tutt 1891); that natural selection favoured
melanics (White 1877; Cockerell 1887). Of particular interest was a debate of the
causes of melanism that was published in the *Entomologist* in 1877. The two sides
of the argument were championed by Mr Nicholas Cooke who argued that
melanism was caused directly by the effects of fumes and other pollutants, acting
chemically on the food and physiology of larvae, and Dr Buchanan White, who
took the view that 'natural selection' and 'protection' were responsible. White
writes of natural selection in this context:

> 'Natural selection' may be defined as the weeding out of all but those individuals
> who are best fitted to survive in the struggle for existence. and this weeding (for
> the most part—like the majority of the universal mother's operations—a gradual
> process) is carried out by many and varied agencies. It may happen that in certain
> districts dark forms of certain species have some advantages over their lighter-
> coloured brethren. By their more obscure colour they may escape detection by
> their enemies, and hence have a greater chance of being the means of continu-
> ing the species than the more conspicuous lighter-coloured individuals; or, in
> another district the very reverse of this may occur, and the advantage be on the
> side of the light-coloured.

Unfortunately, White drew for his examples species in which melanism was not
associated with industrialisation. Indeed, he specifically mentioned 'smoky varieties' of the peppered moth and the engrailed, *Ectropis bistortata* Goeze, as

exceptional cases, which could not 'be considered as throwing much light on the origin of the majority of melanochroic forms'.

On the other side, N. Cooke (1877a,b) specifically denied a role for natural selection, although in his remarks, he appears to confuse natural selection with sexual selection (*sensu* Darwin 1859). Thus it seems that even in the early debates of Darwin's theory of evolution, examples of melanism were seen as possibly providing evidence to endorse it.

Tutt was a strong supporter of selection theory. Although he initially may have put too much emphasis on the role of humidity as a cause of melanism, it is clear that he recognised most of the major classes of melanism. In 1896 he gave a clear indication that he believed that industrial melanism was the result of melanics having increased crypsis against soot blackened surfaces in industrial regions. Singling out the peppered moth he wrote:

> The speckled peppered moth as it rests on trunks in our southern woods is not at all conspicuous and looks like a . . . piece of lichen and this is its usual appearance and manner of protecting itself. But near our large towns where there are factories and where vast quantities of soot are day by day poured out from countless chimneys, fouling and polluting the atmosphere with noxious vapours and gasses, this peppered moth has during the last fifty years undergone a remarkable change. The white has entirely disappeared, and the wings have become totally black. As the manufacturing centres have spread more and more, so the 'negro' form of the peppered moth has spread at the same time and in the same districts. Let us see whether we can understand how this has been brought about! Do you live near a large town? Have you a greenhouse which you have tried to keep clean and beautiful with white paint? If so what is the result?, the paint is put on, all is beautifully white, but a little shower comes and the beauty is marred for ever. But in country places . . . it is not spoily . . . No! near large towns, when rain falls it brings down with it impurities, the smoke and dirt, hanging in the air . . . and so we find fences, trees, walls and so on getting black with the continual deposit on them.

> Now let us go back to the peppered moth. In our woods in the south the trunks are pale and the moth has a fair chance of escape, but put the peppered moth with its white ground colour on a black tree trunk and what would happen? It would . . . be very conspicuous and would fall prey to the first bird that spied it out. But some of these peppered moths have more black about them than others, and you can easily understand that the blacker they are the nearer they will be to the colour of the tree trunk, and the greater will become the difficulty of detecting them. So it really is; the paler ones the birds eat, the darker ones escape. But then if the parents are the darkest of their race, the children will tend to be like them, but inasmuch as the search by birds becomes keener, only the very blackest will be likely to escape. Year after year this has gone on, and selection has been carried to such an extent by nature that no real black and white peppered moths are found in these districts but only the black kind. This blackening we call melanism.

Previously, Tutt had recognised the prevalence of melanics on western coasts, and in the northern isles which he related to humidity. He also knew that coniferous woodlands frequently host melanic forms. Again he saw a role for humidity, in writing: 'Probably no trees have a tendency to hold so much moisture as the different kinds of fir trees'. This tendency he suggests will darken the trees. However, he goes on to write: 'add to this that fir trees, by means of their foliage, shut out an immense proportion of light, and we can readily understand that "natural selection", would leave, to a great extent, only the darker insects in such situations'.

It is clear, then, that the value of melanism in the Lepidoptera for the study of Darwin's theory of evolution by natural selection was already realised by the beginning of the twentieth century. The great problem was in understanding how melanic variants were initially produced, and many writers, including Tutt, gave environmental factors, particularly low temperature or high humidity a role in producing heritable melanic forms. The problems of initial production of melanic varieties and their subsequent inheritance was largely resolved by the rediscovery of Mendel's work on inheritance. The progress of work on melanism in the Lepidoptera during the twentieth century is related in Chapters 5–8. But, it is of interest to realise that the currently accepted evolutionary causes for most types of melanism seen in British Lepidoptera were speculated upon over a hundred years ago. Some knowledge of the early work into melanism is surely of value to modern day workers, if only to save them from needless effort. As Lord Walsingham put it in 1885:

> The mass of ever-increasing literature on all subjects at the present day is constantly rendering more difficult the task of mastering the details of any appreciable part of it. It is only too probable, as has been the case in the history of the world from the earliest dawn of civilisation, that one generation may forget what another knew, and that men may devote their time and energy to lines of research which have long since been followed to their legitimate deductions.

That said, it is also pertinent to note that verification or refutation of the speculations on the causes of melanism have been very slow, largely because little relevant systematic or experimental work on melanism was conducted until Kettlewell's celebrated experiments with the peppered moth in the 1950s.

In many instances, as I shall show, the necessary experimental work to elucidate the causes of melanism in British Lepidoptera has still not been undertaken. In Britain, despite our tradition of study of the Lepidoptera and other wildlife, we have left undone so much that ought to have been done; this is due in part to the lack of any co-ordinated plan of action in the investigations of this phenomenon, and in part to the lack of long-term funding for the necessary research. It is my hope that both these difficulties may be resolved over the next two or three decades, when the study of melanism in general, and industrial melanism in particular, may prove so informative about the fate of disadvantaged forms as many of our industrial melanics decline towards extinction.

2 The diversity of melanics

On a sign by a car park at a distillery in Pitlocherie it says: 'Here you will find more black trees than in the Black Forest'. A quick glance around and the reason for the sign is easy to see. All the trunks, branches and twigs of all the trees and shrubs in the vicinity are as black as soot. These include pines and firs of various species, beeches, birches, rhododendrons, flowering cherry, and mountain ash (Plate 1a). The reason for the apparent melanism of these trees is explained on the notice. The atmosphere around the distillery is so loaded with alcohol, that black yeasts grow all over the woody parts of the trees. Close examination of the trunks does indeed reveal fungal colonies, although the fungi in question are not true yeasts. Rather, they are a sooty mould, *Ayreobaccidium pullulans*, which is sometimes called 'black yeast'. The reason that this fungus is black is that the mycelia are heavily melanised, possibly as a screening defence against ultraviolet (UV) light in high light conditions (Hudson, personal communication).

The black trees of Pitlocherie are an unusual case of melanism, but the array of properties of the colours produced by melanin pigments and of the melanic pigments themselves, mean that black or dark organisms are found in a great range of taxa, from bacteria to humans, for many different reasons. In this chapter I will briefly describe what melanic pigments are, before showing the ubiquity of melanism throughout the animal kingdom and discuss the reasons why so many organisms are dark or have dark forms.

The melanins

The melanin pigments, of which there are a number, are one of the most common groups of pigments in most animal taxa. In the Lepidoptera, they are by far the commonest group (Nijhout 1991). At a simple level, the melanins are divisible into two groups: the eumelanins, most of which are black, and the phaeomelanins which may be tan, brown, or reddish brown. Eumelanins are complex polymers of *o*-diphenols, such as tyrosine or dopamine, or of indoles. Polymerisation is catalysed by enzymes known as phenoloxidases or tyrosinases. The structures of phaeomelanins, are poorly understood. Work on the biochemistry of melanins in general, and phaeomelanins in particular, has been hampered by the fact that these compounds are only soluble in hot alkali and a few strong acids. As these solvents disrupt the chemical configuration of the melanin molecules, it is almost impossible to reconstruct the structural configuration of melanins in the normal way. However, it is thought that the production of the non-black phaeomelanins requires the incorporation of a range of molecules into the polymer complex. For example, the catalytic action of phenoloxidase on mixtures of 3-OH-kynurenine and tyrosine, or of sulphur-containing amino acids and tyrosine, are reputed to produce reddish melanins (Inagami 1954; Prota and Thomson 1976; Riley 1977).

Although the synthesis of melanin can theoretically be achieved by the catalytic action of a single enzyme on a single substrate, in practice melanin synthesis is much more complex. Phenoloxidases are initially produced in an inactive form and they have to be activated by a process called proteolysis before they can act as catalysts. Specific enzymes called proteases cause this activation. The timing of protease action and of phenoloxidase activation thus becomes critical to melanin synthesis. As Nijhout (1991) has described it, it is like a multistep cascade, each step only proceeding once the previous step has been completed (Fig. 2.1). Added to this complexity is the possibility of phenoloxidase action being inhibited, which appears to occur as a result of the action of serine protease on haemolymph phenoloxidase in the Lepidoptera (Sugumaran et al. 1985). Furthermore, in most organisms, although not in butterflies, melanins invariably exist as granules surrounded by a membrane. The exact details of granule size and their distribution in an organism have a major influence on the phenotype produced. Thus, although theoretically simple, in practice, the process of melanisation involves a complex series of chemical interactions mediated by enzymes

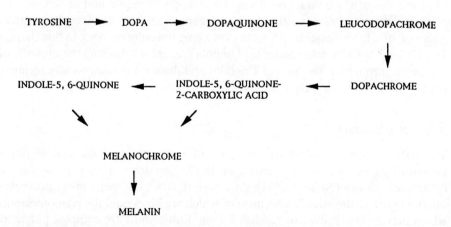

Fig. 2.1 Pathway of melanin synthesis, following a direct sequence, in which each step can only proceed on completion of the previous stage. (After Needham 1974; Kayser 1985.)

and structural controls which have to be triggered at precise stages during the process.

Recent work on the mallard, *Anas platyrhynchos*, has shown that the production of eumelanins and phaeomelanins is strongly influenced by hormonal levels, with testosterone having a particularly marked effect (Haase *et al.* 1995). In mammals, the production of melanins is also mediated by hormones. For example, in mice, a yellow form lacking melanin, is controlled by a recessive mutation of the murine extension gene, which produces an inactive receptor for the melanocyte-stimulating hormone. Conversely, two dominant forms of this gene (alleles), called *sombre* and *tobacco darkening*, both of which have melanising effects, produce hyperactive melanocyte-stimulating hormone receptors (Robbins *et al.* 1993).

Despite the complexity of colour patterns on the wings of Lepidoptera, the most parsimonious hypothesis for the production of these patterns depends on a relatively simple developmental system (Scoble 1992). The colours on the wings result from the colours of their scales, as the underlying cuticular membranes of the wings are generally colourless or brown. The colour of a scale in turn depends on the pigment it contains, which is determined by a series of chemical switches. This is illustrated in Fig. 2.2, which shows that the five scale colours of heliconiid

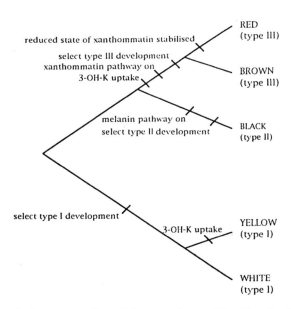

Fig. 2.2 Hypothetical representation of the 'chemical switches' involved in the production of five colour pigments in the scales of heliconiid butterflies. The three types of development are determined at least in part by the fine morphology of the scale in question. (After Gilbert *et al.* 1988.)

butterflies may result from a relatively small number of alternative 'decisions' being made during scale development (Gilbert *et al.* 1988).

The patterns seen on the wings of butterflies and moths can be divided into a few fundamental elements such as bands, spots, and stripes. Nijhout (1991) proposes that each element in a pattern is the product of a chemical signal, called a morphogen, secreted by a small cluster of cells, termed a focus, which causes pigments to be produced in individual scales. As the pigments produced in a particular scale may be influenced by signals from more than one focus, the array of patterns seen is the result of interference between the signals from different foci. The way that this system may result from the action of products of particular genes, is best illustrated with an example. One of the best analysed systems is that producing mimetic forms in the swallowtail butterfly, *Papilio dardanus* Brown, in which many of the female forms are melanics. The genetics and evolution of mimicry in *P. dardanus* are considered in Chapter 8, so it seems appropriate to explain the details of the analysis of pattern development there in the light of these genetic and evolutionary details (see p. 217). Although Nijhout's explanation has not been confirmed for all pattern elements, much evidence in its support has accumulated for some elements, particularly eye-spots, over the last two decades (e.g. Nijhout 1978, 1981, 1985, 1991; Brakefield *et al.* 1996).

Almost all scales possess just a single pigment, the concentration of the pigment in the scale determining the precise colour of that scale. In the case of melanins, the greater the concentration of pigment, the deeper the colour produced. The colour pattern of a moth or butterfly depends on both the colour and the distribution of the individual scales. Few moths and butterflies produce more than three or four basic colours, of which at least one is usually a melanin. The diversity in patterns is a consequence of brightness of colour in each scale, and the juxtaposition of differently coloured scales. If anyone needs to be convinced of this, they might examine the underside of an orange-tip butterfly, *Anthocharis cardamines* (Linnaeus), which although appearing dappled white and green, contains no green scales at all, the green colour resulting from a mixture of black and yellow scales (Plate 1b and c).

The diversity of melanism

It is not possible to review the presence of melanism in all animal taxa here. Rather, I will consider some of those taxonomic groups in which melanism has been most studied, and those in which future study may be valuable.

Micro-organisms

Melanism is common in bacteria. Indeed, bacterial melanins, and fungal melanins, are at least partially responsible for the colours and the structures of some soils (e.g. Paim *et al.* 1990). Although little work has been conducted on the role of melanins in bacteria, or indeed most other micro-organisms in the wild, bacteria which show melanism are proving to be particularly useful in the study of melanin structure and synthesis at the molecular level. For example, Agodi *et al.* (1996) have recently shown, using scanning electron microscopy, that a melanic pigment in *Proteus mirabilis* is organised as rounded aggregates of spherical bodies. Indeed, two melanogenesis systems have recently been defined at the molecular level. Other uses of melanisms in bacteria and fungi are as genetic markers, to aid in the investigation of genetic systems such as transformation and recombination.

Molluscs

Attention to melanism in this phylum has tended to concentrate on terrestrial and littoral species. Melanism is again seen very commonly, many species exhibiting melanic polymorphism. Among the most obvious examples are the banded land snails, *Cepaea nemoralis* and *C. hortensis*, which have been the subject of much study. These species show variation in both shell ground colour, and the number and breadth of dark brown or black bands upon them (Fig. 2.3). The most melanic individuals are those having five bands which are fused together. These species have been studied widely throughout their range in Europe (e.g. Cain and Sheppard 1950, 1954; Cain and Currey 1968; J.S. Jones 1973, 1982; Cowie and Jones 1985; Honek 1995). Various adaptive features, including thermal properties, activity patterns, and avoidance of predation have been identified.

Fig. 2.3 A variable sample of the banded land snail, *Cepaea nemoralis*.

Similar investigations have been conducted on another banded snail, *Helicella candicans* in Czechoslovakia (Honek 1993). Again degree of melanism depends on the number of dark stripes (up to nine), and their width. Honek found that melanics were commonest in habitats with dense vegetation, and in regions where the level of incident light was reduced by fog, cloud, or air pollution.

Melanism in slugs is also common (Williamson 1959). The large slug, *Arion ater*, is a very variable species, having both jet black and paler forms. The black form is most common in the north of the species' range and in dark woodland habitats, paler forms being found in open habitats and in the south of Europe, where a red form predominates. Again it is likely that several factors are important in the evolution of the black form. For example, melanism may warn potential predators that the species is unpalatable, play a part in the thermodynamics of the slug, and act as a barrier against harmful UV light (Kettlewell 1973).

Crustaceans

Melanic pigmentation of the cuticle has been reported in a number of species of cladoceran. In *Daphnia* melanic forms predominate in arctic populations, occur together with pale forms in subarctic populations, and are absent from most temperate populations (Herbert and Emery 1990). In general, melanisation is associated with polyploid clones (Van Raay and Crease 1995).

Arachnids

Melanism in spiders in Britain has received relatively little attention. This is unfortunate because there is circumstantial evidence to suggest that several species exhibit industrial melanism for much the same reason that it is observed in some moths. Kettlewell (1973) cites four species, *Arctosa perita*, *Salticus scenicus*, *Ostearius melanopygius*, and *Drapetiscea socialis*. In each instance, the melanism is recent, and only occurs locally in areas of high pollution. This group would certainly repay increased attention.

In Sweden, the sheetweb spider, *Pityohyphantes phrygianus*, exhibits continuous variation from pale to dark. The variation is controlled genetically. Gunnarsson (1993) has shown that levels of bird predation do not vary between the morphs and, as yet, the factors affecting melanism in this species have not been identified.

Insects

It is among the insects that melanism is perhaps most common. The order in which melanism is best known is the Lepidoptera. Melanism in adult Lepidoptera will be discussed in detail in Chapters 5–8. However, insects of many other orders, from Collembola to Diptera, exhibit melanism. In many species melanics occur only as rarities which probably arise through genetic mutation and are subsequently eliminated from populations by natural selection. Melanic individuals of species such as the wasp, *Dolchovespula media* Retzius, in Belgium (Bruge 1992), or of the 16 spot ladybird, *Tytthaspis 16-punctata* (Linnaeus) (Majerus 1991*a*), are examples. In others, melanic forms make up an appreciable proportion of at least some populations, and the species is said to show melanic polymorphism. Either a single, or many melanic forms may exist in a species. For example, in the Hemiptera, the spittlebug *Philaenus spumarius* (Fallén), a common species in the Palaearctic, shows great variation in its dorsal patterning (Berry and Willmer 1986; Stewart and Lees 1988). This variation can be divided into 13 categories, of which eight are considered melanic and five non-melanic. Similar variation is seen in many populations on the Continent (e.g. Honek 1984).

Melanic polymorphism also occurs in bumble bees. An interesting case is that of the recent rise in melanic frequencies in *Bombus veteranus* (Fabricius) and *B. soroeensis* (Fabricus). Melanic frequencies in the range 48–75 per cent were recorded in southern Finland in 1985. The melanism in *B. soroeensis* is believed to have increased substantially about 1960, while that in *B. veteranus* is thought to have arisen even more recently (Pekkarinen and Teras 1986). Again the causes of the recent increases are not known and would repay deeper investigation.

In some insects, melanic races are known. Such races are particularly prevalent in the Lepidoptera (see Chapter 7); however, they also occur in other orders. So, for example, a melanic population of the damsel-fly, *Coenagrion pulchellum* (van der Linden), has been recorded from Turkey (Dumont *et al.* 1988).

Among beetles, melanic forms of many ladybirds (Coccinellidae) are known. In some, the variation appears continuous, differences in the amount of black being due either to changes in the size or to the number of black spots. In others, distinct melanic forms exist. Thus in many populations of the 2 spot ladybird a range of different melanic forms are present along with the nominate non-melanic form. This species illustrates one of the difficulties of working on melanism. Kettlewell's definition of melanism speaks of an increase in blackness compared with the norm. The normal form of the 2 spot ladybird must be judged to be the nominate form, f. *bipunctata*, which is red with two black spots. This is the form that Linnaeus first described, and in the majority of populations it is one of the commonest forms. Very many other forms of this species have been described (Mader 1926–37; Majerus 1989*a*, 1994*a*). Some of these are obviously melanics, being predominantly black (see Plate 8a and b). However, others, such as many of the '*annulata*' forms are mainly red, although they do have more black patterning on their elytra than the typical form (Fig. 9.8, p. 232). Should these be considered melanic or non-melanic? If we follow Kettlewell's definition to the letter, then almost all forms of the 2 spot would be categorised as melanics, the only exceptions being the nominate form and a small clutch of rather rare forms, such as 'halo', 'mini-spot', and f. *impunctata* (Fig. 9.8, p. 232). Yet, most authors who have discussed polymorphism in this species, including myself, have based designation into melanic and non-melanic forms on whether the majority of the area of the elytra is black or red. This deviation from the normally quoted definition of melanism does not greatly matter as long as the criteria on which designation of a particular form to 'melanic' or 'non-melanic' is made clear. Unfortunately, this has not always been the case, and there are many data sets, particularly on ladybirds, in which samples have simply been split into melanic and non-melanic without details of which forms fall into each category, making the data virtually useless.

Ladybirds are by no means the only beetles which have melanic forms. In New Zealand, the sand-burrowing beetle, *Chaerodes trachyscelides* White, varies from pale to almost black, with a strong similarity being shown between the beetle dorsal coloration and the colour of the sand locally (Harris 1988; Harris and Weatherall 1991).

Among some grasshoppers and locusts (Orthoptera), another form of melanism is seen. Melanism in this order is often very flexible, appearing and disappearing depending on circumstance. For example, many species will change colour if put on backgrounds that are darker or lighter than they are themselves. Thus, if pale migratory locusts, *Locusta migratoria* (Linnaeus), are placed on to fire-burned terrain, they rapidly darken becoming almost black. The colour change results from migration of melanin granules into the upper epidermis, triggered by stimuli from the eyes and ocelli. But this is not the only cause of melanisation. Population density also has an effect. The migratory locust has cycles of solitary and swarm phases. In the former, population densities are low, the locusts relatively inactive, and during both the nymphal and adult instars they are green. In the swarm phase, when population densities are high and the insects become active and restless, the nymphs are black, while the imagines are black and yellow. Wigglesworth (1964) asserts that the melanisation is the result of increased activity of the *corpus allatum* and an increase in the production of juvenile hormone.

The same type of mechanism may explain the increase in melanisation in larvae of some noctuid moths, including the army worms, *Spodoptera exempta* (Walker), and the small mottled willow, *S. exigua* (Hübner), where melanisation increases when larvae are at high density (Faure 1943*a,b*, Simmonds and Blaney 1986). However, in these there is a genetic component (Long 1953) which has not been demonstrated in the migratory locust. Evidence that melanisation of moth larvae at high density has a genetic component was first suggested by the finding by Oertel (1910) of a single bilateral mosaic larva, half green and half brown, of the elephant hawk moth, *Deilephila elpenor* (Linnaeus). Later work by Long (1953) on the silver Y moth, *Plusia gamma* (Linnaeus), demonstrated that crossing adults resulting from either the lightest, or the darkest larvae, produced mainly light or mainly dark progeny, respectively, even when larvae were reared in identical conditions.

There are of course many species of insect in which predominantly black coloration is the norm. In some of these, the dark coloration may serve a camouflage purpose, but in many its function appears to be to advertise the distasteful properties of its wearer. For example, beetles of the genus *Necrophorus* are well defended chemically, secreting an putrid smelling fluid if attacked. Some species, e.g. *N. interruptus* Stephens, are red and black, but others, such as *N. humator* Goeze, are completely black. Many other completely black beetles are foul-smelling and distasteful, such as the common oil beetle, *Meloë proscarabaeus* Linnaeus, the cellar beetle, *Blaps mucronata* Latreille, and the bloody-nosed beetle *Timarcha tenebricosa* (Fabricius). All release acrid secretions when disturbed, the latter releasing a drop of red fluid from its mouth, hence its name. Many solitary bees are black for the same reason, as are some species of fly, caddis fly, and moth. In the latter group, both the European amatidid *Syntomis phegea* Linnaeus, and the red-necked footman, *Atolmis rubricollis* (Linnaeus), are predominantly black and produce repellent secretions.

Fish

Melanism in vertebrates has received relatively little attention, and in only a few species have the adaptive reasons for the phenomenon been well researched. However, melanism is found in some species in every major grouping.

In fish, melanism has been largely ignored. This is partly because of the problem of studying melanism in the field in a medium rather foreign to us, and partly because of the difficulties of interpreting the importance of colours, patterns, and tones to organisms that live in waters where the amount and wavelengths of light that permeate to particular depths are very variable, and until recently have been difficult to measure. However, melanism does occur in fish, and in a few instances some evidence exists for the factors that may be responsible for it. For example, in the three-spined stickleback, *Gasterosteus aculeatus*, males commonly have a red nuptial throat patch. However, in some populations, the red throat patch is replaced by a black patch. McDonald *et al.* (1995) demonstrated that black patches were particularly prevalent in Canadian bog habitats where the aquatic spectrum is shifted to long wavelengths which causes the typically red patch to be 'spectrally masked' thus reducing its effectiveness in nuptial displays. In an elegant series of experiments in which video images of red- and black-throated males were projected on to differently coloured backgrounds, females were found to choose the males whose throat colour showed greatest contrast, whether that be red or black.

Amphibians

Melanic forms of species of frog, toad, newt, salamander, and axolotl are known. In many instances, melanic variation is also seen in tadpoles. Melanism in the amphibians has been most intensively studied in axolotls, where the genetics of melanic forms has been elucidated, and where various drug (e.g. allopurinol) induced melanic forms have been produced (Frost *et al.* 1989) as a means of investigating the biochemistry, physiology, and behaviour of pigment cells. Colour in axolotls is dependent on an interaction among three types of specialised pigment

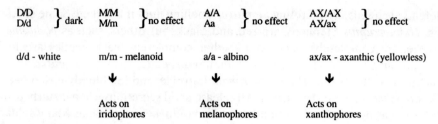

Fig. 2.4   Genetics of colour in axolotls. (After Scott 1981.)

cells, called chromatophores, in the dermal layer of the skin. These are melanophores containing melanin, iridophores containing guanine and producing a shiny iridescence, and xanthophores containing yellow pigment. In essence, the wild axolotl is dark greeny-black which is a product of a dense mixture of melanophores and xanthophores. Virtually no work has been done on the natural variation in axolotls in the wild or the reasons for it. However, several colour forms occur commonly among captive axolotls. These include full melanoid, in which the iridophores are suppressed, 'albino' (which is in fact yellow) where melanophores are absent, and white, in which both black and yellow pigment cells are inhibited. Genes have been identified which cause a lack of each of types of chromatophores individually, or in certain combinations (Fig. 2.4). An interesting parallel can be drawn between the colour of axolotls and Lepidoptera, for, as with the individual scales on the wings of butterflies and moths, an axolotl chromatophore will only contain one pigment type.

Reptiles

Melanic polymorphism is common in terrestrial reptiles, particularly snakes. Snakes of the genus *Vipera* have perhaps received most attention in this context. In Britain, children are taught to recognise our one venomous snake, the adder, *Vipera berus*, by its zigzag pattern. However, across much of its range a melanic form of the adder is also found, sometimes at high frequency. In Britain, melanic forms occur in some populations, such as on the Gower Peninsula in south Wales and parts of the New Forest, but it is on the Continent, in a range of situations, that melanism is most common. For example, in the Black Forest in Germany, where adders are restricted to damp cool habitats, all populations have some melanic individuals, the frequency of melanics varying from 30 to 90 per cent (Lehnert and Fritz 1993). Melanics are also common in many alpine regions. Indeed, melanic frequency reaches 49 per cent in the Bernese Prealps (Monney *et al.* 1996*a*). The maintenance of polymorphisms in this species appears to involve a number of different factors including growth rates, which are more rapid in the melanics, activity rates at low temperatures, and crypsis. Interestingly, the frequency of melanics in Switzerland and Italy was higher in females than males, suggesting that melanism is more advantageous to females than males (Monney *et al.* 1996*b*). This may result from differences in behaviour of the sexes leading to morph-related differences in predation. An elegant study of adder populations on islands in the Baltic Sea and on mainland Scandinavia showed that melanic males are more prone to predation than non-melanics, the reverse being the case for females (Madsen and Stille 1988; Forsman 1995). The opposite is probably true for melanism in Reeve's turtle, *Chinemys reevesii* and in the yellow-bellied slider turtle, *Trachemys scripta*, in which only males are melanic (Garstka *et al.* 1991; Yabe 1994). It is possible that melanism in this latter species is related to reproduction, because melanic males were all more than 6 years old. Melanism is also reported in several other species of turtle and tortoise, but the prevalence is variable. For example, in the European pond turtle, *Emys orbicularis*, melanic adults occur in some subspecies, but not in others (Fritz 1995).

Sexual differences in melanic frequencies are not seen in all reptiles. In the garter snake, *Thamnophis sirtalis*, in the Lake Erie area, melanics made up from 0 to 59 per cent of populations on islands and the mainland, but the level of melanism was similar in males and females (King 1988). Nor is the higher incidence of melanic snakes in cool climes seen in all snakes. In the eastern cottonmouth, *Agkistrodon piscivorus*, melanism increases southwards from the northern most edge of its range (Blem and Blem 1995).

Melanic individuals of some lizards have also been recorded, although in this group melanism is often rare and frequency of melanics may be little above mutation rate. However, in the common wall lizard, *Podarcis muralis*, dark forms are not uncommon (Smith 1964; Kuranova 1989; Sound 1994), and here again, as in the adder, melanism appears to be more frequent in females than in males.

Birds

Many species of bird are predominantly black, or have one sex that is mainly black, or show melanic polymorphism, or show clines in the degree of melanisation. For example, in the New Zealand bellbird, *Anthornis melanura*, the decree of melanisation of plumage characters increases southwards, and may be associated with increased humidity (Bartle and Sagar 1987).

Again the causes of melanism are diverse, and in many instances not clearly understood. Melanic polymorphism occurs in some species. One of the most obvious and best worked examples is the Arctic skua, *Starcorarius parasiticus*, which was the subject of a major research programme led by Peter O'Donald in the 1960s and 1970s. This species has three genetically controlled colour phases: pale, intermediate, and dark (Plate 1d and e). The frequencies of the three forms vary geographically, dark forms generally being commonest in the south (Fig. 2.5), and there being a good inverse correlation between melanic frequency and latitude, in the north-east Atlantic (Fig. 2.6) (O'Donald 1983). The maintenance of the melanics depends mainly on differences in the mating success of the forms,

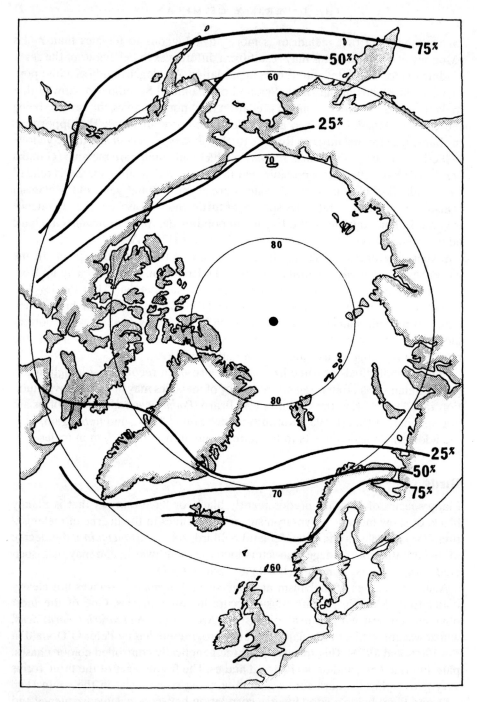

Fig. 2.5 Distribution of melanic forms of the Arctic skua. (Courtesy of Dr Peter O'Donald.)

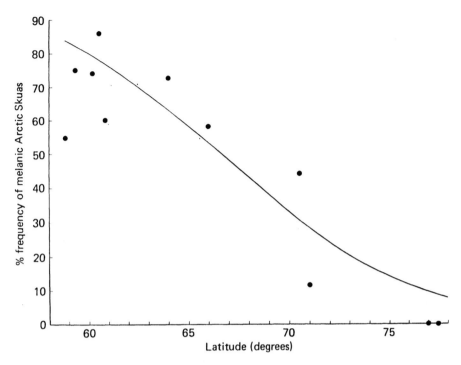

Fig. 2.6   The cline in frequencies of melanic Arctic skuas in relation to latitude.
(Courtesy of Dr Peter O'Donald.)

the age that they breed (p. 94). The ease with which both predators and prey of
the skua see the different colour forms possibly also has some effect. Two other
species of skua, the long-tailed skua, *S. longicaudus*, and the Pomarine skua, *S.
pomarinus*, also have these colour forms, possibly controlled by the same gene.
O'Donald (1983) argues that as the great skua, *Catharacya skua*, does not show
this polymorphism, all birds being dark. The melanic forms of the three *Steroco-
rarius* species are probably ancestral. This highlights another problem with
Kettlewell's definition of melanism (p. 4), for in these species, the dark forms
should probably be deemed the norm.

   A perplexing case is seen in the fulmar, *Fulmarus glacialis*. The melanic form
of this species is rare in its breeding colonies in the south Atlantic, but common
in the north, yet in the Pacific, this pattern is reversed (J. Fisher 1952). The melanic
polymorphisms in this species, and the opposing clines in melanic frequencies in
other sea-birds, well illustrate the dangers of extrapolating interpretations gained
from studies of one population of a species, both to other species, and to other
populations of the same species.

   Melanic polymorphisms are found in a variety of other species of bird. The
snow goose, *Anser caerulescens*, has a so-called blue form that is melanic, and

again mating preferences have been implicated in the maintenance of the poly-morphism (F. Cooke and Cooch 1968; F. Cooke and McNally 1975; F. Cooke *et al.* 1976, 1985; Findlay *et al.* 1985). Melanic great blue herons, *Andea herodias*, are common in the west of the species range in North America, but become rarer to the east, so that in the Florida Keys, the population consists entirely of the white form (Mayr 1955). In the feral pigeon, *Columba livia*, high melanic fre-quencies are associated with urban populations, but this is largely a consequence of high food availability having an effect on the breeding cycle and hormonal levels (Murton and Westwood 1977) (p. 36).

In some species of bird, the reasons for melanism are not so difficult to fathom. For example, a population of melanic partridge, *Perdix perdix*, occurs in north-west Germany on black peaty soils (Sage 1962). Similarly, where desert larks, *Ammomanes deserti*, live on black larval terrain, they tend to be melanic (Sage 1962, 1963). In both these cases, the melanics obviously gain a cryptic advantage.

In cases where melanism is restricted to one sex, either sex may be melanic. The melanism in these cases implies that being dark may have different advan-tages and disadvantages in the two sexes as a result of different factors effecting survival and reproductive success in each sex. In most instances when melanism is confined to males, it is because being black increases mating success, as for example, in blackbirds, *Terdus vulgaris*, black grouse, *Lyrurus tetrix*, and long-tailed widow birds, where the males are strikingly black. Conversely, when melanism occurs in females, as in many ducks and pheasants, the females are cryptically melanic, being dappled brown to avoid predators, particularly while incubating their eggs, while the males are brightly coloured and patterned for sexual reasons.

Among bird species which are all dark, melanism is often associated with unpalatability, as in many species of crow. Cott (1946) working on desert living birds in Africa, offered 350 species to three carnivores: hornets, domestic cats, and humans. He demonstrated that black-coloured birds were in general inedi-ble, and were rejected, while closely related paler, and presumably cryptic, species were devoured.

Mammals

Melanism has been reported from many families of mammals. In Britain, for example, black rabbits, *Oryctolagus cuniculus*, are not uncommon in some popu-lations, particularly in the north-west of Scotland and some of the Scottish Isles. Black grey seals, *Halichoerus grypus*, also occur in some populations around the coast of Scotland, and melanism has been reported in a number of mice, rats, bats, deer, foxes and, of course, in all common domestic animals.

Squirrels, both in Britain and elsewhere, provide some interesting examples. The town of Letchworth is famous for the presence of black individuals in the population of grey squirrels, *Sciuris carolinensis*, that resides there. These black squirrels have survived in the population since early in the twentieth century when Letchworth was founded as the first garden city in England. Indeed, the first, and for many years the only pub in Letchworth, was called the Black Squirrel. In North America many populations of grey squirrel show melanic polymorphism. The current frequencies of forms are partly a consequence of natural processes and partly the result of human activities. In the middle of the nineteenth century, the melanic form was much sought by hunters for its pelt. The result was that the frequency of melanics, which in many populations had exceeded 50 per cent, declined dramatically (Kennicott 1857). In recent years, this pressure has been lifted as fashions have changed, and the frequency of this melanic is now increasing in some localities. Melanic forms are now found mainly in urban and northern populations. Despite much recent experimental research, the adaptive significance of this melanism eludes discovery (Gustafson and Vandruff 1990).

The red squirrel, *Sciurus vulgaris*, also has melanic forms, although these have not been recorded in Britain. However, on the Continent they occur in parts of France, Germany, Switzerland, and Denmark. In Denmark, Shorten (1954) cites a record of a completely black population of approximately 72 000 squirrels in the Fumen region. Many other squirrels have melanic forms, and some, such as the fox squirrel, *Sciurus niger*, have received considerable attention, efforts being made to discover the evolutionary causes of their melanism (Kiltie 1989, 1992*a,b*) (p. 33).

Perhaps the mammal which has been best studied in respect of melanism, apart from humans, is the deermouse, *Peromyscus maniculatus*. Fur colour varies from pale straw to almost black. Many studies have shown that the coat colour in a region is closely correlated to soil colour (Dice 1930, 1947; Blair 1941, 1943).

Two final examples of melanism in mammals, of the many that could be cited, deserve brief mention. Many shrews are dark in colour, and Kettlewell (1973) regards them as showing melanism. The reason for this melanism he considers to be that they are unpalatable, the dark colour being an advertisement of their unpalatability. To support this view, Kettlewell draws attention to the remarkable case of five species of unpalatable tree shrews (Tupaiidae) from Borneo, which are mimicked by at least five species of edible squirrel to such a precise degree that they can only be distinguished by examination of their skulls (Shelford 1916). At the other end of the scale, some species of predatory cat, such as the jaguar and leopard, *Felis pardus*, have melanic forms, as immortalised by Bagheera, the panther, in Kipling's *Jungle Book*. On the open savannahs of Africa, the melanic form of the leopard, which is genetically dominant to the non-melanic form, is very rare. However, in jungle habitats of South-east Asia, the melanic is often the commoner form, and in some populations reaches almost 100 per cent. It is likely that coat colour in these large predators is as much the result of crypsis as it appears to be in many prey species, but here the advantage is that prey cannot see the predator. A non-cryptic predator may not lose as much as a non-cryptic prey that becomes some predator's lunch, but losing lunch by being too easily seen is still a disadvantage.

The causes of melanism

As the brief foregoing examples of melanism in different taxa make clear, the advantages and disadvantages of being darkly coloured are diverse. Reasons include increased activity in cool conditions, being attractive to mates, avoiding predation by crypsis, being more successful in catching prey through crypsis, being difficult to see in specific light conditions, and warning coloration. In addition, melanic pigments have some properties that appear advantageous because of their structure rather than their colour. These include a screening capability against UV light, a strengthening effect in a number of contexts, and, in at least one case, resistance to disease. Additionally, melanism may not always have an adaptive reason. Some melanisms may be a by-product of other biochemical systems, while others may be due to chance. After all, almost every organism is some colour. Although some of these factors have already been implicated in the descriptions of melanism in the examples previously mentioned, it is perhaps valuable to discuss the basis of each of these factors in turn.

Physical properties of being melanic

Thermal melanism

Many species of invertebrate exhibit melanism at high altitude and high latitude. One of the reasons for this only applies to melanism in flying insects, primarily moths, occurring at high latitude. This is the fact that moths which are on the wing in the summer do not have the cover of darkness to fly under, as it is light around the clock. Dark-coloured moths are more difficult to see on the wing than pale ones, and gain an advantage at northern latitudes as a result of reduced predation by birds that hunt moths on the wing. As all reported examples falling into this category are from the northern hemisphere, Kettlewell (1973) termed this phenomenon, northern melanism; although it is likely that examples exist in the southern hemisphere as well. This category of melanism will be discussed in greater depth in Chapter 7. The other reason proposed to explain why melanism is often prevalent at high altitude and high latitude is thermal melanism and applies equally to both situations.

Thermal melanism arises as a consequence of the differences in thermal properties of dark compared with light surfaces. Dark surfaces both absorb and radiate heat more rapidly than pale surfaces. Consequently, melanic organisms warm up more quickly than their light counterparts given a heat source such as sunlight. The reverse is also true, for melanics will cool down more rapidly after dusk. In cool climes, melanics may be at a significant advantage if they can warm up and become active when non-melanics cannot.

Evidence for thermal melanism comes from a diverse array of organisms, including butterflies, ladybirds, snails, and snakes. For example, many studies of butterflies, particularly those involving the Pieridae, have shown that increases in the proportion of melanic scales in the basal third of the wings is correlated to increases in both latitude and altitude. Many of these arctic and alpine butterflies have a characteristic behaviour, basking in the sunshine with their wings held at right angles to the sun to catch as much heat as possible. Other species, such as some of the arctic *Colias* species, which habitually rest with their wings closed, have the undersides of their wings heavily dusted with black scales (Plate 1f).

Correlations between melanism and high altitude have long been known. The prevalence of melanic coloration in invertebrates at high altitude was first reported by Heer as early as 1836. This is a world-wide phenomenon, examples having been observed in almost all the major mountain ranges (Hudson 1928; M. Mani 1968; Kettlewell 1973). In Britain, obvious examples of high-altitude melanism include species such as the black mountain moth, *Psodos coracina* (Esper) (Plate 1g), and the netted mountain moth, *Semiothisa carbonaria* (Clerck) (p. 161). Elsewhere, other taxa show a similar phenomenon. Rapoport (1969) demonstrated a strong correlation between altitude and the percentage of dark or black species of Collembola in the Pyrenees and the Swiss Alps. In the same order, M. Mani (1968) reports that species from several genera (*Isotoma, Proistoma, Hypogastrura*) are commonly dark on the snow fields of the Himalayas. In these cases, species are often monomorphic; however, in some species there is a grading from light to dark, with increasing altitude. Although thermal factors have not been demonstrated to be the cause of melanism in most of these species, in a few the increased rate of heating has been formally demonstrated.

In the spittlebug, *Philaenus spumarius*, a correlation between melanism and altitude in three Scottish glens has been demonstrated. Although other factors have been implicated in the melanic polymorphism in this species (p. 54), Berry and Willmer (1986) showed that the melanics absorbed more heat and had higher temperature excesses than non-melanics, concluding that thermal melanism does play some part in the complex selection regime acting on melanic polymorphism in this species. Other homopterans, such as *Eupteryx urticae* (Fabricius), and *E. cyclops* (Matsumura), also show melanic polymorphism, and again a positive correlation between melanic frequency and altitude has been found, implicating thermal melanism. Indeed, Stewart (1986) suggests that the advantage of thermal melanism is balanced by visual selection against black morphs by entomophagous parasitoids.

Honek (1993) has shown that melanic forms of the snail, *Helicella candicans*, heat up more rapidly under lights in captivity, and attain a higher equilibrium temperature than pale forms. This type of experiment has also been carried out on other snails, various species of ladybird (e.g. Brakefield and Willmer 1985), and reptiles.

Melanin as a barrier against ultraviolet light

Melanic pigments are well known to act as a barrier against the harmful effects of UV light in humans (see Dobzhansky 1962 for review). More recently, because of concerns over the depletion of the ozone layer, which prevents the passage of much UV irradiation, attention has focused upon the effects of UV on other organisms, particularly at high latitudes where UV levels reaching the Earth's surface are exceptional as a result of the much publicised 'holes' in the ozone layer. For example, Herbert and Emery (1990), investigated the causes of the prevalence of melanic clones of *Daphnia pulex* and *D. middenorffiana*, in high arctic and subarctic areas. Having shown that the pigment in the melanics was indeed a melanin, they demonstrated that although the melanin present only comprised 0.03 per cent of body weight, it prevented the transmission of over 90 per cent of the incident UV light. Furthermore, they showed that melanic clones of both species survived twice as long as unpigmented clones when exposed to high levels of UV irradiation. Finally, using reciprocal transplant experiments into natural ponds, they showed that while melanics could survive in either clear or cloudy ponds, the unpigmented forms only survived in the latter. Similar results have been obtained by Zellmer (1995).

Kettlewell (1973) has speculated that the presence of melanic forms of many species of slugs is also a consequence of melanins acting as a barrier against UV, although he presents no data to support this view.

A discussion of the costs and benefits of melanic skin pigments in humans is outside the scope of this book. The situation is not simple, for in some ways the distribution of differently coloured races is anomalous. On the theory of thermal melanism, one might except darker-skinned races to occur primarily at high latitude, with paler-skinned races frequenting the tropics. Yet precisely the reverse is the case. However, as we have some control over our environment, thermal considerations may not be the main concern. The question, why are some peoples melanic, should be coupled with the question, why are some peoples pale skinned? Indeed, the latter question is perhaps the more crucial in evolutionary terms, for melanism is the ancestral condition in humans. Although several hypotheses have been put forward in answer to these two questions, none is completely satisfactory.

There is a considerable body of evidence that points to melanins being important as a barrier against UV. This evidence can be broadly split into two categories: that from the correlation between degree of melanisation and distance from the equator, and that from medical evidence showing a correlation between melanisation, UV light, and the incidence of some types of skin cancer. In general, the most highly melanised races come from the lowest latitudes, and in such peoples, skin carcinomas are virtually absent. However, when paler-skinned peoples have emigrated to equatorial regions, many have suffered from such conditions.

Everyone who ever sunbathes to get a tan, should recognise that the production of melanins is one of our body's defences against the harmful effects of the suns rays, for a naturally obtained tan is no more than an increase in the amount of melanin in the skin. The role of melanins as a defence against UV irradiation has been highly publicised in recent years, partly in recognition of the dangerous consequence of our disruption of the ozone layer. In addition, the amount of research into the effects of lights of different wavelengths, and the variance in tanning abilities of different skin types, particularly in relation to basic skin content has been increasing recently (e.g. see Chardon *et al.* 1991). Sadly, it is common now to hear warnings against prolonged exposure to the sun on weather forecasts even in the British summer. Furthermore, sun tan lotions now are sold as much on the basis of their effectiveness at blocking UV, as on their ability to aid the tanning process. Despite this, the incidence of skin cancers is rapidly increasing, this increase being partly attributed to increased exposure to UV irradiation.

Yet, cancer is certainly not the whole story. Even in light-skinned peoples, malignant melanoma is a rare disease that occurs mainly in the old. Those that die from it will have already passed on the genes for their skin colour. But prevention of skin carcinomas may not be the only benefit of possessing a barrier against UV. There is some evidence that UV can destroy certain vitamins known to be essential in growth and pregnancy (J.S. Jones 1993). In this case, the effect would certainly influence the passage of genes to the next generation.

The question of why peoples living in higher latitudes have pale skins, is possibly also related to vitamins. Lack of vitamin D produces a disease in children known as rickets, in which bones do not harden properly, often leading to deformity. Most European children now get sufficient vitamin D in their diet. But vitamin D can also be made in the skin by the action of UV light, and pale-skinned races manufacture vitamin D more efficiently than those with dark skin. When food availability was less than now, this greater ability may have been crucial in higher latitudes.

Resistance to injury or as a response to injury

Melanic pigments in most organisms are present in the epidermis in the form of pigment granules within cells. These have the property of strengthening the epidermis. In many organisms, melanic pigments are laid down around wounds in the epidermis. This type of response is seen in both vertebrates and invertebrates, and an example of each will suffice to make the point. In humans, it has been reported that one of the consequences of skin grafting procedures is often an

Fig. 2.7   Abnormal melanin laydown around an injury in a 7 spot ladybird.

increase in melanisation of the region from which the skin graft has been taken. In ladybirds, the result of injury to pupae is abnormal asymmetric melanin production in the ensuing adult (Fig. 2.7). .

The moth, *Ephestia kuhniella* Zeller, presents an intriguing case, for in this species melanics were found to have a greater ability to encapsulate the eggs of parasitoid wasps and their larvae (Verhoog *et al.* 1996). This suggests that melanism may be a defence against certain types of pathogenic attack. The possibility that melanism may have a role in disease resistance is strengthened by Gershenzon's (1994) finding that a melanic strain of the oak silkmoth, *Antheraea pernyi* Guérin-Ménéville, from Manchuria was substantially more resistant to nuclear polyhedrosis virus than the typical form. Whether the resistance observed is a consequence of melanism itself, or is simply linked to the melanism gene(s) is not yet known.

Resistance against abrasion

One of the paradoxes in the insect world is that many beetles living in desert regions are black. This seems counteradaptive from the point of view of crypsis and thermal properties, and has led to the so-called 'black beetle paradox' (Cott 1940; Hamilton 1973). One possible reason for the colour of these beetles is that they are unpalatable and the black coloration which stands out against the pale desert sands advertises this unpalatability (Cloudsey-Thompson 1964; Rettenmeyer 1970). However, a second possibility is that the melanin granules in the surface of the exoskeleton of the beetles makes them more resistant to sand abrasion.

Resistance against abrasion has been widely reported as a property of melanic, as opposed to non-melanic, feathers of birds (Averil 1923; J. Harrison 1947; Prater *et al.* 1977). In parallel to the situation in beetles, Burtt (1979, 1986) has shown that melanic feathers wore less when submitted to bombardment by sand particles, and proposed that this was the main reason why most desert birds are melanic (Burtt 1981). Burtt also found that non-melanic barbules, which hold the feather barbs together to form the feather vane, were more likely to break than melanic barbules.

Bird feathers are made up of keratin, the same basic material as mammalian hair, although the structure of avian and mammalian keratins are quite different. Feathers, particularly flight feathers, have to be extremely strong for their weight to withstand the tremendous stresses placed upon them. Although melanic feather keratin has long been considered to be more resistant to wear than non-melanic feather keratin, that this is the case has only recently been demonstrated. Bonser (1995) showed that the indentation hardness of melanic keratin is 39 per cent greater than that of non-melanic keratin. As the indentation hardness of a material is inversely proportional to its rate of wear (Lipson 1967; Barwell 1979), melanic keratin will wear more slowly than non-melanic keratin under the same abrasive conditions.

As Bonser (1996) notes, melanin production must have a cost, so it is likely that in birds there is a trade-off between the cost of melanin production and the benefit of increased flight efficiency arising from the reduction in wear of melanic feathers.

Once again caution must be applied when discussing such matters, because it is unlikely that these are the only factors that should be considered in a cost–benefit analysis. For example, considering melanic desert birds, Burtt's (1981) deduction that melanic plumage is common in such birds as defence against the abrasive effects of sand should be taken alongside Cott's experiment's showing that melanic desert birds are less palatable than non-melanics. Both features are presumably beneficial to dark or black desert birds now, but one is left wondering which developed first. Did the birds first become melanic for the aerodynamic advantages in a habitat where abrasive bombardment was of regular occurrence, and then, being black and conspicuous, did unpalatability evolve because dark feathers are of little value for crypsis in a generally pale habitat? Or conversely, did unpalatability and conspicuousness evolve initially, the protection against abrasion being an incidental fringe benefit? Here are a pair of possibilities that a comparative phylogenetic analysis might easily resolve.

Melanism in defence

The external colours of most organisms are thought to have evolved, at least in part, for reasons of defence against predators. Colour, by itself or coupled with other characteristics, can imbue an organism with defensive properties in any of a variety of different ways. As the defensive strategies of organisms are often considered to be influential in the evolution of melanism, it is worth considering the characteristics and constraints of the various defensive strategies.

Camouflage and crypsis

Camouflage and crypsis when applied to organisms both conjure up the idea of animals, or more rarely plants, that are difficult to see. For some organisms this impression is partly true if the organism hides away in secretive situations. However, the art of camouflage is not in not being seen, it is in not being recognised. In practice there are two commonly used strategies to avoid being recognised. An organism may blend in with its surroundings so it is not easily detected as an entity in itself. Alternatively, it may be obvious to the eye, but be of a form that is not easily recognised as to what it is. In this context the many examples of arthropods that resemble bird droppings come to mind.

The mechanisms used to achieve camouflage or crypsis are very varied. Organisms can simply match their background, blending in with the habitat or specific elements of it. Some organisms, such as chameleons, cuttlefish, and many flat-fish, change colour. The colour changes in these species are achieved rapidly by the migration in the epidermis of pigments, including melanins in every case studied, in response to light stimuli. However, not all colour changes are so rapid. The colour patterns of many late instar lepidopterous larvae and pupae are determined by light, or other stimuli, acting on larvae during a critical period in their development (p. 67).

Some species camouflage themselves by incorporating parts of the environment into themselves. Thus early instar larvae of many moths, such as the angle-shades, *Phlogophora meticulosa* (Linnaeus), have a transparent body wall, so that their colour is that of their gut contents, which in turn is the colour of the vegetation that they are sitting and feeding upon (Plate 2a–d). Conversely, other species, such as many caddis-fly larvae, lacewing larvae, and crabs, practice self-adornment, sticking objects on to their exterior.

Various mechanisms exist to avoid the problems of casting shadows. Adult moths of many species 'clamp down' on to the substrate, the fringes of their wings flush against the substrate so that there is no gap. Some butterflies and grasshoppers tilt themselves towards the sun so that the shadow produced is minimised. Yet perhaps the commonest method used by prey species to reduce the risk that their shadow will give away their presence, is to be countershaded. This means that the surface that is usually directed towards the light, is darker than that away from the light, usually due to melanic pigments. The result is that the normal cylindrical body shape appears flattened and unicolorous. Countershading occurs in many taxa, fish and antelope providing some of the best examples.

In addition, organisms may have dummy heads or dummy eyes to deflect predator attacks to non-vital parts. They may become smaller, as in the case of wing-less moths, and they may suffer reduced predation simply by being different.

The evidence that melanism has often evolved as an anti-predator device is overwhelming. Cases involving adult Lepidoptera will be discussed in Chapters 5–8, but there are many other animals in which the cryptic advantages of being melanic have been demonstrated.

One of the most elegant studies demonstrating a correlation between melanism and background colour was conducted by Harris and Weatherall (1991) in New Zealand. They collected samples of the sand-burrowing beetle, *Chaerodes trachyscelides*, from 11 beaches in New Zealand. They also collected samples of sand from each beach. By measuring the colour of the dorsal surface of the beetles and the sand, using a reflectance spectrometer, they demonstrated a very strong correlation (correlation coefficient = 0.961) between the colour of the sand and the mean dorsal colour of the beetles. They also demonstrated that the degree of variation in the colour of the beetles from a particular beach closely paralleled the amount of variation in the colours of the sand on that beach.

A rather more complex situation has been revealed in the fox squirrel, *Sciurus niger*. Populations of this squirrel are polymorphic around the Mississippi delta and three forms are present: pale-agouti, dark-agouti, and non-agouti black (Kiltie 1992*a*). Tests of the relative ease with which these forms could be seen against a variety of tree trunks suggested that the pale-agouti form would gain a cryptic advantage compared with the darker morphs when motionless on the trees (Kiltie 1992*b*). However, the melanic squirrels matched burned trees better, and indeed, Kiltie (1989) found a positive correlation between the frequency of wild fires and melanism. Kiltie (1992*b*) considered it unlikely that the advantage gained by the dark squirrels on burned trees would, on its own, be sufficient to maintain them in the population. However, by observing the time taken for captive red-tailed hawks, *Buteo jamaicensis*, to respond to moving squirrel models of different colours, he found that the intermediate and black squirrels elicited a slower response than the pale models. In this work, Kiltie also showed that having patterned coloration may also be beneficial, because the hawks responded to models with black heads and white noses and ears, as exhibited by the squirrels from the south-eastern coastal plain, more slowly still. Here the head colours may act as a disruptive pattern, breaking up the body outline.

It is worth noting here that associations between melanism and burned vegetation have been observed in other taxa. Several lepidopteran examples are known (p. 182), and so-called fire melanism has been found in the grasshoppers on the savannahs of West Africa (Fishpool and Popov 1981).

One intriguing case is that of the larvae of the elephant hawk moth. The facts are these. In their early instars, elephant hawk moth larvae are green. However, a few change colour at the third moult, becoming dark brown, while the majority make this transition at the fourth moult, with only a few remaining green throughout. Experiments have shown that larvae reared singly remain green at the third moult. However, if five larvae are reared together the majority turn brown on entering the fourth instar. The question is why do the larvae change

colour, and why does the timing of the transition vary. This problem has enter-
tained entomologists for many decades. Denis Owen (1980) questions how this
density-dependent colour change could be related in any way to camouflage, and
points out that the colours of animals such as caterpillars may have other functions. Then, noting that larvae kept in groups tend to feed up more quickly, Owen
speculates that the development of dark coloration by crowded larvae may be a
form of communication between larvae competing for food. Whether this is the
case or not awaits investigation. However, the colour change certainly could be
for protection from predation.

It is known that many birds form searching images for types of food that they
find commonly (p. 92). Over a period, birds foraging regularly in an area where
elephant hawk moth caterpillars occur at high density are likely to form a searching image for green larvae, as all the larvae are initially this colour. If larvae then
change colour, the birds will have to form a new searching image. Tinbergen
(1960), working on great tits, *Parus major*, has shown that switching searching
images takes time, there being a lag of several days after a new type of prey
becomes common before the birds began to search actively for it. The reason that
larvae at low density do not change colour may simply be because predators may
not find it profitable to form a searching image for scarce types of prey. One question remains: Why do some larvae not change brown at the third ecdysis when
at high density? This may be a question of the balance between the advantage
of changing colour to avoid the bird's old searching image, and the advantage of
retaining the old colour once bird's have switched their attention to larvae of the
new colour. The colour change that occurs in the majority of larvae of the angleshades moth, where about 80 per cent of larvae change from green or olive to
brown at the end of the fourth instar, has previously been explained in this way
(Majerus 1978).

Warning coloration

While being difficult to see is often an advantage for both prey and predators,
some prey species also gain an advantage by being easily seen, and being recognised. Many organisms have an armoury of chemical and sometimes physical
defences against their predators. These defences may be manifested in distastefulness, by sting or bite, by bearing irritating hairs, or by spines. Organisms bearing
one or more of these defences often advertise their unpalatability by being memorably coloured. In turn, such species have become the object of copying by
edible species that dupe potential predators into leaving them alone believing
them to be harmful. These deceitful mimics, and the honest models they resemble, frequently use black coloration as part of their colour pattern, for black is a
very memorable colour.

Several examples of truthful warning coloured melanism have already been
mentioned (e.g. black beetles, black desert birds, and black tree shrews). Further

examples from the Lepidoptera and ladybirds will be discussed in Chapters 8 and 9, respectively.

Sexual selection

Darwin (1859, 1871) recognised that in addition to struggling for survival, which he termed natural selection, organisms would have to struggle to reproduce. This he termed sexual selection. To Darwin, this type of selection had to exist to explain the secondary sexual characteristics that he observed in many species. The conspicuous male breeding plumage of many birds, the antlers of deer, the bright colour patterns on some male butterflies, could not be explained by natural selection, for were they an advantage in the struggle for survival, females should surely also bear these features. Thus, Darwin conceived that they had evolved as a result of one of two mechanisms to increase the likelihood of being successful in the struggle to secure mates and reproduce. On the one hand, males (or occasionally females) may compete among themselves for access to mates (male competition). On the other hand, females (or more rarely males) may choose between potential mates on the basis of their characteristics (female choice) (see also p. 93).

In many cases of both male competition and female choice, the sexually dimorphic traits that have evolved are melanic and monomorphic in the sex bearing them. Thus, in blackbirds and the long-tailed widow bird, all adult males are predominantly black. However, in a few cases polymorphism within the sexually selected sex results. The case of the three-spined stickleback is instructive in this regard. Here, female choice has selected for the colour of male throat patches (Semler 1971), males with red throats being favoured over males without a red patch. In many populations all males develop red throats when reproductively mature. However, in some populations male throat colour is polymorphic, and in others all males have black throats. Polymorphism in throat colour results from a balance between the reproductive advantage males with red throats gain from being preferentially chosen as mates, and a predatory disadvantage of being more obvious to predatory fish such as trout. Black-throated males result from the fact that the mating preference appears not always to be for a specific colour, but for contrast between the colour of the throat and the rest of the male, as seen under water. When this is the case, black or dark patches of colour may prove an attractive trait, as occurs in the murky waters of some boggy ponds in Canada (McDonald et al. 1995) (p. 19).

Other cases of melanic males in sexually dimorphic species gaining a mating advantage are known. For example, Garstka et al. (1991) have shown that melanic males of the yellow-bellied slider turtle, court females more frequently than non-melanics.

Interestingly, in some species, mate recognition mechanisms prevent the evolution of melanism in one sex. Thus, in the polymorphic mimetic butterfly, Papilio *dardanus*, melanic mimetic forms are confined to females, the males having to retain their ancestral yellow and black colour pattern for females to recognise them.

Not all sexually selected melanic traits are limited in their expression to one sex. Sexual selection by female choice has been implicated in the maintenance of melanic polymorphism in snow geese, Arctic skuas, the scarlet tiger moth, *Callimorpha dominula* (Linnaeus), 2 spot ladybirds, and the ladybird, *Harmonia axyridis* Pallas, and in none of these is melanism limited to one sex.

Melanism as a by-product

In the foregoing discussion of the causes of melanism, the examples cited have involved melanism evolving because it was beneficial, at least in some circumstances. However, that being dark may in itself be neither advantageous or disadvantageous is also a possibility. In such a case, being melanic may simply be the by-product of other physiological considerations. For example, in some bacteria, melanin production may be a method of getting rid of excess amounts of tyrosine, that is itself a by-product of biochemical reactions (Oliver, personal communication).

Melanism in feral pigeons may serve to explain this idea. As mentioned previously, melanism in feral pigeons is particularly associated with urban populations which benefit from high food availability throughout the year. To understand the connection, one has to consider the length of the reproductive season of populations of pigeons with different food availabilities. When pigeons have a virtually unlimited food supply, the necessity to breed only during a short period of the year, when sufficient food can be found for the young, is removed. Thus the day-length cues by which this period is determined lose their adaptive relevance. Indeed, a gene that causes pigeons to relax the limits of their breeding season will be an advantage where food is plentiful, for these birds will produce more young per year. Murton and Westwood (1977) have argued that not only has such a gene arisen and spread in many urban pigeon populations, but that this gene, by affecting the photoperiodic response of pigeons and the length of their breeding season, has also dramatically affected the hormonal systems of the birds which have had a knock-on effect on melanism production. The effect on melanin production is thus not of itself adaptive, but a by-product of an extension to the breeding season. Indeed, in a considerable proportion of melanics, the gonads do not diminish at the end of the normal breeding season, and these birds continue to breed throughout the winter (Murton *et al.* 1973).

Interestingly, melanism in urban pigeons is not always a by-product. Kettlewell (1973) relates an interesting tale of a dramatic increase in fully dark pigeons in Glasgow, birds with significant amounts of white having disappeared over a very short period. Investigations revealed that due to the high population density of pigeons, and the damage that they and their excrement do to buildings, the authorities had taken steps to control numbers. However, due to public concern over the activities of pigeon catchers, it was felt politic to only catch pigeons under cover of darkness while the birds were at roost. At such times, fully black birds were much more difficult to see and destroy than those that had white markings. Thus here is another case of melanism as an anti-predator device, in this case humans being the predator.

The complexity of melanism

In this chapter, we have seen that melanism is a common phenomenon in many and diverse taxa. The evolutionary causes of melanism are also diverse. The colours and patterns of organisms are the evolutionary result of a number of different factors, which may often not be in complete harmony, and in some instances are undoubtedly in conflict. These factors include, defensive, and more rarely aggressive strategies, the need to attract, find, recognise and excite a mate, the physical attributes of the pigments involved and the colours that they produce, the energetic costs of colour production, and the past history of the organism in respect of both mutational events and the previous selective pressures to which the species has been exposed. The evolution of melanism in a species will rarely be the result of just one of these factors. Although one factor may have a greater influence than others, in virtually every instance studied in any detail, several factors have been shown to play a part. Indeed, in the species in which melanism has been studied in most detail, the greatest complexity of interaction between a variety of factors is seen. This suggests that the complexity found is the result of the amount of study, and indicates that complexity is likely to be the rule rather than the exception.

3 The principles of genetics

Introduction

Some species of organism have melanic forms, while other morphologically similar and closely related species lack them. Furthermore, within a species, some populations contain no melanics, others are all melanic, and some have a mix of both melanics and non-melanics. If we are to understand the reasons for the presence or absence of melanic forms within a species, or why the frequencies of melanic forms vary in populations in either time or space, we must understand how melanic forms arise, how they increase or decrease temporally or spatially, and why in some species the increase of a melanic form stops before it completely replaces non-melanic forms. This is a long-winded way of saying that we have to understand the evolution of melanism, and to do that we need a knowledge of inheritance. How are traits passed from one generation to the next? Why do progeny have the characteristics of one or other of their parents in some features, appear to be a mix of their parents in other features, and are unlike either of their parents in still others? The study of these questions of inheritance is the field of genetics.

This chapter outlines the principles of genetics that are relevant to an understanding of the evolution of melanism. Within this, it is necessary to use a small number of technical terms. This is unavoidable, and indeed desirable, for any reader wishing to investigate the matter in more depth will find a knowledge of these terms is generally assumed in most of the source material I cite. Furthermore, the meanings of many genetic terms that were known to few who were not specialists in the field a decade or two ago, are now filtering into common usage as genetics begins to impinge more and more on our daily lives.

From generation to generation

The characteristics of organisms can broadly be split into three groups: those that are the product of hereditary resemblance; those that are a product of the environment to which the individual has been exposed; and those that are partly a product of both. In the main, biological evolution, as we currently understand it, is concerned with traits that can be passed down lineages, from parent to offspring or to direct descendants in later generations. The transmission of the vast majority of such traits from one generation to the next involves one or other of two huge molecules, deoxyribonucleic acid (DNA) or ribonucleic acid (RNA). These are the basic molecules of inheritance and carry the so-called 'blueprint of life'.

I say only that the vast majority of traits are the product of DNA or RNA because some characteristics, particularly in humans and some other higher vertebrates, may be passed down the generations by learning. So, for example, I have the habit of holding my chin between my thumb and index finger of my left hand, rubbing my chin intermittently when I am thinking. My father had this same habit. Yet I do not believe that I carry a specific piece of DNA, inherited from my father, which determines that I have this habit. Rather, I picked up this behaviour by copying my father when I was a child, and I have retained it as a habit for more than 20 years since his death. It will be interesting to see whether any of my children 'inherit' this habit. However, examples of traits passed down in this way are rare and need not concern us here.

To understand the principles of genetics, it is first necessary to understand how genetic material is passed from generation to generation during reproduction. Most organisms are composed of a huge number of minute units, called cells. In general, cells consist of two main parts, the nucleus and the cytoplasm. The nucleus lies within the cytoplasm and is surrounded by a semi-permeable membrane. Most genetic material resides in the nucleus, where it is arranged in strands, known as chromosomes. Humans, for example, have 46 chromosomes, 22 almost identical pairs called autosomes and a single pair of sex chromosomes (p. 58). Other species may have different numbers of chromosomes, and there may be considerable variation between quite closely related species, so while the brindled beauty moth, *Lycia hirtaria* (Clerck), has 14 pairs (Harrison and Doncaster 1914), the pale brindled beauty moth, *Apocheima pilosaria* (Denis & Schiffermüller), has 112 pairs (Regnart 1933). But the number found in any particular species is relatively invariable. In animals, the nuclei of most cells contain two of each different chromosome and are said to be diploid. However, in the specialised germ cells, eggs in females and sperm in males, there is only one copy of each chromosome, so the cells are haploid. This means that when a new organism is formed through the fertilisation of an egg by a sperm, the fusion of the two haploid nuclei brings together two copies of each chromosome, so diploidy is restored. This newly formed cell is the zygote.

Cell division

A zygote that is the result of sexual fertilisation gains one chromosome set from each parent. It is on the chromosomes that the hereditary units, genes, exist. These genes, of which there are many thousands, collectively make up a plan for the development of the newly formed zygote. The zygote grows by a process of cell division, known as mitosis. Without going into detail, during mitosis each chromosome duplicates itself, so at this stage a cell will contain four copies of each chromosome, two originals and two new copies. The membrane around the nucleus then disappears. For each chromosome, one copy migrates to one end of the cell, while the other moves to the other end, so that two complete sets of chromosomes end up in groups, one at each end of the cell. The cell then cleaves down the centre, and new membranes form around each of the chromosome groups forming two new nuclei. The result is that one cell splits into two which, in respect of their genetic material, are virtually identical to one another, and, of course, to the parental cell which gave rise to them. Consequently, all cells in an organism contain the same genetic blueprint, with the exception of a few specialised cells which lack nuclei, and the more crucial exception of the haploid gametes which are produced by a different type of cell division, called meiosis.

Genes and gene products

The genes on a chromosome are frequently likened to a string of beads. However, to understand patterns of inheritance and subsequently the mechanisms of evolution, a more precise description of genes and their action is necessary. The basic genetic material in all higher organisms is DNA. This enormous molecule comprises two similar, but not identical, long, thin strands, running parallel to one another like a spiral staircase. This is the celebrated 'double helix' discovered in Cambridge by James Watson and Francis Crick in 1953. The strands consist of a backbone of sugar molecules with phosphate groups attached, and are linked by hydrogen bonds between the nucleotide bases that are arranged along each strand. There are four types of nucleotide base in DNA: adenine, thymine, cytosine, and guanine. These bases are at the centre of the 'genetic code'. The chemical properties of the bases mean that a particular base in one strand virtually always has the same base opposite it in the other strand. This means that one strand is effectively a negative of the other, with cytosine pairing with guanine and adenine pairing with thymine. It is the order in which these bases lie along the DNA molecule that is important, for they determine how proteins and enzymes, the primary structural and catalytic biological molecules, are made up. Proteins and enzymes consist of chains of amino acids. The type and order of amino acids in a chain is determined by the order of nucleotide bases of DNA, and this arrangement in turn determines the structure and function of the protein or enzyme.

A gene is a plan of one amino acid chain. It has three important components: (a) a special short sequence of bases, called a start sequence, that says this is where the plan begins; (b) the central reading frame, which is a series of sets of three bases, called codons, each of which corresponds to an amino acid; and (c) a second specialised short sequence of bases, called a stop sequence, which says this is the end of the plan for this chain. It is the reading frame that concerns us most in understanding how proteins and enzymes are produced.

The translation of the reading frame into a protein is a complicated process. Initially, an enzyme, RNA–DNA polymerase, makes a negative copy of the reading frame of the gene in question. That is to say, it produces a molecule, messenger RNA, with a sequence having guanine wherever cytosine occurs and vice-versa, adenine wherever thymine occurs, and a fifth base, uracil, very similar to thymine, wherever adenine occurs. The messenger RNA separates from the DNA and migrates through the nuclear membrane into the cytoplasm. Here it becomes attached to small organelles called ribosomes. The cytoplasm of a cell is a kind of chemical soup, containing among other things, amino acid molecules and a second type of RNA called transfer RNA. The codons of nucleotide bases in the transfer RNA molecule are specific for a particular amino acid, and are a negative of part of the messenger RNA code. The transfer RNA molecules attach to the amino acid molecules their code specifies, and then with their accompanying amino acids, these attach in sequence to the messenger RNA codons, so that a specific series of amino acids link up into a chain forming a protein or enzyme.

The production of germ cells

Owing to the process of mitosis, almost all cells in an organism, with the important exception of the germ cells, or gametes, have the same complement of genes. The germ cells are of course critical to both inheritance and evolution, for it is through these cells that the characteristics encoded by genes are passed from generation to generation. Germ cells are produced by a specialised type of cell division called meiosis. This differs from mitosis in that it involves two cell cleavages rather than one. However, there is still only one chromosomal replication. The result is that from one original diploid cell, four haploid cells are produced.

The way in which this happens is crucial, for it is during this process of meiosis that recombination takes place, allowing new combinations of genes to arise. As in mitosis, the two chromosomes of a particular type pair up and produce copies of themselves by a similar process to that which occurs in mitosis. Again the DNA divides down the middle and each of the two strands produces a negative copy of itself, each negative copy being identical to the strand its partner strand separated from. The chromosomes have now become a quadruplet and it is at this stage that the genetic phenomenon known as recombination, or crossing-over, occurs. Here, the chromosomes of a pair may break and reattach to one another, so that nucleotides are exchanged from one chromosome of a pair to the other. As we shall see later (p. 57), this process is important in the generation of variation because it allows genes close to one another on the same chromosome to become separated, and new genes that were separate to be brought into close proximity. The rest of meiosis is similar to two mitotic cell divisions. First, two of the four chromosomes of a set migrate to each of the opposite ends of the cell, and the cell divides. Then, in each of these daughter cells, one of the two chromosomes of each type migrates to each end of the cell and the cell divides again, in all giving rise to the four haploid cells. These are the germ cells, eggs in females and sperm in males.

Genetic mutation

During the replication of DNA in meiosis, the genes generally remain unaltered. This is of great importance because it means that the characteristics that the genes code for get passed down from one generation to the next without change. I say the genes generally remain unaltered, because occasionally mistakes occur during DNA replication, or at some other time in meiosis. During replication, the wrong nucleotide base may be inserted, or one or more bases may be missed out, causing a shift in the reading frame of a gene. During recombination, when chromosomes break, a number of mishaps may occur. The wrong broken ends of the chromosomes may become reattached, or whole sections may flip over and so lie in reverse order, or a fragment from one chromosome may fuse to a chromosome of a different pair. Occasionally, a piece of chromosome or a whole chromosome may be lost completely from a gamete, or conversely a gamete may receive an extra copy of part or of a whole chromosome. All these types of errors during cell division may be included under the heading of mutations.

It is not necessary to go into the mechanisms of mutation here. All such mutations occur only rarely, but they are extremely important in evolution, for ultimately all genetic evolution is based on heritable variation, and the source of heritable variation is mutation.

It is clear then that the vast majority of characteristics of an organism that are passed from one generation to the next, are encoded in the DNA in the form of genes. Genes are usually passed down unchanged, but through the various processes of mutation, variations in genes do arise through time. We may now consider how genes and the characteristics that they produce are inherited within the so-called 'laws of inheritance'.

Mendelian inheritance

The inheritance of single gene differences

To begin, we should consider the simplest situation. Imagine that there is a character which varies in respect of its colour, being either pale or dark. The difference in the colour of this character is controlled by a single pair of genes, one derived from the mother, the other from the father. Each of the two genes will be on the same chromosome type, and in a similar position. This position is termed the gene locus, and the two genes are called allelomorphs, or alleles for short. In this case we may imagine a pale allele and a dark allele. These will differ in respect of one or more of the nucleotide bases in the gene. A particular individual may receive two copies of the same allele, one from each parent, or two different alleles. When the alleles are identical, the individual carrying them is said to be homozygous for the gene; if the alleles are different, the carrier is heterozygous for the gene.

In the scarlet tiger moth, the colour pattern differences between the nominate form, f. *dominula* and the variety f. *bimacula* Cockayne are controlled by different alleles of the same gene (Cockayne 1928*a*). Individuals homozygous for the *dominula* allele have about eight white and two yellow spots on the dark forewing, the red hindwing being rather lightly marked with black, while those homozygous for the *bimacula* allele have less, and in some cases, just a single white and a single yellow spot on each forewing, and are much more strongly marked with black on the hindwing. The heterozygote, that is an individual with one *dominula* and one *bimacula* allele, has an intermediate pattern, called f. *medionigra* (Plate 7a). If we mate two *dominula* individuals, the progeny are all *dominula*. Similarly, if we cross two *bimacula* individuals, only *bimacula* progeny result. When, however, we cross a *dominula* with a *bimacula*, all the progeny are of the intermediate *medionigra* form, for all will receive a *dominula* allele from one parent and a *bimacula* allele from the other parent. If two of these heterozygotes are crossed, all three types of progeny should result (Fig. 3.1), for some will receive a *dominula* allele from each parent, others will receive a *bimacula* allele from each parent, and some will get a *dominula* allele from one parent and a *bimacula* allele from the other. The number of each form among progeny of this type of cross is predictable: they should conform to a 1 *dominula*: 2 *medionigra*: 1 *bimacula* ratio, because there are two ways of producing intermediates and only one way of producing each type of homozygote. In this case, the heterozygote *medionigra* is dissimilar from both homozygotes. This is because both alleles are partially expressed in the heterozygote; one does not completely mask, or dominate, the other.

There are many other examples in which melanic forms are controlled by a single allele and show a similar type of inheritance, with the heterozygote being distinguishable from either homozygote. Just among the Lepidoptera several examples come to mind from my own work. For example, two of the melanic forms of the alder moth, *Acronicta alni* (Linnaeus), are controlled by a single pair of alleles on one gene. The more extreme melanic, f. *melaina* Schültze, is homozygous, while the less extreme melanic form, *steinerti* Caspari, is heterozygous (Majerus 1982*a*). Here the melanic allele could be said to be incompletely dominant to the typical allele as the heterozygote is much more similar to the melanic homozygote than to the non-melanic form, yet it is always distinguishable except in very worn individuals. A similar instance is seen in the case of the *griseovariegata* Goeze and *grisea* Tutt forms of the pine beauty, *Panolis flammea* (Hübner), the former being heterozygous and the latter homozygous (Majerus 1982*b*).

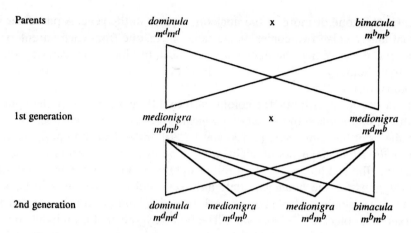

Parents

dominula   x   bimacula
$m^d m^d$ $m^b m^b$

1st generation

medionigra   x   medionigra
$m^d m^b$ $m^d m^b$

2nd generation

dominula   medionigra   medionigra   bimacula
$m^d m^d$ $m^d m^b$ $m^d m^b$ $m^b m^b$

Fig. 3.1 The genetic control of the *dominula*, *bimacula*, and *medionigra* forms of the scarlet tiger moth (m^d and m^b are the *dominula* and *bimacula* alleles, respectively). Neither allele is completely dominant, with the result that the heterozygote is distinguishable from both homozygotes.

An interesting case involving incomplete dominance was that of two melanic forms of the grey arches moth, *Polia nebulosa* (Hufnagel), that occurred in Delamere Forest, Cheshire, in the early part of the twentieth century. One was an almost black form with white outer margins, called f. *thompsoni* Arkle, while the other, f. *robsoni* was a little less dark and lacked the white outer margins. Extensive breeding data on these forms showed that f. *thompsoni* is homozygous and f. *robsoni* heterozygous for a melanic allele of a single gene (A. Harrison 1908, 1909; A. Harrison and Main 1910; Mansbridge 1909*a*,*b*; Newman 1915, 1916). Subsequently, this melanic allele appears to have become extinct, being replaced by another melanic, f. *bimaculosa* Esper (Kettlewell 1973).

Other single gene examples of melanism involving no dominance or incomplete dominance could be cited from a diverse array of animal taxa. However, this situation of no dominance or incomplete dominance is less common among examples of melanism than the situation where the heterozygote is indistinguishable from one or other of the homozygotes.

The inheritance of the melanic form of the waved umber moth, *Menophra abruptaria* (Thunberg) f. *fuscata* Tutt (Fig. 3.2), is controlled by a single allele. In crosses between known homozygote f. *fuscata* and typical (i.e. non-melanic) individuals, all offspring are *fuscata*. These progeny, despite their melanic appearance, all carry a non-melanic allele from their typical parent. In other words, all are heterozygous for the *fuscata* and typical alleles. When two of these offspring are crossed, their heterozygosity is revealed, for one-quarter of the progeny are typical (Fig. 3.3). This 3:1 ratio of *fuscata* to typical is then a variant of the 1:2: 1 ratio, for a third of the *fuscata* are homozygous, the other two-thirds being het-

Fig. 3.2 Typical (above) and melanic (f. *fuscata*) forms of the waved umber.

erozygous. This situation, when one allele masks the presence of the second allele in the heterozygote, is called genetic dominance. Here the *fuscata* allele is completely dominant to the typical allele. Conversely, we may say that the typical allele is recessive to the *fuscata* allele.

Mendel's laws

These two examples highlight a number of important principles of inheritance. First, when two different alleles occur together in a heterozygote they remain

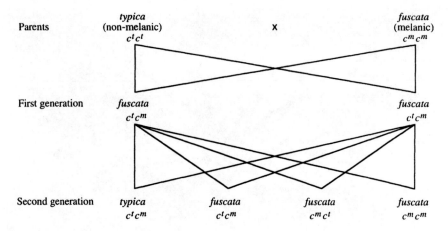

Fig. 3.3 Genetic control of the melanic (f. *fuscata*) and typical forms of the waved umber (c^m and c^t represent the melanic and typical alleles, respectively). The melanic allele is completely dominant to the typical allele.

independent. They do not contaminate one another. Their effects may be additive, so that the heterozygote is intermediate in form between the two homozygotes, or one allele may dominate so that the heterozygote looks the same as one of the homozygotes. But each allele will be passed to the next generation as a separate entity: the genetic material does not merge. This separation of the members of a pair of alleles, when they pass into germ cells during meiosis, is called segregation. The principle of segregation, and the lack of merging of alleles, was proposed by Father Gregor Mendel, Abbot of the Augustinian monastery of St Thomas at Brünn (now Brno), in Moravia, as early as 1862 (although his results were not widely recognised until 1900).

The basic rules of inheritance are generally referred to as Mendel's laws, and the normal inheritance of nuclear genes is generally termed Mendelian inheritance. The law of segregation, Mendel's first law, says that when homozygous individuals, exhibiting a pair of contrasted characters are crossed, the original types separate out in definite proportions in the second filial (F_2) generation, that is among the grandchildren.

Mendel's second law, the law of independent assortment, states that when two or more pairs of contrasted characters are involved in the same cross, they segregate independently of each other. This second law, while holding true for the majority of characters, has been modified for characters that are controlled by genes that are close together on the same chromosome, for some of those positioned on the specialised sex chromosomes, and for the relatively few that are found in the cytoplasm of cells rather than in the nucleus. Each of these categories will be discussed later in this chapter. First, however, we should consider examples in which the law of independent assortment does apply.

Two gene systems

In a cross between a *bimacula* form of the scarlet tiger moth and an individual homozygous for a mutant form named f. *rossica* Kolenati, which has the red of the hindwing replaced by yellow (Plate 7a) and which is inherited as an unifactorial recessive, the first generation were all *medionigra* ($n = 42$). A second generation was produced using three pairs of these progeny. These crosses produced a total of 145 offspring (Table 3.1). Of these, 30 were *dominula*, 53 were *medionigra*, 26 were *bimacula*, 11 were *dominula-rossica*, 17 were *medionigra-rossica*, and eight were *bimacula-rossica*. These numbers approximate to a 3:6:3:1:2:1 ratio. This is the expectation when a 1:2:1 ratio (the expected ratio when crossing two individuals heterozygous for a pair of alleles showing no genetic dominance, i.e. the *bimacula* locus) is superimposed on a 3:1 ratio (the expected ratio when crossing two individuals heterozygous for a pair of alleles of which one is dominant to the other, i.e. the *rossica* locus). If the alleles are notated as follows: *bimacula* = m^b, *dominula* = m^d, red hindwing = h^r, yellow hindwing = h^y, then the genetic composition, or genotype, of the first cross would have been:

$$bimacula \times rossica$$
$$m^b m^b h^r h^r \times m^d m^d h^y h^y$$

The gametes from one parent would all have carried one m^b and one h^r allele, while those from the other parent would each carry one m^d and one h^y allele. The genotype of all first generation offspring must then be $m^b m^d h^r h^y$. That these offspring were all *medionigra* patterned with red hindwings, confirms that m^b and m^d show no dominance and that h^r is dominant to h^y. These double heterozygotes will produce germ cells, be they eggs or sperm, of four types, $m^b h^r$, $m^b h^y$, $m^d h^r$ and $m^d h^y$. Figure 3.4 shows the combinations of genes that will occur in progeny of a cross between two of these double heterozygotes. Twelve of the 16 zygote combinations have the h^r allele occurring at least once, and, as h^r is dominant to h^y, will have red hindwings. Of these 12, three are homozygous for m^b, three homozygous for m^d, and six heterozygous $m^b m^d$. These thus give the 3:6:3 ratio of *bimacula*: *medionigra*: *dominula* among the red hindwinged individuals. The other four

Table 3.1 Details of progeny from three families produced by crossing F₁ progeny of a cross between a *bimacula* female and a *rossica* male of the scarlet tiger moth

| Family | Progeny numbers | | | | | | |
	dominula	*medionigra*	*bimacula*	*dominula-rossica*	*medionigra-rossica*	*bimacula-rossica*	*Totals*
1	15	29	11	5	8	5	73
2	9	14	8	2	4	2	39
3	6	10	7	4	5	1	33
Totals	30	53	26	11	17	8	145

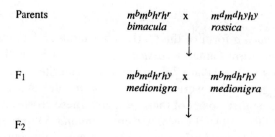

gametes ♂♂ ♀♀	m^dh^r	m^dh^y	m^bh^r	m^bh^y
m^dh^r	$m^dm^dh^rh^r$ dominula	$m^dm^dh^yh^r$ dominula	$m^bm^dh^rh^r$ medionigra	$m^bm^dh^yh^r$ medionigra
m^dh^y	$m^dm^dh^rh^{py}$ dominula	$m^dm^dh^yh^y$ dom/rossica	$m^bm^dh^rh^{py}$ medionigra	$m^bm^dh^yh^y$ med/rossica
m^bh^r	$m^dm^bh^rh^r$ medionigra	$m^dm^bh^yh^r$ medionigra	$m^bm^bh^rh^r$ bimacula	$m^bm^bh^yh^r$ bimacula
m^bh^y	$m^dm^bh^rh^{py}$ medionigra	$m^dm^bh^yh^y$ med/rossica	$m^bm^bh^rh^{py}$ bimacula	$m^bm^bh^yh^y$ bimac/rossica

Fig. 3.4   A 'Punnett square' showing the genotypes and phenotypes expected among the progeny of a cross between two offspring from mating pure-breeding (i.e. homozygous), *bimacula* and *rossica* forms of the scarlet tiger moth. The allelic notation is m^d = *dominula* (dom), m^b = *bimacula* (bim), h^r = red hindwings and h^l = yellow hindwings (i.e. *rossica*), med = the *medionigra* phenotype.

are all homozygous for the h^y allele, and so have the yellow *rossica* hindwing. Among these there is again a 1:2:1 ratio of *bimacula*: *medionigra*: *dominula*.

This small series of crosses then demonstrates both of the laws that Mendel formulated from his work on pea plants, for both original parental forms are present in the grandchildren, and the genes for colour and pattern assort independently of one another.

A range of genetic ratios occur commonly. If both of two pairs of independently assorting alleles showed complete dominance, a 9:3:3:1 ratio results, while if both pairs of alleles showed no dominance, nine phenotypes would result in a 1:2:1:2:4:2:1:2:1 ratio. With three pairs of characters, each pair controlled by two alleles of a different gene, with one dominant to the other, when two triple heterozygotes are crossed, the expected ratio of progeny will be 27:9:9:9:3:3: 3:1. As dominance relationships change, so expected ratios change. Moreover, when the number of loci under consideration increases, the ratios expected become more complex. So, for example, the control of melanic larval colour in the angleshades moth involves three biallelic loci with complete dominance, one biallelic locus with no dominance and one triallelic locus with one allele dominant over two co-recessives. The situation is complicated by a number of interactions between the alleles of different genes in the system. But the expected

ratio of the 12 phenotypes produced by crossing two quintuple heterozygotes is
72:39:36:36:24:13:12:12:6:3:2:1 (Majerus 1978, 1983a,b)!

Multiple alleles

Many genes have more than two different allelic forms. For example, in the pale
brindled beauty, *Apocheima pilosaria* (Denis and Schiffermüller), three forms
occur, the typical non-melanic form, a full melanic form f. *monacharia*, and an
intermediate between the two called f. *pedaria*. Form *monacharia* is inherited uni-
factorially, the *monacharia* allele (M) being dominant over the typical allele (m)
(Ford 1937). Form *pedaria* is controlled by a third allele (m^p) at the same genetic
locus, this being dominant to m, but recessive to M (Lees 1974).

In the moth, *Aglia tau* Linnaeus, melanic polymorphism involves three alleles
of a single gene, these alleles showing incomplete dominance to one another
(Standfuss 1910). This incomplete dominance means that all six of the possible
genotypes are phenotypically different, making analysis of the system simple.
However, this type of situation is rare. When alleles in a multiple allelic series
show full dominance, their analysis must be conducted with care, for it is not
always easy to determine whether the genes controlling different forms are truly
allelic, or occur at different loci.

This difficulty is well illustrated by analysis of the inheritance of forms of the
10 spot ladybird (*Adalia 10-punctata* Linnaeus). The three common forms of this
ladybird are f. *decempunctata* (orange or red with a variable number of small dark
dots, sometimes called typical), f. *decempustulata* (orange or red with a dark grid
pattern, called chequered) (Plate 8d), and a melanic form, f. *bimaculata* (black or
dark brown with a yellow, orange or red flash at the outer front margin of the
elytra) (Plate 8d). Analysis of the inheritance of these forms posed one specific
problem. While it was not difficult to demonstrate that *decempunctata* was dominant to both *decempustulata* and *bimaculata*, and that *decempustulata* was dominant to *bimaculata*, it was not easy to say whether three alleles at a single genetic
locus, or two pairs of alleles at different loci, were involved. Two hypothetical
models were feasible to explain early crosses, these being: (a) a single gene model
with three alleles, one for each form, with the *decempunctata* allele (A^{typ}) dominant to the *decempustulata* allele (A^{che}) which was in turn dominant to the *bimaculata* allele (A^{mel}), or (b) a model with two unlinked genes each having two
alleles, where the alleles of the first gene were typical (T^+) dominant to non-
typical (T^-), and the type of non-typical was determined by the two alleles of the
second gene with the chequered allele (B^{ch}) being dominant to melanic (B^{mel}).

To distinguish between these alternatives, a cross between individuals of specific hypothetical genotypes was necessary. One parent had to be a *decempunctata* from a cross between *bimaculata*, and a *decempunctata* known to be carrying
the *decempustulata* allele. The other parent had to be *decempustulata*, from a
cross between a *decempustulata* and a *bimaculata*.

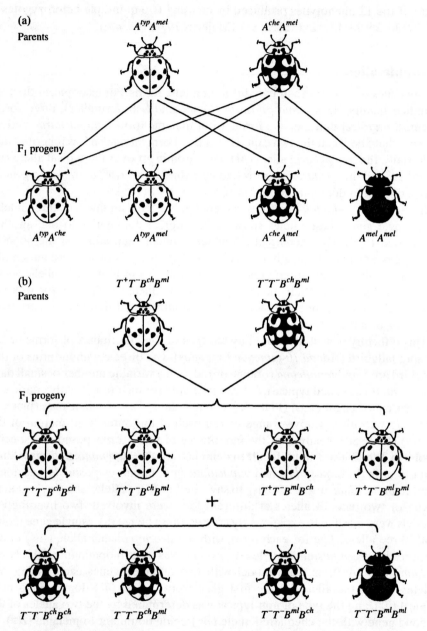

Fig. 3.5 Diagram of expected ratios of the test cross, to distinguish between different models (see text), for the control of melanism in the 10 spot ladybird: (a) a single gene model with three alleles; (b) a two gene model, each having two alleles (see text for allelic notation). (From Majerus 1994a.)

If the one gene model were correct, the genotypes of these parents would then be heterozygous A^{typ}/A^{mel} and heterozygous A^{che}/A^{mel}. Figure 3.5a shows that the expected progeny would be half *decempunctata*, a quarter *decempustulata* and a quarter *bimaculata*.

If the two gene model were correct, the genotypes of the parents of the cross would have to be as shown in Fig. 3.5b, which results in an expected progenic ratio of 4 *decempunctata*: 3 *decempustulata*: 1 *bimaculata*. Results from 13 crosses of this type gave a total of 826 *decempunctata*, 396 *decempustulata*, and 416 *bimaculata*, there being no significant difference in the relative numbers of the forms between families. These data are significantly different from a 4:3:1 ratio, but not from a 2:1:1 ratio. In this the results are consistent with a one gene three allele system.

The cases discussed above have involved three alleles at a single locus. However, many examples are known of multiple allelic series involving many more alleles than this. In *Philaenus spumarius* at least seven alleles of a single locus control a total of eight melanic and five non-melanic forms (Stewart and Lees 1996). Other examples of large multiple allelic series controlling melanism are those of the swallowtail butterfly, *Papilio dardanus*, and the 2 spot ladybird, discussed in Chapters 8 and 9, respectively.

Gene interaction: epistasis and hypostasis

When genes at different loci assort independently, they produce genotypic ratios in accord with Mendelian predictions. How these genotypic ratios convert into phenotypic frequencies depends on two factors: the dominance relationships between the alleles at each locus, and the interactions between the alleles at the different loci. The basic ratio produced by crossing individuals heterozygous for each of two biallelic genes showing full dominance is 9:3:3:1 of the four phenotypes. However, other ratios will occur if the alleles of the two genes interact. One of the most common types of interaction occurs when an allele of one gene blocks or masks the expression of an allele of the other gene. The terms used to describe this masking are epistasis and hypostasis depending on which way round the system is perceived. The blocking allele is said to be epistatic to the blocked allele, while the allele whose expression is prevented is said to be hypostatic to the allele which interferes with its expression. The ratios produced from double heterozygote crosses will depend on the genetic dominance of the epistatic and hypostatic alleles involved. To illustrate this, we can consider two examples.

First, a case in which a dominant allele at one locus masks the expression of a recessive allele at another. Williams (1931, 1933) conducted breeding experiments on two abnormal forms of the willow beauty moth, *Peribatodes rhomboidaria* (Denis and Schiffermüller), one a melanic form, f. *rebeli* Aigner, the other a yellow-grey form, f. *haggardi*. These forms are controlled by alleles at different loci, the *rebeli* allele, *R*, being dominant to its typical allele, *r*, and the

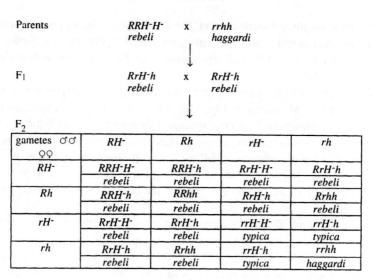

Parents   RRH⁻H⁻   x   rrhh
                       rebeli   haggardi

F₁   RrH⁻h   x   RrH⁻h
                       rebeli   rebeli

F₂

gametes ♂♂ ♀♀	RH⁻	Rh	rH⁻	rh
RH⁻	RRH⁻H⁻	RRH⁻h	RrH⁻H⁻	RrH⁻h
	rebeli	rebeli	rebeli	rebeli
Rh	RRH⁻h	RRhh	RrH⁻h	Rrhh
	rebeli	rebeli	rebeli	rebeli
rH⁻	RrH⁻H⁻	RrH⁻h	rrH⁻H⁻	rrH⁻h
	rebeli	rebeli	typica	typica
rh	RrH⁻h	Rrhh	rrH⁻h	rrhh
	rebeli	rebeli	typica	haggardi

Fig. 3.6 The effect of epistasis in the willow beauty. The expression of the *haggardi* allele, *h*, is prevented if the *rebeli* allele, *R*, is present.

haggardi allele, *h*, being recessive to its typical counterpart, *H⁻*. The F₂ progeny were 134 *rebeli*, 28 or 29 typical and 6 or 7 *haggardi*. This is close to a 12:3:1 ratio which would be expected if the *rebeli* allele masks the expression of the *haggardi* allele as shown in Fig. 3.6.

The second case, involves two indistinguishable melanic forms of the small dusty wave, *Sterrha seriata* (Schrank) (Fig. 3.7), one form controlled by the allele *At* which is dominant to its typical allele, the other being due to the *ni* allele at a different locus, which is recessive to its typical counterpart. Of the progeny from the double heterozygote cross, 285 were melanic and 62 typical. This is very close to a 13:3 ratio, the only genotypes that produce typical phenotypes being those that are homozygous for the typical allele at the *At* locus and carry at least one typical allele at the *ni* locus (Fig. 3.8). In this case, the *At* allele is epistatic to the typical allele at the *ni* locus, while the *ni* allele, when homozygous, is epistatic to the typical allele at the *At* locus.

Other ratios such as 9:6:1, 9:3:4 or 10:6 are produced with different dominance and epistatic relationships between pairs of alleles of two genes. If the number of alleles or the number of gene loci involved is increased, more complex ratios result. In these cases, the problem is generally one of understanding how dominance and epistatic interactions affect the relationship between the genotype and the phenotype. However, the various possible genotypes occur in the progeny in ratios consistent with Mendel's laws.

One other point should be made in respect of gene interactions before moving on. The genetic interactions which affect the relationship between genotype and

Fig. 3.7 Non-melanic (above) and melanic (below) forms of the small dusty wave. There are two genetically distinct melanic forms, which are phenotypically indistinguishable.

phenotype discussed above involve large effects. However, the phenotypic expression of many genes is affected to very small degrees by a variety of other genes in the organism. In other words, the exact expression of a particular gene often depends on the genetic environment in which it is operating. Two examples will suffice.

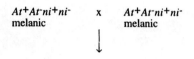

$At^+At^-ni^+ni^-$ x $At^+At^-ni^+ni^-$
melanic   melanic

gametes	At^+ni^+	At^+ni^-	At^-ni^+	At^-ni^-
At^+ni^+	$At^+At^+ni^+ni^+$ melanic	$At^+At^+ni^+ni^-$ melanic	$At^+At^-ni^+ni^+$ melanic	$At^+At^-ni^+ni^-$ melanic
At^+ni^-	$At^+At^+ni^+ni^-$ melanic	$At^+At^+ni^-ni^-$ melanic	$At^+At^-ni^+ni^-$ melanic	$At^+At^-ni^-ni^-$ melanic
At^-ni^+	$At^+At^-ni^+ni^+$ melanic	$At^+At^-ni^+ni^-$ melanic	$At^-At^-ni^+ni^+$ non-melanic	$At^-At^-ni^+ni^-$ non-melanic
At^-ni^-	$At^+At^-ni^+ni^-$ melanic	$At^+At^-ni^-ni^-$ melanic	$At^-At^-ni^+ni^-$ non-melanic	$At^-At^-ni^-ni^-$ melanic

Fig. 3.8 Control of melanism in the small dusty wave. The non-melanic form is only produced when At^- is homozygous, and ni^+ is present.

In Finnish populations of *Philaenus spumarius*, some of the melanic patterns are restricted to females. The reason for this is that the alleles producing these patterns are not expressed in males. In British populations, the same melanic patterns are found, and it is assumed that the same alleles are responsible for their control. However, in these British populations, the strict limitation of these melanic patterns to females is relaxed, leading to a 20-fold increase in the proportion of melanic males. Indeed, in populations from the Cynon Valley in south Wales, where melanics reach their highest frequency known, melanic alleles have the same level of expression in males as in females. Because the melanic alleles in Britain and Finland are thought to be the same, the difference in their expression in the two countries must be a consequence of the different cohorts of genes that make up the genetic environments in which they lie.

As a second example we may consider melanic polymorphism in the 2 spot ladybird, in which the melanic form, f. *quadrimaculata* (Plate 8c), is usually controlled by a single allele that is fully dominant over the typical form. If a male f. *quadrimaculata* from Glasgow is crossed to a typical female from the same population, it produces either only *quadrimaculata* progeny, or a 1:1 ratio of the parental types, depending on whether the *quadrimaculata* male was homozygous or heterozygous for the *quadrimaculata* allele. Crosses between *quadrimaculata* progeny from this cross produce a 3:1 ratio of *quadrimaculata*: typical. However, when the same melanic male is crossed to a Cambridge typical female, the progeny produced show a range of phenotypes varying in their degrees of melanism (Fig. 3.9), and this is also the case if an F_2 generation is produced. The *quadrimaculata* allele which shows full genetic dominance over the typical allele in Glasgow is not expressed in precisely the same way when placed on the Cambridge genetic background. This is because the other genes that have minor effects on the expression of the Glasgow *quadrimaculata* allele, in particular

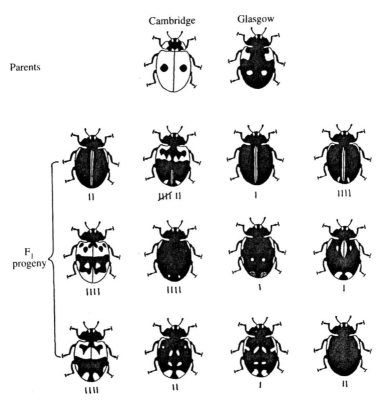

Fig. 3.9 F_1 progeny from a cross between a Cambridge typical and a Glasgow *quadrimaculata*, showing breakdown of dominance of the melanic allele over the typical allele. (From Majerus 1994a.)

those promoting its expression over that of the typical allele in heterozygotes, are not present in the Cambridge population.

Genes of this sort, which affect the expression of other genes in small ways, are called modifiers. Modifiers are important in a range of evolutionary contexts, most relevantly for melanism in their impact on genetic dominance, the penetrance of melanic alleles in the sexes and the evolution of precise mimetic colour patterns. Consequently, they will be discussed further when these topics are considered in detail (pp. 90, 119, 217, 238).

Lethal genes and reduced viability

One group of alleles appear to lead to the production of progeny in numbers at odds with the ratios expected from Mendelian genetics, but only if perceived at the phenotypic level. These are the lethal or partially lethal alleles. When fully lethal and dominant, a cross between two heterozygotes leads to an apparent 2:1 ratio of the dominant and the recessive trait, rather than a 3:1 ratio, because the homozygous dominant class all die. When death occurs late in development, establishment of the lethality of dominant homozygotes may be apparent from the death of one-quarter of progeny. However, if death occurs early in embryo development, say before birth, or before hatching from eggs, it may be far from obvious that a quarter of the zygotes formed have died. Then, the lethality of the homozygous dominant is only revealed when the progeny of F_1 crosses are analysed. If being homozygous for a dominant allele is not fully lethal, any ratio between 2:1 and 3:1 may result from a cross between two heterozygotes.

If a recessive allele is fully lethal when homozygous, its existence is even more difficult to ascertain, because its effects will only be manifest by the mortality of a quarter of progeny in crosses between two heterozygotes, and of course all those that survive will appear normal. This difficulty in ascertainment means that few fully lethal recessive melanic alleles have been revealed, although a few are known. One possible example will suffice to demonstrate that such genes have, until recent advances in molecular genetics, only come to light as a result of exceptional circumstances. Only one example of a melanic grey birch moth, *Aethalura punctulata* (Denis and Schiffermüller), has been recorded, and this was a bilateral mosaic, with one side melanic and the other typical. The mechanism by which this type of oddity arises need not concern us here, but the deduction from this record is that a melanic allele does occur in the species. This allele is recessive and fully lethal when homozygous, except in the unusual case in point when the typical half of the zygote appears to have allowed survival (Murray 1928).

Cases of partial lethality or reduced viability of recessives are far more common and easy to identify, for in these a proportion of homozygotes survive. Again, the progenic ratios produced by matings between two heterozygotes deviate from those expected from Mendel's laws, with the homozygous recessive class being deficient. Examples in the Lepidoptera include the melanic forms of many warningly coloured species of tiger and ermine moths (Arctiidae) and burnet moths (Zygaenidae) (see Chapter 8). Another interesting case is the melanoid allele, *m*, of axolotls, which when homozygous has a tendency to produce scoliosis, or curvature of the spine (Scott 1981).

These instances, in which reduced viability of one particular genotypic class produces progenic ratios which appear on the surface to deviate from Mendel's laws, do not in fact do so, for zygotes are formed in ratios consistent with these laws. However, there are a number of genetic systems which do contravene the rules of inheritance discovered by Mendel.

The breaking of Mendel's laws

Some breeding experiments throw up progenic ratios which do not conform to the expectations based upon Mendel's laws. There are a variety of causes of such deviations.

Genetic linkage

Mendel's law of independent assortment does not hold true for all genes. When two genes are located close together on a chromosome they will pass together into germ cells, and so to the next generation, unless a cross-over event takes place between them during meiosis. The likelihood of a cross-over occurring between two particular gene loci that lie on the same chromosome depends mainly on the distance between them. The closer together they are, the less likely a cross-over will separate them and the more tightly linked they are said to be.

An understanding of genetic linkage is important for two reasons. First, anyone carrying out genetic analysis must be aware of its existence, and have some appreciation of what happens when recombination unlinks the alleles of two previously linked genes. Second, linkage allows alleles of genes which are advantageous when together, but not when associated with other alleles, to be passed on together, so that the advantageous traits are held in concert. Several examples of this type are known in cases where melanism is controlled by so-called supergenes. A supergene is a group of two or more gene loci, controlling different sets of characters which lie so close together on a chromosome that they usually behave as a single switch mechanism in controlling different forms. Many of the various forms of the banded land snails, *Cepaea nemoralis* and *C. hortensis*, and the melanic forms of the 2 spot ladybird are controlled by supergenes. The latter will be discussed further in Chapter 9.

A simple example of the inheritance of two genetically linked melanic traits is seen in the small psocid bug, *Mesopsocus unipunctatus* (Müller), which displays

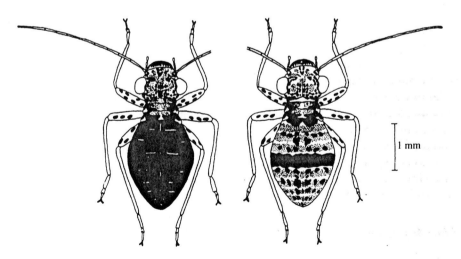

1 mm

Fig. 3.10 Melanic and non-melanic forms of *Mesopsocus unipunctatus*. The markings on the thorax are controlled by the alleles of a gene that is linked to another which controls melanism on the abdomen. (From Popescu *et al.* 1978.)

industrial melanism in Yorkshire (Fig. 3.10). One of the two genes involved affects the colour of the head and thorax, while the other affects the abdomen. Each of the genes has two alleles, one melanic, the other non-melanic, with the melanic alleles dominant over the non-melanic alleles. The two genes are loosely linked (Popescu *et al.* 1978), with the rate of recombination between the two genes being 25 per cent. This means that in a cross between a bug that is homozygous for the non-melanic alleles at both loci and one that is heterozygous for both genes, with, for example, the melanic alleles being on one chromosome, and the non-melanics on the other, then a quarter of the progeny will be the result of cross-overs, with the two melanic alleles becoming disassociated.

A more complex case involves melanism in both the larvae and adults of the oak eggar moth, *Lasiocampa quercus* (Linnaeus). The allele for one of the melanic forms of the larvae, called 'chocolate' is inherited as a unifactorial recessive to that for typical larvae, and is linked to one of the genes that produces a melanic form of adult called f. *olivacea* (Cadbury 1969; Kettlewell *et al.* 1971). However difficulties arise in analysis, because both melanic larvae and adults may be produced by other genes. For example, the recessive allele of a different gene produces both melanic (silky black) larvae, and *olivacea* adults, while other genes which produce melanic adults do not affect larval colour at all (Kettlewell *et al.* 1971; Kettlewell 1973).

Sex determination and sex linkage

I said previously that chromosomes are present in pairs, one of each pair being derived from each parent. The members of a pair are generally very similar in length and structure for all pairs expect one. The exceptional pair are the sex chromosomes. The longer of the sex chromosomes is the X chromosome, the shorter is the Y. In humans, females carry two X chromosomes and no Y chromosome, while males carry one X and one Y chromosome. This is the case in the majority of organisms, although in two major groups, the Lepidoptera and the birds, the male is XX while the female is XY. This reversal is also seen in some fish and amphibians. The sex that carries two similar X chromosomes is referred to as the homogametic sex, as all the gametes it produces carry an X chromosome. Conversely, the sex with one X and one Y chromosome is called the heterogametic sex because it produces gametes of two different types, either carrying an X or a Y chromosome. In species in which males are the homogametic sex and females are heterogametic, the sex chromosomes are sometimes given different letters for distinction, the X being notated by Z and the Y by W. However, for the purposes of this book, it is perhaps less confusing simply to use X and Y throughout.

It is the sex chromosomes that carry the genes that determine sex. In most insects, sex determination does not depend on the presence or absence of the Y chromosome. The Y chromosome generally carries few genes, and from some insect species it is completely absent. Species which lack the Y chromosome are said to be X0. In such cases it is difficult to envisage how sex determination could depend on the Y chromosome. Rather, in most insects, sex determination depends on the ratio of the number of X chromosomes to the number of sets of autosomes (A). In fruit flies (*Drosophila* spp.), an investigation of the sexual traits of flies with a variety of chromosomal arrangements, including flies with abnormal chromosome complements, has shown that if this ratio is 2X:2A (i.e. two X chromosomes and two sets of autosomes) the result is a female. If the ratio is 1X:2A the result is a male. Abnormalities with three X chromosomes and the normal two sets of autosomes (i.e. a 3X:2A ratio) are referred to as super-females, for they have all the female traits, but in extra measure; a fruit-fly equivalent of Pamela Anderson! Conversely, individuals with either one or two X chromosomes and three sets of autosomes, giving ratios of 1X:3A and 2X:3A, are super-males, drosophilid Schwarzeneggers perhaps, and intersexes, respectively. The basis of this system is a balance between female determining genes on the X chromosomes, and male determining genes on the autosomes. When the number of X chromosomes equals (or exceeds) the number of sets of autosomes, the female determining genes are expressed and the male determining genes are suppressed. When the number of X chromosomes is equal to (or less than) half the number of sets of autosomes, the autosomal male genes are expressed.

The X chromosome/autosome balance mode of sex determination, found in insects and many other groups of organisms, is not relevant to mammals including ourselves, where genes on the Y chromosome play a critical part in determining maleness.

The rules of inheritance for genes carried on sex chromosomes will obviously be different from those for genes on autosomes. A large portion, usually the majority, of the X chromosome has no pairing equivalent on the Y chromosome. Consequently, any gene that lies on this non-pairing part of the X chromosome can only have one allele present in the heterogametic sex. Such genes are said to be sex linked.

To illustrate the difference in the inheritance of a sex linked and an autosomal character we can consider some of the many melanic forms of the black arches moth, *Lymantria monacha* Linnaeus. Because the phenotypic expression of melanic alleles varies somewhat between males and females in this species, and because the forms produced by different melanic alleles, either individually or in combination, are often not phenotypically distinguishable, the names used for the forms by lepidopterists would be confusing in this illustration. Consequently, I will consider just two of the gene loci controlling melanism in this species (the B and C loci of Goldschmidt (1921)), and simply use the phenotypic classes melanic, intermediate, and typical, ignoring the complexity of the full range of melanism in this species.

Considering first the B locus. This is autosomal, and exhibits incomplete dominance. A series of crosses initially involving a cross between a homozygous

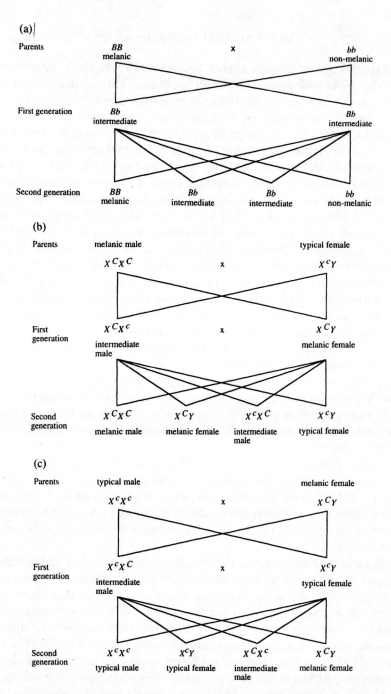

Fig. 3.11 The inheritance of melanism in the black arches moth, showing the difference in patterns of inheritance between autosomal and sex-linked traits: (a) inheritance due to an autosomal locus, B = melanic allele, b = non-melanic allele; (b) and (c) the alternate patterns of inheritance of a sex-linked trait depend on which sex shows which phenotype in the parental generation, where X^c = melanic allele, X^c = non-melanic allele. (Note: in Lepidoptera, females are the heterogametic sex.)

melanic (BB) and a homozygous non-melanic (bb), and subsequent crosses between the progeny, gives normal Mendelian ratios: i.e. all intermediate progeny in the F_1, and a $1:2:1$ ratio of melanic: intermediate: typical in the F_2. Crucially for this illustration, the sex of the original parents does not matter, and the ratios of the forms in the F_2 are the same in males and females (Fig. 3.11a).

The C locus behaves in a very different way, for it lies on the non-pairing portion of the X chromosome. If we consider a similar series of crosses to those described for the B locus, it is essential to define the sex of each parent. Take first the case in which a male homozygous for the melanic C allele is crossed to a typical female. The male progeny from such a cross are all intermediate, reflecting their heterozygosity. However, the female progeny are all melanic. This is because they carry just one allele of the gene on their single X chromosome, and they always receive this from their father, never from their mother (Fig. 3.11b). When the progeny of this cross are themselves crossed, the male progeny show a $1:1$ ratio of melanics: intermediates, all receiving a C allele from their mother, while half receive a C allele and half a c allele from their father. The female progeny will also show a $1:1$ ratio, but this time of melanics:typicals because they will either receive a C or a c allele from their father, and they receive no allele of this locus from their mother. This illustrates that females cannot be intermediate as they carry only one copy of the gene, and so can never be heterozygous for this locus. If the initial cross is performed the other way round, i.e. a homozygous typical male × a melanic female, the F_1 males will again all be intermediate, while the F_1 females will all be typical. In the F_2, males will show a $1:1$ ratio of intermediates to typicals, while females again will be half melanic and half typical (Fig. 3.11c).

A similar system is seen in domestic cats where an X-linked gene with two alleles, one producing black, the other orange, is responsible for the production of the tortoiseshell coat. However, here females are homogametic and males heterogametic. Tortoiseshells which are heterozygous for the black and orange alleles are always female because males only have one copy of this gene.

In the above cases, the relationship between the sex-linked alleles under consideration was one of incomplete or no dominance. When one allele of a sex-linked gene is fully dominant to the other, the result is that the recessive character is expressed phenotypically much more commonly in the heterogametic sex than in the homogametic sex. This is because, in the heterogametic sex, the recessive allele can never have its expression masked by the dominant allele, as happens in heterozygotes in the homogametic sex.

While the Y chromosome (when present) is known to lack counterparts of most of the genes on the X chromosome, in some organisms it does carry some genes. Such genes may be unique to the Y chromosome, in which case they will always and only be transmitted down the heterogametic sex line. Alternatively, genes on the Y chromosome may be on the short region that pairs with the X chromosome, so that they have a counterpart on the X. Such genes will give patterns of inheritance similar to those of autosomal genes. Y-linked genes of both types are rather rare, but a few are known. As an example of a trait being controlled by a gene on the non-pairing portion of the Y chromosome, we may consider a melanic form of the pale brindled beauty moth. This is produced by a gene that is transmitted from mother to all daughters, but not to any sons, which are all normal. This pattern is repeated in all following generations, the males having no effect on the inheritance of the trait. (Note, another much commoner autosomal melanic locus also exists in this species.)

Other examples of Y-linked melanic genes are seen in the tiger swallowtail butterfly, *Papilio glaucus* Linnaeus, and in the eastern mosquitofish. In the former, melanic, as opposed to yellow, females give rise to melanic daughters and yellow sons. In the eastern mosquitofish, a Y-linked allele produces black spotting. However, this case differs slightly, because the penetrance of the allele (i.e. the proportion of individuals in which it is expressed when present) is variable, depending on temperature. Angus (1989) demonstrated that in fish raised at 22°C, penetrance was almost complete. However, in fish raised at between 26°C and 29°C, the penetrance was reduced to 42 per cent.

Sex-limited inheritance

A second sex-related phenomenon warrants mention. This is sex-limited or sex-controlled inheritance, in which genes carried on autosomes are expressed in one sex only. Here the expression, or lack of expression, of a gene is dependent on its environment, in the form of the genetic setting provided by each of the two sexes. Sex-limited genes occur in both sexes, and are transmitted equally by both. It is just their effects which are confined to one sex. This contrasts with X-linked traits which appear in both sexes, although often more commonly in one sex than the other, but are not transmitted equally by the two sexes.

An example is provided by the silver-washed fritillary, *Argynnis paphia* Linnaeus. In this species, males are all similar in respect of their upperside ground colour which is orange-brown. Typical females resemble males but are somewhat duller in colour. However, a second type of female, f. *valesina*, occurs in many populations (Plate 5a). This form is controlled by a single dominant allele of an autosomal gene whose expression is suppressed in males. That the gene responsible is sex limited, rather than on the non-pairing portion of the Y chromosome, is most easily demonstrated by collecting normal females from a population which contains *valesina* at reasonable frequencies, and rearing progeny from them. Some of these females should produce *valesina* offspring because they will have mated with males carrying the *valesina* allele. In the New Forest, between 1972 and 1989, the frequency of *valesina* females was 10.4 per cent (*n* = 396). Broods from 28 wild-mated phenotypically, normal females were reared at various times during this period. Of these, three produced roughly half normal and half f. *valesina* female offspring (Majerus, unpublished data). The rest produced all normal offspring. These data demonstrate that three of the females had previously mated with males that were heterozygous for the *valesina* allele.

Cytoplasmic inheritance

One final group of genes that break Mendel's laws warrant a brief mention. These are those genes that occur not in the nucleus, but in the cytoplasm, either located on the limited amount of DNA in mitochondria, or in micro-organisms that live inside some cells, and are inherited. In most animals, such genes are inherited only from the mother. This is because the sperm of most species is essentially just a nucleus with a tail; it contains virtually no cytoplasm, unlike the large female gametes. The result is that traits coded for by genes in the cytoplasm are passed from mothers to all progeny, but it is only their daughters that pass the trait further. This unusual mode of inheritance is of considerable use in studying evolution, for the inheritance of these genes is unisexual, from mother to daughter, to daughter, to daughter, etc. There is no mixing resulting from recombination or sex.

I know of no example of a melanic trait that is cytoplasmically inherited, although the male killing bacteria of some ladybirds, which may have an influence on melanic polymorphism (p. 258), are inherited in this way (Hurst *et al.* 1992). If a cytoplasmically inherited melanic gene could be found, this would open a considerable range of possibilities for analysis because of the relative simplicity of cytoplasmic genomes.

Continuous variation and polygenic inheritance

Individual variation is a conspicuous phenomenon whenever organisms of the same species are carefully examined. We find little difficulty in recognising different people, however, many people we know, through facial features, skin pigmentation, hair colour, shape and style, body configuration, height, weight, and so on. We notice differences between ourselves more than variation in other organisms. But morphological variations may be detected, almost invariably, when individuals of one species are carefully scrutinised. Much of this variation involves small subtle differences in characteristics, rather than the somewhat coarse and obvious differences I have discussed so far. When the individuals of a population show small variations in a character, the trait is said to vary continuously. As I said at the beginning of this chapter, variation may have a genetic basis, or it may be the product of differences in the environment to which individuals have been exposed, or it may be due partly to both. With continuously varying characters, it is not always easy to discern whether a particular type of variation has a genetical, or an environmental basis. This, of course, is the crux of the 'nature versus nurture' controversy, in respect of some human traits, such as intelligence, or aggressive behaviour.

Variation is considered continuous when a range of forms exist, merging, more or less, one into another, with a full range of intermediates between the extremes. For example, the twin-spotted wainscot moth, *Archanara geminipuncta* (Haworth), exhibits a range of forms from pale brown to almost black (Fig. 3.12). The question is then, is the observed variation in colour due to environmental factors, or to 'polygenes', where each of a large number of genes have a small effect on the ground colour of individuals, or to some combination of both? What we are really asking is, how much of the observed variation is attributable to genetic, and how much to environmental factors?

The heritability of a trait is defined as the proportion of total variation that is under genetic control. Perhaps the simplest way of estimating heritability is to conduct what is termed a mass selection experiment. The procedure involves assessing the darkness of a random sample of moths against some colour standard (such as the *Munsell Book of Colour* 1966), to assess the total variation present in the population. From this sample, individuals of more or less the same colour, from near one of the extremities of the distribution, i.e. either relatively light, or relatively dark, are selected. These are allowed to breed among them-

 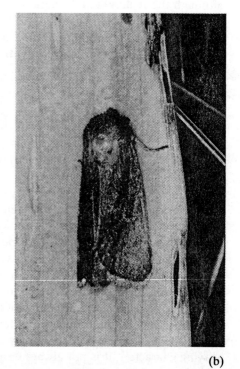

(a)　　　　　　　　　　　　　　　　　　　　　　　　　　　　　　　　　(b)

Fig. 3.12　The twin-spotted wainscot shows continuous variation from very pale (left) to deep brown (right). It is not known whether this variation is controlled genetically or environmentally.

selves and produce progeny. All the progeny are then assessed against the standard. Once the relevant data (the average darkness of moths in the whole original sample C(Tot.), the average darkness in the selected breeding sample C(Par.), and the average darkness of their progeny C(Prog.)) has been collected, an estimate of heritability is obtained by a simple calculation. Theoretically, we are concerned with two measures, the 'selection differential', which is simply the difference between C(Par.) and C(Tot.), and the 'selection gain', which is the difference between C(Prog.) and again C(Tot.). It is intuitively obvious that if the average colour of progeny is no different from that of the original sample (i.e. the selection gain is zero), despite all the parents of these progeny being either very light, or alternatively very dark, the variation must be environmental, for selecting parents from a particular part of the distribution has no effect on offspring tone. Similarly, if the average of progeny is equivalent to the average of the selected parents, i.e. the selection differential and selection gain are the same, the variation must all be genetic. Such gains, either negligible, or conversely equivalent to the selection differential, are rare. In most cases, some selection gain will be observed, but this value will be less than the selection differential. The heritability is then obtained simply by dividing the selection gain by the selection differential. In terms of the averages above, the calculation is:

$$\text{Heritability} = C(\text{Tot.}) - C(\text{Prog.})/C(\text{Tot.}) - C(\text{Par.})$$

Melanism has been investigated in this way on rather few occasions considering the number of cases in which the darkness of the colours of organisms vary continuously. However, one such study, by Bengt Gunnarsson (1987), showed that the heritability of darkness in the spider, *Pityohyphantes phrygianus*, from a Swedish population was 0.43. In other words, almost half the variation in the darkness of the spiders was due to genetic factors.

Similar studies on species such as the November moth, *Oporinia dilutata* (Denis and Schiffermüller), or the winter moth, *Operophtera brumata* (Linnaeus), which show continuous variation in tone over much of their range, would certainly be valuable.

Behavioural genetics

The traits I have been discussing so far have been morphological, with the phenotypic traits involved usually being fairly easy to score and relate to genotype. However, the study of melanism may also involve considering behavioural traits associated with melanism. For example, in many Lepidoptera, melanism appears to be associated with behavioural preferences to rest on appropriate backgrounds, or actively select certain types of habitat. Similarly, in scarlet tiger moths and some ladybirds, mating preferences are implicated in the maintenance of melanic polymorphisms. To understand the evolution of such behavioural traits and the melanisms they affect, the genetic systems underlying these behaviours need to be understood. Yet the study of the genetics of behaviour is usually complex and time consuming. The problem is that most behaviours are not absolute, so assigning a phenotype to an individual is often difficult. Let me give an example. Rory Howlett (1989) attempted to discover whether there was a genetic basis to the background choices made by melanic and non-melanic peppered moths. The problems were, first, that peppered moths only live for a few days, so only chose resting backgrounds at dawn five or six times, and second that the trait under consideration involved preference rather than absolute choice. This meant that unless a moth always consistently rested on the same type of background, it was not easy to determine whether a moth was showing preference or choosing its background randomly.

Howlett resolved this problem to some extent by inventing his 'dawn box' (p. 143), which allowed four dawns to be simulated each 24 hours, and by analysing choices made by all the progeny in a family as a single unit. He thus showed that there was a genetic basis to the resting site preferences shown (Grant and Howlett 1988; Howlett 1989). However, the exact nature of the genetic system was not determined, despite the enormous amounts of work involved.

A similar approach was used in analysing the genetics of the mating preference some female 2 spot ladybirds show for melanic males. In this case, the analysis involved looking at the mating choices made by female progeny of just 21 families. As the underlying genetic system proved to be uncomplicated, the preference is controlled by a single dominant allele (O'Donald and Majerus 1985; Majerus *et al.* 1986), this apparently simple regime was sufficient. Yet the experiment still involved over 10000 ladybirds and 4000 observed matings.

The interactions between an organism's morphology and its behaviour are crucial to its fitness. If we are to understand the complex arrays of interactions between melanic forms of species and their environments, we have to understand how they behave, and how they have evolved to behave in a particular way. As yet, only a handful of studies have attempted to discover the genetics of behavioural systems that impinge on the evolution of melanism, largely because of the difficulties inherent in such studies. However, it is to be hoped that advances in molecular genetics will allow genetic markers of behavioural traits to be identified. Markers of this type would certainly make the analysis of behaviour easier, for it would at least be possible to know whether an individual had the trait or not.

Environmental variation

Finally, in this chapter on the principles of genetics as they concern melanism, a few words on the influence of the environment are necessary. Although a great many melanic forms of species are under predominantly genetic control, there are many in which the external environment, particularly in the form of either temperature or surrounding colour, has a critical role to play. In some, environmental factors appear to have the major role. In others, such features play a subordinate part, simply modifying the phenotypic expression of particular alleles to a small extent.

It has been known that environmental factors influence melanin laydown in the Lepidoptera for over a hundred years. Professor E.B. Poulton (1885, 1886, 1887, 1892, 1893) examined the possible influence of surroundings upon larval coloration in many species of Lepidoptera. Most of the species he used had arbivorous larvae, which were found to be very sensitive to the background colours of their immediate surroundings. Larvae of the brimstone moth, *Opisthograptis luteolata* (Linnaeus), and the peppered moth produced the most striking results. In the main, larvae of these species reared just on green leaves were of a green colour, while those reared with brown twigs became brown. Larvae of several other species which have brown larvae, were also found to be susceptible, the shade of brown depending on the darkness of the twigs which the larvae rested upon. These species included: the scalloped oak, *Crocallis elinguaria* (Linnaeus); the August thorn, *Ennomos quercinaria* (Hufnagel); the lunar thorn, *Selenia tetralunaria* (Hufnagel); and the silver ground carpet, *Xanthorhoe montanata* (Denis and Schiffermüller); as well as several species of *Catocala*. Poulton also produced lichen marked larvae of the scalloped hazel, *Odontoptera bidentata* (Clerck), and the lappet, *Gastropacha quercifolia* (Linnaeus), by rearing larvae in cages with bits of white paper mixed among their food.

Cockayne (1928b), notes that these experiments are borne out by field observations of larvae of the green brindled crescent, *Allophyes oxyacanthae* (Linnaeus), which often have a mottled pattern when feeding on blackthorn in the New Forest. Cockayne found that the proportion of lichen patterned larvae was closely correlated to the profuseness of lichens growing on the blackthorn. On the other hand, in London and other polluted regions where lichens were rare or absent, Cockayne failed to find any lichen marked larvae of this species. Similar field observations of brimstone and scalloped hazel larvae in Scotland, also corroborate Poulton's findings (Cockayne 1928b).

Poulton found that larvae were most sensitive to the colours of their background in the third, and to a lesser extent the second instars. The effect was cumulative: the longer a larva was exposed to particular surroundings, the greater the effect produced. More recently, experiments conducted in the same way on larvae of the scalloped hazel from several localities in England, have shown that the potential to produce lichen marked larvae in this species has a genetic basis. Some families of larvae have this potential, others do not (Majerus 1983c).

A similar effect of the environment is seen in the pupae of many Lepidoptera. In the European swallowtail butterfly, *Papilio machaon* Linnaeus, pupae are almost invariably either green or brown. Pupae that are destined to hatch in the year that they are formed are usually green, while those that overwinter are mainly brown. The same type of situation is seen in the small white butterfly,

Pieris rapae Linnaeus, and in many tropical swallowtail species, where green pupae are commoner in the wet season and brown in the dry season. The reasons for the differences are easy to understand and may be attributed to crypsis. However, the mechanism by which some larvae become green while others become brown is more difficult. Analysis of many families has shown that it is not possible to predict the colour of progenic pupae from that of their parents. Clearly, the production of green and brown forms is not under direct genetic control. The variation of pupal colours produced under a wide array of environmental conditions has been studied in many species of butterfly and some moths. Although no two species appear to be exactly alike, substrate colour, colour of the surroundings, substrate texture, day length, temperature, and humidity have all been shown to influence the colours of pupae in some species. For example, in a study of melanisation in the pupae of 13 species of nymphalid butterflies, Koch and Bueckmann (1982–83) found that in nine the level of melanin deposition depended on the background colour at pupation sites.

Although most melanic forms of moth are thought to have a genetic basis, in a few cases exposure of larvae or pupae to abnormal temperatures can affect the degree of melanisation of the adults. In the map butterfly, *Araschnia levana* Linnaeus, temperature triggers the switch between the predominantly orange spring brood, and the melanic summer brood. In many other butterflies, exposure of pupae, or occasionally larvae, to cold temperatures produces exceptionally dark imagines. This may be an adaptive response for the dark forms produced may have a thermal advantage in cool climes. The same response has been observed in some moth species. For example, Kettlewell (1943–44) showed that prolonging the pupal period of the bordered straw, *Heliothis peltigera* (Denis and Schiffermüller), by keeping pupae under cool temperatures, caused a general darkening of the resultant adults. In the horse chestnut moth, *Pachycnemia hippocastanaria* (Hübner), a phenotype indistinguishable from the genetically controlled melanic form *nigrescens* can be produced by rearing larvae with a non-melanic genotype at low temperature. Rearing individuals with the *nigrescens* genotype at low temperatures does not affect the colour of the imagines produced (Majerus 1981).

In cases of this type, melanism is a product of an interaction between genetic factors and the environment. Some or all members of a population will carry the necessary genes to produce melanism, but these genes are only expressed if triggered by some environmental cue.

Opportunities and the need for original research

That the laws of genetics are more or less the same in all living things, with the exception of some micro-organisms, means that any organism may be used for genetic investigations. Lepidoptera with melanic forms have had a place of prominence in genetic research throughout the twentieth century for a number of reasons. First, most species are easy to breed and rear. Second, they produce large numbers of progeny. Third, some species will produce several generations per year. And fourth, the melanic varieties of many species are obvious, so that scoring phenotypes of progeny is easy. Although melanic forms of many species of Lepidoptera, and of some other animals, have been subjected to genetic analysis, those of many more have not been investigated in this way. Even in Britain, where melanism in moths has received more attention than anywhere else, the hereditary control of melanic variation in the majority of species is still unknown. This variation provides any careful observer, who is prepared to apply logically the simple principles outlined in this chapter, the opportunity to make original and valuable genetic discoveries. Indeed such work would be timely, for, as we shall see, in the cases in which melanic forms occur as a result of changes in the environment brought about by industrialisation and pollution, the melanics are declining, and there are predictions that some of these forms will become extinct within the next 50 years or less.

4 Evolutionary processes

Introduction

Our Earth is 4.5 billion years old. Latest evidence suggests that life developed on Earth at least 3.8 billion years ago. By the passage of hereditary material, from parent to progeny, generation upon generation, and as a result of changes in and mixing of this material, there are now thought to be over 30 000 000 species sharing the planet with us. They range from single-celled microbes to blue whales and giant redwoods. The cumulative change in the hereditary material of organisms, compared with their antecedents, is called biological evolution. At the phenotypic level, such changes range from the invisible, involving minute changes in non-coding hereditary material which has no obvious discernible effect on the organism, through the apparently insignificant, such as small changes in the precise pattern of black flecking on the wings of a typical peppered moth, to more overt changes, such as those that produce the distinct melanic forms of many organisms. Evolutionary change may result from a number of different processes: natural, sexual, and artificial selection, a variety of biases in the behaviour of the hereditary molecules DNA and RNA, and chance. In considering the reasons underlying the evolution of melanism in animals, an understanding of the principles of selection, be it natural or sexual selection, and chance, in the form of random genetic drift, is essential, for these processes determine changes in the frequencies of the genes responsible for melanism in populations. Whether biases in the behaviour of the genetic molecules influence the evolution of melanism to a significant extent has been little considered to date, although there are some systems in which molecular genetic biases may have played a part (e.g. the evolution of the colour pattern supergene in the 2 spot ladybird, p. 237).

Two mechanisms of evolution

To illustrate the essential difference between random genetic drift and selection, consider a clutch of moth eggs on the leaves of an oak tree, somewhere in industrial northern England. The tree is cut down for timber, and the branches, leaves, and the clutch of eggs are incinerated. The death of this clutch is independent of the genetic constitution of the embryos in the eggs. The death of the eggs would have occurred whatever the genes carried by the embryos. In respect of the genotype of the eggs, this fatal event was simply due to chance.

In a second clutch, on another tree that does not fall victim to the forester's saw, the eggs hatch, and, in the fullness of time, some survive to emerge as adult moths. Some of these are black forms, while others are lichen patterned, with the difference between the two types being genetically controlled. One of the lichen patterned moths resting during the daylight hours on the branch of a tree is spotted by a bird and is eaten. The reason that the bird saw this moth is that it was resting on a branch denuded of lichens and blackened by soot, and that on this branch the freckled patterning on the moth's wings was not an effective camouflage. On another branch of the same tree, a black moth from the same clutch was not seen by the bird.

In the first scenario, a chance event causes the destruction of a group of individuals and the genes that they carry. As the event is independent of the genes carried by these individuals, any change in the frequency of alleles in the population resulting from this event will be directionally random. The process is thus called random genetic drift. This contrasts with the second case, where the genes carried by individuals strongly affect the likelihood of their survival. Here for instance, individuals carrying and expressing a melanic gene are more likely to avoid predation, as a result of improved crypsis, in a polluted habitat. Thus the frequency of the black allele is more likely to increase than decrease. The change is not random. Rather it has a direction that is a product of the process identified by Charles Darwin in 1859, which he named natural selection. Together the processes of drift and selection combine to mould the genetic constitution of future generations and, ultimately, species.

Genetic drift

Random genetic drift affects all genes, whatever their effect on their bearer. Conversely, natural selection affects only a proportion of genes: those that have a significant effect on the phenotype of their carriers.

The process of random genetic drift, acting on a polymorphic gene, that is to say a gene with two or more different allelomorphs, is analogous to someone drawing coloured balls out of bag, without looking. For a gene having two alleles, imagine a large bag containing equal numbers of white and black balls. Drawing 10 balls, we might find seven white and three black. This may not be the most likely combination, but it is certainly a plausible one. Now let these balls determine the proportions of a new large bag population, containing 70 per cent white and 30 per cent black. Repeating this procedure again and again, each time drawing 10 balls to start a new large population, it is easy to envisage the proportion of white and black balls oscillating up and down until on one occasion all 10 balls drawn are of one colour.

Random genetic drift works in exactly the same way. Some genes make no contribution to the next generation because, by chance, their carriers fail to survive, or fail to mate, or they are carried in gametes that are unsuccessful during fertilisation. At each genetic locus, alleles which happen to be carried more frequently by successful gametes will increase in number, while those that are poorly represented will suffer a decline. As a consequence of this random sampling of genes each generation, alleles will drift up and down in frequency until, eventually, one is lost and the other becomes fixed. A gene, previously polymorphic for two alleles, would then be monomorphic. In theory, if random genetic drift was the only process affecting genes, variants would be lost one after the other until, given enough time, all members of the population would become, genetically, the same.

Despite about 3.8 billion years of evolution, this obviously has not happened. While drift gradually eradicates genetic variation, a second process generates new variation. That process is mutation, the ultimate source of all genetic variation (p. 42).

The dynamics of genetic drift are fairly simple, being determined by three main factors: the size of the population concerned, the substructure of the population, and the mutation rate supplying the system with new variations. For example, it is easy to see that drift will have a proportionately greater effect in small populations than in large ones, by returning to the bag of equal numbers of white and black balls, and imagining taking out 10000 balls rather than 10 for each generation. While seven white/three black is not unlikely, 7000 white to 3000 black seems highly improbable.

The relationship between the proportional magnitude of gene frequency changes due to random genetic drift and population size is crucial. Random genetic drift is far more important in small populations than in large ones. For most of the genes that we are considering in this book which have a significant phenotypic expression, it is probably reasonable to say that in populations of 1000 or more, the changes in allelic frequency produced by random genetic drift will be negligible. This statement needs some corollaries. First, the population size refers to the number of individuals that reproduce, which may be only a small proportion of the number that survive to reproductive maturity. Second, the population size referred to is the minimum population size, it is not the mean population size over a period. Third, the population in question must mix freely for the purposes of mating. Substructuring of a population may reduce an apparently large population to a series of sub-populations of small size. The effect of random genetic drift on the population as a whole will then be a combination of the size of each sub-population and the amount of movement of individuals between the sub-populations (p. 74).

The Hardy–Weinberg law

A concept central to the understanding of the behaviour of genes in populations was formulated independently by the biologists Hardy and Weinberg in 1908.

This concept, which has become known as the Hardy–Weinberg law, states that given that a number of assumptions apply, then, after a single generation of mating, the frequencies of each homozygote genotype of a particular gene will be the square of the frequency of the allele involved while that of a heterozygote genotype will be twice the product of its component alleles. Thereafter, the frequencies of alleles and genotypes will be stable. Seven assumptions underlie this law:

(a) the organisms in question must be diploid;

(b) they must reproduce sexually;

(c) they must be in a population of infinite size;

(d) they must mate randomly;

(e) there must be no mutation;

(f) there must be no migration into or out of the population;

(g) there must be no selection.

Obviously, no such population of organisms can exist. Yet the Hardy–Weinberg law is the mathematical bed-rock of population genetics, the study of genes in populations. The reason for its importance is that it allows the theoretical frequencies of genotypes to be estimated for a large population in which nothing of biological import is happening. These genotype frequencies can be compared with data from real populations. Any differences between the frequencies expected as a consequence of the Hardy–Weinberg law and the observed frequencies implies that one or more of the Hardy–Weinberg assumptions is being contravened.

When observed genotype frequencies agree with Hardy–Weinberg expectations, the population is said to be in Hardy–Weinberg equilibrium. This may mean that the assumptions of the Hardy–Weinberg law are effectively being met. However, this is not always the case, for, as we shall see, in some instances the frequencies of genotypes may be in equilibrium not because nothing is happening, but because several opposing things are happening, which together hold the population in equilibrium, as in the case of balanced polymorphisms (p. 83).

In practice, investigating deviations from Hardy–Weinberg expectation involves asking a series of questions relating to the Hardy–Weinberg assumptions. It is usually fairly easy to ascertain whether an organism is diploid and if it reproduces sexually. In practice, mutation is such a rare event that it can be ignored in most studies. Migration is a different matter. All organisms disperse to some extent, and measuring migration rates into and out of a large population is not a trivial matter. However, here we need only be concerned with migration that is not random with respect to the genotypes under consideration. This can often be tested by multiple mark–release–recapture experiments, using grids of traps or collection sites, over a period of time.

The size of a population can also be estimated using multiple mark–release–recapture techniques. However, this assumes that one has some idea of the limits of the population under consideration. This may be possible for a colonial species living in an isolated wood, but defining the limits of a population of moths with a wide and continuous distribution is much more problematic. In deciding whether a population is large enough for the effects of random genetic drift to be negligible, the difficulty in defining the limits of the population under consideration may not matter. It may be sufficient to say that the population size is simply large. However, there is a second consideration, which relates to the amount of mixing in the population, and whether mating is random.

As an illustration, consider the brindled beauty moths that live in and around my garden and the three adjacent to it. These four gardens together comprise about three acres of mixed deciduous woodland surrounded for 2 km in each direction by agricultural land with no suitable habitat for this species to breed. I know that occasionally immigrant individuals reach these gardens from Madingley Wood some 2 km away. However, I suspect such migrants are relatively rare. While the population of brindled beauty moths around my garden is undoubtedly part of a much larger population, its partial isolation means that it is almost certainly not mating randomly, for individuals in this small woodland are far more likely to mate with one another than with moths that emerge anywhere else. This leads to the likelihood of some degree of inbreeding within the population.

As a general rule, inbreeding increases the number of homozygotes in the population. This is easy to envisage by considering the extreme case of self-fertilisation. Consider an animal that is heterozygous for a gene carrying two alleles, A and a. The F_1 progeny will be in the Mendelian ratio of $0.25AA:0.5Aa$ $:0.25aa$, or half homozygotes and half heterozygotes. In a self-fertilising organism, homozygotes can only breed true and hence can be thought of as being fixed. Thus, there will be a steady loss of heterozygotes. The basic outcome is that with full selfing the proportion of heterozygotes declines by a half each generation.

In cases where the degree of inbreeding is less extreme, the reduction in heterozygotes is less each generation, but it is still in the same direction. In most animals the sexes are distinct, and selfing is not possible. The closest matings are either between full siblings or between parent and offspring. In both these cases only half the genetic material of the mating pair is shared. Consequently, the decrease in heterozygotes resulting is half of that expected in the case of self-fertilisation, i.e. a quarter per generation. Matings between more distant relatives produce a smaller decline, so from half-sibs, which share a quarter of their genes, the decline in heterozygosity is an eighth, and so on.

In fact, for the population geneticist, the term inbreeding may involve matings between individuals that are not known to be related at all, for it covers just the type of situation I described with the brindled beauties in my garden. The problem is that not only are real populations not infinite, but any given individual is only ever likely to encounter a small number of all possible partners, because most natural populations have some degree of substructure. The result is that the majority of brindled beauties in my garden do share at least some of their genes, as a result of common descent, because they are at least distantly related. Again the result is a decrease in heterozygosity compared with Hardy–Weinberg expectation. The deficiency of heterozygotes as a result of population subdivision is known as the Wahlund effect.

Inbreeding, whether as a result of mating between close relatives or as a result of population subdivision, is only one contravention of the random mating assumption of the Hardy–Weinberg law. There are a number of other types of non-random mating. For example, some female ladybirds prefer to mate with melanic males rather than non-melanics, regardless of whether they themselves are melanic or non-melanic (Majerus *et al.* 1982*a,b*; Osawa and Nishida 1992). In the case of scarlet tiger moths from one colony in Berkshire, there is a preference for moths of particular genotypes with respect to the *bimacula* locus, to mate with partners that have a genotype different from their own. This is the reverse of inbreeding, which is essentially non-random because the partners are more alike than two randomly chosen individuals would be, for here the partners are specifically different. Not surprisingly, the result of this so-called disassortative mating between 'un-alikes' is the opposite of inbreeding, i.e. an increase in the proportion of heterozygotes. Many of the different types of non-random mating were initially suspected because of differences between genotype frequencies in natural populations and Hardy–Weinberg expectations, the exact nature of the mating biases then being revealed by careful mating preference experiments (pp. 86 and 255). However, not all deviations from Hardy–Weinberg expectations involving a deficiency or excess of homozygotes are the result of non-random mating. Other explanations are possible, such as selection acting in favour of, or against, heterozygotes.

Finally, if all the above assumptions of the Hardy–Weinberg can be shown to hold, with reasonable probability, the final question can be addressed—Is there selection? I have left this subject, and the full explanation of what natural selection is, until last, because, although easy to understand, it is often the most difficult to investigate.

Natural selection

July the 1st, 1858, marked perhaps the most important development in biological thought, for it was on that date that the Darwin/Wallace lecture was delivered to the Linnaean Society of London. The lecture was entitled: 'On the tendency of species to form varieties; and on the perpetuation of varieties and species by natural means of selection'. This lecture contained the essential elements of Darwin's theory of evolution, formulated 20 years earlier. Darwin had long been put off publication of his ideas because of the controversy and criticism he expected his theory to provoke. However, his hand was forced when he received a letter, on 20 June 1858, from Alfred Russell Wallace. In this letter,

Wallace asked Darwin's opinion on a theory suggesting that natural selection played an essential part in shaping the development of living species. Darwin commented to a colleague: 'If Wallace had had my manuscript written out in 1842, he could not have made a better short abstract!'.

In 1859, Darwin published *The Origin of Species* which explained his theory in great depth and provided a plethora of examples to illustrate and support his views on evolution by natural selection. This book was an immediate popular success. In it, Darwin extends his selection theory to include sexual selection, a phenomenon which he developed in the ensuing years, and which was to be one of the central themes of his second great book on evolution, *The Descent of Man*, published in 1871.

Darwin's theory of evolution through natural selection is based on four propositions which Darwin held to be true, and three logical deductions.

(a) Organisms produce many more reproductive cells than ever give rise to mature individuals.

(b) The numbers of individuals in species remain more or less constant.

(c) Therefore, there must be a high rate of mortality.

(d) The individuals in a species are not all identical, but show variation in virtually every character.

(e) Therefore, some variants will succeed better than others in the competition for survival, and the parents of the next generation will be naturally selected from among those members of the species that have traits which best allow them to survive the rigours of their environment.

(f) There is an hereditary resemblance between parents and offspring.

(g) Therefore, subsequent generations will maintain and improve on the degree of adaptation of previous generations by gradual change.

Despite considerable opposition from many quarters, particularly the clergy, many Victorian scientists readily embraced Darwin's theory of evolution. Its basis was so simple that T.H. Huxley remarked that the most remarkable thing about it was that no one had thought of it before.

Lack of communication between scientists during the nineteenth century was one of the main factors which slowed scientific advancement. Darwin's evolutionary theory is an excellent example. One of the facts on which Darwin's theory is based is that there is hereditary resemblance between parents and their offspring. The mechanism underlying this heredity evaded Darwin, and troubled him in his writings on evolution. He toyed with ideas of blending inheritance and, later, with Lamarckian inheritance, the inheritance of acquired characteristics. Yet while he was working on the *Descent of Man*, and the revised editions of *The Origin of Species*, Gregor Mendel was growing peas and unravelling what we now call Mendel's laws of inheritance (Chapter 3).

Knowledge of Mendel's findings would have made Darwin's deliberations over the pathways of evolution easier. His theory, when first made widely public, was highly controversial. The orthodox view of the time was that life on Earth had been formed by The Creator as it is now. The species of organism had been formed by The Creator in all their perfection and did not change. They were immutable. Darwin's theory challenged this view. Species evolved in response to the harsh conditions of their circumstance. Those with characteristics which made them more likely to survive and reproduce in the conditions they were exposed to, would produce offspring with these successful characteristics. Individuals with novel deleterious traits died or failed to reproduce. Such traits were thus lost. However, individuals with novel and beneficial traits survived and reproduced, at the expense of those without these traits. Species could thus change. They were not immutable.

That species change, becoming ever better adapted to their environs under the auspices of natural selection, was the prime deduction of Darwin's theory of evolution. However, Darwin had no examples of such a change actually observed in nature. The best demonstrative examples Darwin could muster were of domesticated animals and crop plants which had changed under selective breeding programmes. Yet, first in Manchester, and soon afterwards, in industrial regions across England, evidence that species could and did change under natural selection could have been available to Darwin.

In 1864, only 5 years after *The Origin of Species* was first published, Edleston published the first report of a black form of the peppered moth. By the time of Darwin's death, in 1882, the melanic form of the peppered moth was in the majority in many industrial parts of Britain (Chapter 5). This would have included the Potteries which Darwin knew well from the Wedgewood side of his family (Josiah Wedgewood who founded the Wedgewood pottery was Darwin's great uncle). However, the link between the increase in the frequency of the black form of the peppered moth in industrial areas and Darwin's evolutionary ideas was not made until after his death, by which time the frequency of the melanic form was high in most industrial areas. That Darwin did not appear to be aware of the changes affecting the peppered moth is a scientific tragedy, for as Professor E.B. Ford (1975) pointed out a century later, the failure to monitor the early rise in the frequency of the melanic form of the peppered moth was one of the great lost opportunities in science.

The Neo-Darwinian theory

Since Darwin's death, his evolutionary theory has been modified and amended in a number of ways. One of the most important steps was the merging of selection theory with Mendelian genetics. This 'New Synthesis', often called the Neo-Darwinian theory, can be summarised in a number of points.

(a) The origin of heritable variation is mutation.

(b) The combinations of genes which can occur together depend on a variety of factors, such as the type of reproduction, the level of inbreeding and outbreeding, the position of genes with respect to one another, the existence of chromosomal inversions, and so on.

(c) The relationship between genotypic variation (the organism judged by its genetic constitution) and phenotypic variation (the organism judged by its appearance) depends upon factors such as genetic dominance relationships between alleles, gene interactions, and linkage effects.

(d) Some individuals will be better adapted to a particular environment than others, due to greater viability or fecundity, which will lead to differences in Darwinian fitness.

(e) The genetic composition of a species will change through the mechanism of selection.

(f) In some circumstances, a range of genotypes will be maintained within a species by selection producing balanced polymorphism.

(g) Selection may act on different populations within a species in different ways, leading to divergence and speciation.

The Darwinian and Neo-Darwinian theories have been amended in various ways, for specific cases or sets of cases. For example, as we have seen in Chapter 3, not all genes are inherited in a Mendelian manner. Some are inherited from just one parent, usually the mother, because they occur in the cytoplasm, which is usually not transmitted in sperm (p. 63). But for our purposes here, we need only be concerned with the basic phenomena of Darwinian selection and random genetic drift.

Natural selection operates whenever three conditions are met. First, that there is variation in the particular character in question. Second, that some of this variation is genetically controlled. Third, that some variants produce, on average, either more or fewer offspring than others. In other words, the genotype must be correlated with fitness.

Any trait that is affected by natural selection will also be affected by random genetic drift, because drift affects all genetic traits. The question of whether drift or selection has the greater influence on a particular trait depends crucially on two factors: the differences in fitness of the variants, and the population size. If fitness differences are great, selection overwhelms drift, except in very small populations. Conversely, if a mutation is only slightly beneficial or detrimental to the individuals that bear it, except in very large populations, drift may play a significant part in the fate of the mutation. The spread of such a mutation is like a delicate moth trying to fly on a stormy night. Although there is a preferred direction of travel, this is often overwhelmed by the directionless vagaries of the wind. Only when the wind drops does the moth's flight become controlled and directional. In evolutionary terms, the still night is equivalent to a population which is large enough for the effect of drift to be negligible.

However advantageous a new mutation may be, it will be most vulnerable to chance loss from the population in the first few generations after it arises, because initially only a few copies of it will be present. Return to our imaginary tree with the clutch of moth eggs. Even if say half of these eggs contain a novel beneficial mutation inherited from their mother, the catastrophe of the tree being cut down and burned will cause all to be lost. If the mother moth also happens to be eaten by a bird before laying a second clutch of eggs, the new beneficial mutation will have been eradicated. Of course, once the eggs hatch and begin to disperse, the danger gradually reduces. When some reach maturity, fly to other trees, and begin to mate themselves, the risk of chance loss is dissipated still further, for figuratively the eggs have been spread around a number of different baskets. As the frequency of the new allele increases, the importance of drift diminishes.

To study the way that natural selection causes a change in the genetic constitution of a population of organisms, we need to consider how selection acts on genes and genotypes. This is essential, for the precise genetic control of the variation within a trait can have a profound influence on its evolution. This can be illustrated by considering three basic genetic systems: an advantageous dominant allele; an advantageous recessive allele; and a polygenic system controlling a trait showing continuous variation.

Selection for a dominant allele

Our starting point is the production, by mutation, of a new allele, in one individual in a population. Let us assume that this allele, A, is genetically dominant to the old form of the gene, a, and that it produces a melanic phenotype in a moth. We will also assume that this phenotype is selectively advantageous to the old non-melanic form. Assuming that this first moth with the melanic mutation survives to reproduce and that more than one of its progeny reach reproductive age, in the following generation there will be the potential for three genotypes: AA, Aa, and aa; the former two being melanic because of the dominance of A over a, and the latter being non-melanic. Because the melanic form is at an advantage over the non-melanic, both the AA and Aa genotypes will be selectively favoured. In the initial stages of the spread of a gene, this is important, because initially the majority of the mutant alleles with be found in heterozygote genotypes. This is easy to show by using the Hardy–Weinberg frequencies.

Let the frequency of the melanic allele be p, and that of the non-melanic be q, then the Hardy–Weinberg law says that the frequencies of the three genotypes should be $AA = p^2$, $Aa = 2pq$, and $aa = q^2$. If the frequency of the A allele, at some point early in its spread is 0.01 (i.e. 1 per cent of the alleles), then p = 0.01 and q = 0.99, so that $p^2 = (0.01)^2 = 0.0001$, $2pq = 2 \times 0.01 \times 0.99 = 0.0198$ and $q^2 = (0.99)^2 = 0.9801$. Converting these frequencies into the numbers of moths of each genotype in a population of say 10 000, we would have one homozygous melanic, 198 heterozygote melanics and 9801 non-melanics. Of the 200 melanic alleles in this population (two in the single homozygote melanic, and one in each of the 198 heterozygotes), 99 per cent are present in heterozygotes, and only 1 per cent in homozygotes. This can be compared with the situation later when the melanic allele has increased in frequency to say 50 per cent, i.e. both p and $q = 0.5$. Now if the same calculation is performed, we get 2500 homozygote melanics, 5000 heterozygote melanics, and 2500 non-melanics in our population of 10 000. This means that there are 10 000 A alleles in the population, 5000 in the homozygotes and 5000 in the heterozygotes, and the proportion of A alleles in homozygotes has risen to 50 per cent.

The importance of the dominance relationship between the alleles now becomes evident. When a new mutation arises, it will initially occur mainly in heterozygotes. Consequently, for selection to act upon it, the beneficial effects of the allele must be expressed, and they will only be fully expressed if the allele is dominant. Put another way, when a new beneficial mutation arises, it will be in the heterozygous state, and will only be subject to selection if it is expressed in this state.

Selection for a recessive allele

We can compare this situation with the case in which the new advantageous mutation is recessive, as in the case of the form, f. *nigra* Cockayne, of the brindled beauty moth. Again when the mutant allele first arises it will be in a heterozygote, but now it will not be expressed. It only has a selective advantage when it is homozygous, because only then is it expressed. Going back to our population of 10 000 with the melanic allele at a frequency of 0.01 (which is already much higher than the mutation rate), only one individual in the population will be melanic and so gain a selective advantage, if the melanic allele is recessive. This compares with the 199 that were melanic in the case when the melanic allele was dominant.

The take home message is that the spread of an advantageous mutation is much faster if the allele is dominant than if it is recessive, at least in the early stages.

At the other end of the spread, once the new mutation has spread to high frequency, and the original allele has become rare, the situation is reversed. Now further increase in the dominant allele is slow, because the increasingly rare recessive allele that it is replacing is generally found in heterozygotes. Here, in heterozygotes, the disadvantageous allele is protected from selection by the favoured dominant allele. Conversely, once an advantageous recessive mutation becomes common, its disadvantageous dominant allele is eradicated swiftly because it is always exposed to selection whether in the heterozygous or homozygous state.

Once a new mutation has spread right through a population, it is said to have become fixed. The original allele that it has replaced can then only arise as a result of back mutation.

The eradication of deleterious mutations

Of course not all novel mutations are advantageous. Indeed, quite the reverse, for the majority of mutations that have any appreciable effect on the phenotype are selectively detrimental. Here again their fate depends on their genetic dominance. If a disadvantageous new mutation is dominant, it is quickly eradicated by selection, because it is expressed in the phenotype whenever it occurs. The case is very different if the mutation is recessive, for now, while rare, the allele will usually occur in heterozygotes, where it is not expressed, and so the deleterious allele is protected from selection by the dominant allele that accompanies it. The outcome is that deleterious recessive mutations can hang around in populations, at low frequency, for a considerable time.

One final set of cases involving the spread of advantageous mutations must be considered. Some beneficial mutations that arise and start to spread because they are selectively advantageous, do not go to fixation. Their spread is arrested at some particular frequency so that both they and the allele they were replacing are maintained in the population. The situation when two or more forms are maintained in a population at particular frequencies is known as balanced polymorphism. The study of the ways in which balanced polymorphisms are maintained, in particular the relative roles of drift and selection, has been central to the development of evolutionary thinking. Many of these studies have involved melanic polymorphisms and so the basic features of genetic polymorphism, as defined by Ford (1940a) will be considered in some detail towards the end of this chapter.

Hitch-hiking

Up to now I have talked effectively as if selection and drift act on one gene locus at a time and that each locus can be considered independently with regard to the evolutionary pressures upon it. This is not realistic in all instances. Consider, for example, a case where two gene loci, A and B, are linked on the same chromosome. Each has two alleles, those of locus A being selectively neutral, while the genotypes at the B locus have different fitnesses. Because of the linkage between the genes, the frequency of the A locus alleles will be influenced by selection acting on the B locus. The allele of the A gene that is most often linked to the beneficial allele at the B locus will increase in frequency, in effect hitch-hiking upon its success. In this way, an allele that is of no particular benefit itself may attain high frequency because it is linked to a successful allele close by.

Selection on polygenic traits

Many heritable traits are controlled not by single genes that have a large effect on their bearer, but on sets of genes each of which has a small effect on the phenotype. These are called polygenes, and they produce continuous variation. To look at the way that selection acts on a polygenic trait, it is easiest to consider an example. Surprisingly, very little work has been conducted on the inheritance of polygenic melanism in moths. Therefore, I will use the case of melanism in the 14 spot ladybird, *Propylea 14-punctata* (Linnaeus), as an example. The elytra of the 14 spot ladybird varies continuously from all yellow to all black (Fig. 4.1). In Britain, most individuals have between 30 and 60 per cent black, with the numbers of ladybirds with less or more black decreasing outside this range. Heritability studies using the mass selection method (p. 64) have demonstrated the variation to have a high genetic component ($h^2 = 0.78$, Majerus 1994*a*) and to be inherited polygenically.

Let us consider three situations in which selection favours ladybirds in different parts of the distribution. First, if the fittest form is that with the average amount of black, with those towards the extremes of the distribution being increasingly less fit, the selection is said to be stabilising. As more extreme forms are likely to be eliminated, this type of selection reduces genetic variation.

The influence of changes in population size on polygenic systems under stabilising selection is interesting. Most populations of organisms are stable in number. When a population begins to increase in size, this is usually a result of conditions

Fig. 4.1　The amount of black patterning on the elytra and pronotum of the 14 spot ladybird varies both within and between populations. The majority of British individuals have amounts of black within the limits of the two individuals shown.

being better than average in some way. The stresses of life are thus reduced and we may assume that selection pressures have relaxed. This will mean that more of the extreme forms survive, with a resulting increase in genetic variation. Conversely, if a population is diminishing numerically, genetic variation decreases, because only the fittest individuals, i.e. those close to the optimum for the trait, will survive. Perhaps counter-intuitively, stabilising selection only influences the amount of heritable variation while population size is actually changing. If population size is constant, the level of stabilising selection will be the same whether the stable population size is large or small, although the loss of alleles by drift in a small population may reduce variation (p. 72).

The second situation is when selection favours forms towards one end of the variation distribution. The selection is then said to be directional. Individuals from the favoured end of the distribution, say the black end, will produce more of the next generation than those producing more yellow patterns. Consequently, genes producing more black will increase in frequency, and the shape of the distribution will change, becoming skewed towards the black end.

The final situation is that in which two or more phenotypes, for example, very yellow or very black, are fit, with other forms being unfit by comparison. This is known as disruptive selection. Disruptive selection may occur, for example, when different forms are favoured in different habitats which occur very close together. The expected outcome of disruptive selection is that some mechanism to reduce the number of intermediates produced should evolve. This may be by the evolution of assortative mating, that is to say predominantly yellows prefer to mate with predominantly yellows, and blacks prefer to mate with blacks. Such assortative mating may lead to speciation. Alternatively, a polygenic system may evolve to become polymorphic. If some of the genes in the polygenic system which together produce one of the fittest phenotypes become tightly linked, they will be inherited together as a unit, or supergene. Melanic polymorphism in a number of species is controlled by supergenes that may have evolved in this way.

Polymorphism

Ford (1940a) provided a very precise definition of genetic polymorphism as:

> the occurrence together, in the same locality, at the same time, of two or more discontinuous forms of a species in such proportions that the rarest of them cannot be maintained merely by recurrent mutation.

Ford then explains the limits set by this rather exact definition. It excludes continuous variation, geographic races, seasonal variations, and rare aberrations which arise from time to time by mutation.

Ford goes on to split polymorphisms into two types: transient polymorphisms and balanced polymorphisms. A polymorphism is transient if the frequencies of the alleles are changing in one specific direction; as for example, while an advantageous mutation is spreading through a population, eventually reducing the original allele to the status of an occasional mutant. Conversely, a balanced polymorphism involves the frequencies of forms being stable or changing in a predictable cyclical manner.

As we shall see, the melanic polymorphism in the peppered moth has, at different times, been both transient and balanced. During the initial spread of the melanic form, f. *carbonaria*, in the nineteenth century, it was transient, *carbonaria* increasing in frequency in industrial areas. For much of the twentieth century, the frequencies of forms in a particular locale have been stable, being the result of a balance between selection and migration (p. 132). During this period the polymorphism was balanced. Since about 1970, the polymorphism has again been transient in many areas as the frequency of *carbonaria* has declined following anti-pollution legislation.

The maintenance of genetic polymorphism: the selection-drift controversy

That Ford thought it valuable to imbue a term that simply means 'many inherited forms' with such a precise definition, is perhaps indicative of the important place that such polymorphisms had assumed in evolutionary study. Ever since Darwin's time, polymorphisms posed a problem for selectionists, because it was difficult to see how 'the survival of the fittest' could lead to the persistence of two or more discrete forms within an interbreeding population. Darwin himself discussed polymorphisms in *The Origin of Species*, and carefully divorced them from natural selection, writing:

> This preservation of favourable variations, and rejection of injurious variations, I call natural selection. Variations neither useful nor injurious would not be affected by natural selection and would be left a fluctuating element as perhaps we see in the species called polymorphic.

Darwin's view, that the different forms in conspicuous polymorphisms were not affected by differential selection, was the pervading view held by most evolutionary biologists up to the 1940s. They saw changes in the frequencies of the different forms as the result of random genetic drift. However, two eminent scientists, Sir Ronald Fisher at Cambridge and Professor E.B. Ford at Oxford, strongly disagreed with this view, believing that polymorphisms were the product of a balance of selective factors, some favouring, some acting against, each of the forms. The selection-drift controversy was the most heated and enduring debate in biology this century (Provine 1986). It has its origins around 1930, and continued apace until the early 1960s. Even now, in the 1990s, many research papers contain echoes of the conflict between the Fisher school and those that supported Professor Sewall Wright, who championed the drift side.

The first treatments in the argument were theoretical. As early as 1922, Fisher showed mathematically that two or more alleles of a gene could be maintained in a population by a balance of selective forces. At equilibrium, the selective forces acting on the alleles would be precisely the same. Fisher envisaged two mechanisms that would maintain such polymorphic equilibria, these being heterozygote advantage, in which the heterozygote genotype is fitter than either homozygote (p. 89), and negative frequency dependent selection, in which the fitness of a form is negatively correlated to its frequency in the population (p. 91). Various theoretical treatments relating to the evolution of dominance, why heterozygotes might be expected to be fitter than homozygotes and the effects of substructuring populations followed. In most of these treatments, it was assumed that any selection pressures acting on the forms in a balanced polymorphism would be very small because of the lack of changes in frequencies over time. However, Ford questioned this view. Having considered Fisher's theoretical studies of how polymorphism might be maintained, Ford reasoned, with remarkable insight, that change in balanced polymorphisms might be slow, not because selection pressures were weak, but because they were very strong, and acted in opposition. Thus, he argued, those polymorphism which showed the greatest stability would be those that were maintained by the strongest balancing selection.

Despite this argument, and because of the lack of any empirical evidence to support it, this view was still in the minority into the 1940s. For example, Ernst Mayr, in *Systematics and the Origin of Species*, wrote in 1942:

> Neutral polymorphism is due to the action of genes 'approximately neutral as regards survival value'. Ford (following Fisher) believes that this kind of polymorphism is relatively rare, because 'the balance of advantage between a gene and its allelomorph must be extraordinarily exact in order to be effectively neutral'. . . . There is, however, considerable indirect evidence that most of the characters that are involved in polymorphisms are completely neutral, as far as survival value is concerned. There is for example, no reason to believe that the presence or absence of a band on a snail shell would be a noticeable selective advantage or disadvantage.

And later:

> Even more convincing proof for the neutrality of the alternating characters is evidenced by the constancy of the proportions of different variants in one population. The most striking case is that of the snails *Cepaea nemoralis* and *C. hortensis*, in which Diver (1929) found that the proportions of the various forms from Pleistocene deposits agree closely with those in colonies living today.

It was precisely this type of enduringly stable polymorphism that Ford believed would be the subject of the strongest selection pressures. In the 1930s Ford began to plan an extensive series of research projects to collect evidence to resolve the question of the relative roles of selection and drift on balanced polymorphisms.

The case of the scarlet tiger moth

Working with Fisher, Ford's first target was the melanic polymorphism in the scarlet tiger moth colony at Cothill in Berkshire. This colony is the only one known that is naturally polymorphic for the three genetically controlled forms, *dominula*, *medionigra*, and *bimacula* (p. 43) (Plate 7a). In 1939 and 1940, Fisher and Ford visited the colony to assess the frequency of the *bimacula* allele. Thereafter, they used novel multiple mark–release–recapture techniques to estimate the frequency of the alleles, the average survival rate, and the population size for the years 1941–46 (R. Fisher and Ford 1947).

Fisher (1930) had already shown that it was possible to calculate the extent to which the frequencies of a pair of alleles can alter from generation to generation, through random genetic drift, if population size and initial allelic frequencies are known. The Cothill data (Table 4.1) was the first to which these calculations could be applied, so that the alternative explanations of balanced polymorphisms, selection and drift, could be tested.

Without going into the mathematical details, Fisher and Ford deduced that the changes in the frequencies of the *bimacula* allele were too large to be consistent with a drift theory. The acrimonious debate that followed has been widely reported (e.g. Provine 1985, 1986). Suffice to say that Wright (1948) challenged this deduction on several grounds, which were in turn forcefully answered by Fisher and Ford (1950), and later by Sheppard (1951). The matter was eventually resolved by work undertaken by various ecological geneticists at Oxford, including Philip Sheppard, Lawrence Cook, Bernard Kettlewell, David Lees, and Ford himself. Using a range of techniques, many of which were novel, two principal selective factors were shown to influence the polymorphism. By marking larvae with radioactivity and releasing them, Kettlewell showed that the *dominula* form had a considerable survival advantage over the *medionigra* form (Kettlewell and Cook 1960). Conversely, using mating tests, Sheppard (1952) demonstrated that the different forms had a preference to mate with partners that had a different genotype from their own (Table 4.2). This gives the rarer forms (*medionigra* and *bimacula*) an advantage, because they are the preferred mating partners of the common *dominula* form which usually comprises over 90 per cent of the population. Furthermore, by setting up artificial colonies or introducing the *bimacula* allele into colonies from which it was absent, Sheppard demonstrated that the frequencies of the forms usually converged towards those found at Cothill, whether the *medionigra* allele was initially at a higher or a lower frequency (Sheppard and Cook 1962; Ford and Sheppard 1969) (Table 4.3).

Many later investigations on melanic and other polymorphisms have vindicated the selectionist view. It is not necessary to go into the details here further than saying that the majority of balanced polymorphisms in which the forms are conspicuously different are maintained primarily by balancing selection. However, it is worth briefly considering some of the variety of ways that selection can main-

Table 4.1 The occurrence of three forms of the scarlet tiger moth, at Cothill, Berkshire, from 1939 to 1988. (After D. Jones 1989.)

Year	Numbers captured dominula	medionigra	bimacula	Total	Estimated population size of colony	% Frequency of bimacula allele
1939	184	37	2	223	?	9.2
1940	92	24	1	117	?	11.1
1941	400	59	2	461	2000–2500	6.8
1942	183	22	0	205	1200–2000	5.4
1943	239	30	0	269	1000	5.6
1944	452	43	1	496	5000–6000	4.5
1945	326	44	2	372	4000	6.5
1946	905	78	3	986	6000–8000	4.3
1947	1244	94	3	1341	5000–7000	3.7
1948	898	67	1	966	2600–3800	3.6
1949	479	29	0	508	1400–2000	2.9
1950	1106	88	0	1194	3500–4700	3.7
1951	552	29	0	581	1500–3000	2.5
1952	1414	106	1	1521	5000–7000	3.6
1953	1034	54	1	1089	5000–11000	2.6
1954	1097	67	0	1164	10000–12000	2.9
1955	308	7	0	315	1500–2500	1.1
1956	1231	76	1	1308	7000–15000	3.0
1957	1469	138	5	1612	14000–18000	4.6
1958	1285	94	4	1383	12000–18000	3.7
1959	460	19	1	480	5500–8500	2.2
1960	182	7	0	189	1000–4000	1.9
1961	165	7	0	172	1200–1600	2.0
1962	22	1	0	23	216	(2.2)
1963	58	1	0	59	470	(0.8)
1964	31	0	0	31	272	—
1965	79	2	0	81	625	1.2
1966	37	0	0	37	315	—
1967	50	0	0	50	406	—
1968	128	3	0	131	978	1.1
1969	508	38	0	546	5712	3.5
1970	444	31	0	475	4493	3.4
1971	637	9	0	646	7084	0.7
1972	335	5	0	340	3471	0.8
1973	230	1	0	231	1000–2000	0.2
1974	836	11	0	847	2000–3000	0.7
1975	50	2	0	52	<1000	1.9
1976	167	3	0	170	1000–2000	0.9
1977	12	0	0	12	<500	—
1978	39	1	0	40	<1000	1.3
1988	611	11	0	622	4000–6000	0.9
Totals	19979	1338	28	21345		

Note: from 1941 to 1961 population size was estimated by mark, release, and recapture. The estimates for 1962–72 were obtained from an unweighted regression of population size on the number of moths captured in those years when seven or more were recaptured. From 1973 the earlier method was used.

Table 4.2 Results of Sheppard's mating tests, showing a
preference in the scarlet tiger moth to mate disassortatively with
respect to the genotype at the *bimacula* locus. Three moths, two
of one sex, and one of another, were put in a mating cage, and the
phenotypes of the moths that mated were recorded. (From
Sheppard 1952.)

Phenotypes of the three months in the cage	Result of first mating		Totals
	Like genotypes pairing	Unlike genotypes pairing	
dominula male *dominula* female *medionigra* female	8	20	28
medionigra male *dominula* female *medionigra* female	12	14	27
dominula male *medionigra* male *medionigra* female	13	14	26
dominula male *medionigra* male *dominula* female	11	22	33
medionigra male *medionigra* female *bimacula* female	2	0	2
bimacula male *medionigra* female *bimacula* female	0	1	1
medionigra male *bimacula* male *bimacula* female	3	15	18
medionigra male *bimacula* male *medionigra* female	2	10	12
dominula male *dominula* female *bimacula* female	2	1	3
Totals	53	97	150

tain polymorphisms, for I shall describe empirical evidence supporting these
mechanisms in later chapters.

Broadly, balanced polymorphisms may be maintained by any one of three
processes: heterozygote advantage, a balance between selection and migration,
and negative frequency dependent selection.

Table 4.3 Data of phenotype frequencies from Sheppard's artificial colony of scarlet tiger moths at Hinksey, and the manipulated colony at Sheepstead Hurst. (a) At Hinksey, a new colony was established by releasing 4000 eggs from *dominula* × *medionigra* crosses. (b) At Sheepstead Hurst, eggs from 50 *dominula* × *medionigra* crosses were released into an existing colony from which the *bimacula* allele had previously been absent. (Adapted from Majerus *et al.* 1996.)

Year	*dominula*	*medionigra*	*bimacula*	% of *bimacula* allele
(a)				
1951	2000 eggs	2000 eggs	—	25.00
1952	21 larvae	9 larvae		15.00
1960	269	32	3	6.25
1961	217	35	1	7.31
(b)				
1949–53	all	—	—	0.00
1954	introduction of *bimacula* allele			
1955	873	2	0	0.11
1960	398	9	0	1.11
1961	405	9	0	1.09
1964	739	13	0	0.86
1969	509	25	0	2.34

Heterozygote advantage

A balanced polymorphism will be maintained if, in a single locus biallelic model, the heterozygote has greater fitness than either of the two homozygotes. The fitness of the heterozygote is of benefit to both alleles present in this genotype, and protects them from eradication. The equilibrium frequency, that is the level at which the frequencies of the genotypes are stable, will depend on the relative fitnesses of the two homozygotes, and the heterozygote.

The classic example of heterozygote advantage is sickle cell anaemia in humans (Allison 1954, 1956). However, melanic polymorphisms which are maintained, at least in part, by heterozygote advantage, are also known. For example, Ford demonstrated, in rearing experiments in which larvae were placed under stress, that the heterozygote for the gene inducing melanism in the mottled beauty moth, *Alcis repandata* (Linnaeus), is fitter than either homozygote (Ford 1937) (Table 4.4). Furthermore, in a wild sample of these moths, the heterozygotes were more common than expected on the basis of Hardy–Weinberg expectations, although not significantly so.

For a long time many biologists, following Ford, thought that heterozygote advantage was the commonest mechanism by which balanced polymorphisms were maintained. This was because there was a theoretical evolutionary reason to explain why heterozygotes might be fitter than homozygotes.

Table 4.4 Viability differences between the *typica* and *nigra* forms of the mottled beauty, in backcrosses (heterozygote *nigra* × *typica*) which should produce a 1:1 ratio of the parental types. (a) Families reared in optimum conditions (The first family represents the progeny of a wild female, and contains the parents of the other families.) (b) Families submitted to suboptimum conditions, in the form of reduced food for larvae. (Data from Ford 1937.)

Family	f. *nigra*	f. *typica*	Totals
(a)			
1	43	38	81
2	27	21	48
3	31	32	63
Totals	101	91	192
(b)			
4	3	0	3
5	18	12	30
6	10	6	16
7	4	3	7
8	17	10	27
Totals	52	31	83

Heterozygote advantage is the expected outcome of Fisher's theory of the evolution of dominance. This theory depends on the presumptions that:

(a) major genes have multiple effects;

(b) one allelomorph is only dominant to another with respect to one, or a number of, particular phenotypic effects, while for other characteristics the dominance may be incomplete or reversed;

(c) that a character controlled by a particular allelomorph may vary slightly due to the action of modifier genes.

If we consider a scenario where a novel advantageous mutation is beginning to spread through a population, there will be a considerable period when the new allele occurs mainly in heterozygotes. Selection will then act on the heterozygotes that express the new allele most fully. If the slight variations in these heterozygotes are the result of modifier genes, those modifiers that promote the expression of the new mutation will be selected for. Should the new mutation affect more than one characteristic of its bearers, selection will tend to make the favourable effects of an allele dominant and the unfavourable effects recessive. The heterozygote will then have nothing but selectively advantageous characteristics and will be superior to the homozygotes which will have both advantages and disadvantages.

Despite this theoretical basis, and evidence to indicate that melanic alleles have indeed evolved dominance in a number of cases, there are relatively few known examples of heterozygote advantage, and the evidence that this is a widespread phenomenon is still lacking.

Polymorphism maintained by a balance between selection and migration

The conditions in which a species lives vary across geographic space. Consequently, different populations are exposed to different selection pressures, with the result that the frequencies of forms often vary spatially. Such variations in form frequencies across geographic distances are known as clines. Clines are well known in species showing melanism. In many of the moth species in which melanics have increased since the industrial revolution, melanics are common in industrial areas and rare in rural regions, with intermediate frequencies being seen in between. In such cases, the fact that melanic frequencies rarely reached 100 per cent, even in the most industrialised cities, is usually attributed mainly to migration of non-melanics into industrial areas from more rural settings. Similarly, the finding of melanics at appreciable frequencies in relatively unpolluted parts of the country has been explained in some, although not all cases, by migration of melanics from polluted regions.

The dynamics of clines have received considerable theoretical attention, and are often not simple mathematically. However, for our purposes it is sufficient to say that the frequency of forms of a species at a specific point along a cline will be a function of the relative fitnesses of the forms in that locale, the frequencies of the forms in surrounding areas, and the rate of migratory interchange between the locality and surrounding areas. The greater the interchange is, the less chance there is for local selection pressures to match the frequencies of the forms precisely to local conditions. Consequently, clines in species which migrate considerable distances tend to be smoother than those in species that are relatively static (p. 132).

Polymorphism as a result of negative frequency dependent selection

In the above cases, I have treated fitness as though it has a fixed value in a particular time or place. The melanic form of a moth in an industrial area has a specific fitness advantage, or a heterozygote is fitter than each of its homozygotes by fixed amounts. However, there are some situations in which the fitness of a form depends on variable biological parameters such as its frequency in the population, or the density of the population relative to some other species. When the fitness of a form is negatively correlated to its frequency, selection is said to be negatively frequency dependent, and this type of selection maintains many polymorphisms.

Search image formation

The different ways in which fitness may be negatively correlated to frequency are best illustrated by examples. Perhaps the easiest case to understand is that involving the development of search images by a predator (Tinbergen 1960). We may imagine a population of moths that has melanic and non-melanic forms, with the melanics being much commoner than the non-melanics. This population is at risk from birds that hunt prey visually. It is known that birds will form search images for prey items that they encounter commonly. If the birds preying on the moths at first do so randomly, most will first encounter the melanic form, assuming the two forms of moth to be equally conspicuous. Birds are thus much more likely to begin to search actively for the common melanics than for the rare non-melanics. The non-melanics will thus gain a selective advantage and begin to increase in frequency at the expense of the melanics. As the non-melanics become commoner, this advantage will diminish as it becomes increasingly profitable for birds to form a search image for them. At the same time the selective disadvantage of the common melanics will decrease as they become rarer and birds become less likely to search actively for them. The result is that the frequencies of the two forms will oscillate around a stable equilibrium.

It is not known how widespread this type of phenomenon is, although it has been implicated in maintaining many polymorphisms. Some of the best evidence comes from Cain and Sheppard's (1950, 1954) work on predation of the banded snail, *Cepaea nemoralis*, by thrushes, and on the predation of mock or real caterpillars by a variety of passerine birds (Allen and Clarke 1968; Allen 1972, 1974, 1976; Majerus 1978). For example, the angleshades moth has a variety of larval colour forms, including a green form and a melanic brown form (Plate 2d). Experiments were conducted in which larvae were placed out for birds to feed upon, in a 9 green:1 brown ratio, or the reverse, for 5 consecutive days in both the winter and the summer. The results (Table 4.5) show differences in the ease with which birds initially found the two colour types, in winter and summer, green being more conspicuous in the winter, and brown in the summer. This difference presumably reflects changes in the colour of low growing vegetation between the seasons. However, more interesting is the comparison of the number of the common and rare forms taken though the sequence of 5 days. In every one of the four tests, the number of the common species taken increases, while that of the rarer species consumed declines almost to zero. The only reasonable interpretation of these results is that the birds are forming a search image for the common colour type and eventually ignore larvae of the rarer colour almost completely (Majerus 1978, 1980). Green versus brown, or black, polymorphisms are found in many species of Lepidoptera (Plate 2e and f). It is probable that in many of those cases, in which the forms are genetically controlled, the polymorphisms are maintained by this type of negatively frequency dependent selection, with the equilibrium frequencies being a product of the relative crypsis of the forms.

Table 4.5 Test of the effect of morph frequency of a prey species, presented at low density, on predation rates by wild birds. Fresh killed green or brown larvae of the angleshades moth were offered at a density of two per square metre, on rough grassland, in either winter (December) or summer (August), over 5 consecutive days. The main predators were starlings. (After Majerus 1978.)

Day	No. green put out	No. brown put out	No. green eaten	No. brown eaten	Total	Ratio between number of common and rare morph eaten
Winter
1	180	20	48	6	54	8.0:1
2	180	20	52	3	55	17.3:1
3	180	20	69	1	70	69.0:1
4	180	20	56	0	56	
5	180	20	69	0	69	
1	20	180	8	39	47	4.9:1
2	20	180	4	43	47	10.8:1
3	20	180	0	41	41	
4	20	180	2	56	58	28.0:1
5	20	180	0	62	62	
Summer
1	180	20	36	13	49	2.8:1
2	180	20	41	9	50	4.6:1
3	180	20	50	4	54	12.5:1
4	180	20	56	2	58	28.0:1
5	180	20	50	1	51	50.0:1
1	20	180	48	5	53	9.6:1
2	20	180	58	5	63	11.6:1
3	20	180	63	5	68	12.6:1
4	20	180	56	1	57	56.0:1
5	20	180	71	1	72	71.0:1

Non-random mating

Negative frequency dependence arising from non-random mating has already been alluded to. In the case of the scarlet tiger moth, mating is disassortative with respect to the genotypes at the *bimacula* locus, and this gives an advantage to the rarer forms, while they are rare (p. 86). Other types of non-random mating may also lead to negative frequency dependent selection. For example, the preference of some female 2 spot ladybirds for melanic rather than red males may help to maintain the colour pattern polymorphism (Majerus 1986a; Majerus *et al.* 1986). If we assume a constant proportion of the females of this highly promiscuous species have this preference, it is easy to see that melanic males benefit more when melanism is rare than when it is common, for when rare each melanic male will gain relatively more additional matings from the preferring females than when melanics are common (O'Donald 1967, 1980; O'Donald *et al.* 1984).

The same selective advantage of rarity applies to melanism in the Arctic skua, although as this bird is monogamous, that the selection is frequency dependent is not so obvious. The problem is that although females prefer to pair with melanic males, on the assumption of a 1:1 sex ratio, all males will eventually secure a mate. The problem of mating preferences in monogamous species was actually considered by both Darwin and Fisher. Both resolved the question by speculating that timing of breeding may be important, and that preferred males may be chosen early in the breeding season, thereby gaining optimum conditions for their offspring. This theory was tested and verified by Peter O'Donald in an exhaustive and elegant series of studies of the Arctic skua on the Fayre Isles and Foula between 1959 and 1979 (see O'Donald 1983 for review). This bird is polymorphic, having three forms controlled by a single biallelic gene with incomplete dominance (Plate 1d and e), the heterozygote being only marginally paler than the homozygote melanic. The essence of the system is that the non-melanic form generally starts breeding a year earlier than the melanics, thus having the potential to reproduce for more years (no longevity differences between the forms have been detected). However, females prefer to pair with melanic males, and melanics breed earlier in the year. This means that their offspring have more resources available, longer to fledge, and more time to put on reserves before the winter. Thus the earlier breeding of melanics within a year, resulting from the mating preference for melanics, balances the younger mating age of the pale birds.

Environmental heterogeneity

Natural environments are not homogeneous; they vary, being made up of a mosaic of different habitats. Under certain conditions this environmental heterogeneity can lead to the maintenance of polymorphism. With respect to a particular organism, an environment may be either coarse grained (that is an organism spends most of its life in one habitat patch), or fine grained (if the organism moves from one patch to another, and covers all the different habitats through its lifetime). In both cases polymorphism will be maintained if over all the niches the average fitness of heterozygotes is higher than that of homozygotes. This is simply a variant of heterozygote advantage and need not be frequency dependent. However, if there is competition in a course-grained environment, then rarer morphs will gain an advantage as the niches in which they do best may be less crowded, leading to a negative frequency dependent aspect in the overall selection. Such a mechanism is only powerful enough to produce polymorphism in a very limited range of cases (Levene 1953; Maynard Smith and Hoekstra 1980). But the maintenance of balanced polymorphisms may be facilitated if environmental heterogeneity is coupled with habitat selection, that is to say, if genotypes spend most of their lives in those habitats in which they are fittest. Melanic polymorphisms involving habitat selection are known, and will be discussed in detail in Chapters 6 and 7, along with the evolution of other similar behaviours, such as resting site selection.

Co-evolution between species

The final category to be considered is when negative frequency dependent selection results from the evolution and counter-evolution between different species, often referred to as evolutionary 'arms races'. Obvious examples of arms races include those between parasites and their hosts, or between predators and prey. In the former, rarer forms of the host species may gain an advantage because it will only be profitable for a parasite to be adapted to the common host types. Consequently, the rare form will be less prone to attack by parasites. The case of search image formation by predators for specific prey types is a similar phenomenon in predator–prey systems.

A less obvious example of an arms race is the interaction between a Batesian mimic and its model. Batesian mimics are organisms that are relatively unprotected from predation, except that they closely resemble another species, the model, that is well protected (p. 212). Although the majority of Batesian mimics are monomorphic, polymorphism is maintained in some because the fitness of a mimetic form depends in part on how common it is relative to its model. That this should be so is easy to understand. If a mimic is rare compared with its model, the chances are that naive predators will come across the model before the mimic, and so learn of its unpleasant characteristics. As a mimic becomes more common relative to its model, the chance of it being encountered before its model will increase, and the predator may, at least initially, associate the colour pattern of the mimic with edibility, to the detriment of both the mimic and the model. Thus fitness of a mimic will decrease as it becomes more common. The initial advantage that a novel mimetic form gains, compared with a non-mimetic ancestor, by resembling an inedible species, thus diminishes as its frequency increases. This is relevant to many examples of melanism, because black is commonly a component of warning colour patterns, and many Batesian mimics are essentially melanic forms (p. 212).

How do adaptations arise?

All organisms have characteristics which seem to enable them to live in a particular environment. Darwin's (1859) theory of evolution by natural selection is based upon variation in such characteristics: 'those forms of a species which are more well suited to a particular way of life or to a particular habitat, would be more likely to survive to reproduce than less well suited individuals'.

Such characteristics are termed adaptations. One final question must be posed in respect of the evolution of adaptations. Do adaptations usually arise as a result of selection (in a changing environment) favouring chance mutations which arise concurrently with environmental change, or as a result of selection favouring novel phenotypes which are the result of recombination, transposition, translocation, inversions, and other genetic reorganisation mechanisms?

It is certainly true that in some cases evolutionary change results from selection favouring a beneficial mutation. The celebrated case of industrial melanism in the peppered moth is a case in point (Chapter 5). However, other evidence suggests that this type of case may be the exception rather than the rule. For example, the most melanic forms of the common marbled carpet, *Chloroclysta truncata* (Hufnagel), are controlled by melanic alleles of three different loci with additive effects. Examination of old collections leads to the conclusion that the melanic alleles of all three loci pre-date widespread industrialisation, but that they rarely occurred together. Only as a consequence of heavy pollution has it become selectively advantageous for this species to carry the melanic alleles of all three loci and be almost completely black in some areas. Rural areas in southern Britain support populations that show considerable variation in darkness, but fully non-melanic individuals, which carry none of the melanic alleles, are now confined to the Highlands and Islands of Scotland.

Further evidence favouring the genomic reorganisation theory is provided by the changes that occur in the variance of traits with population size changes. Most polygenically controlled adaptive traits are maintained within limits by stabilising selection. If a population is exposed to harsh conditions, so that the intensity of selection increases, the variance of the trait decreases. Conversely, if selection is relaxed, variance increases. This was dramatically demonstrated in 1976, when a combination of favourable climatic conditions led to great increases in the populations of many species of moth in Britain, and an even greater increase in records of abnormal forms, which survived because of the favourable conditions.

Perhaps the best evidence supporting genetic reorganisation is seen in the wealth variation in many species of crop plants and domestic animals. Here widely divergent forms have been created by artificial selection, over periods of time that are too short for new mutations to have had much effect. Rather, the differences between breeds of dog, or types of cabbage, *Brassica oleracea* (including cauliflower, broccoli, calabrese, and brussel sprout), are the product of the polarisation of naturally existing variation, through selective breeding. Taking dogs as an example, dogs are domesticated wolves. It is thought that they were first domesticated about 13000 years ago. The diverse array of breeds we see today, from daschunds to Dobermans and from chiwuawuas to great danes, vary greatly in size and anatomy, as well as coat colour, hair type, and temperament. Yet molecular genetic analysis has shown that all are still essentially pet wolves, the differences between them being the result of the needs and caprices of their breeders.

5 The peppered moth story

Introduction

One species of moth, the peppered moth, has dominated the study of melanic polymorphism. The reasons for this are easy to appreciate. First, the change in some populations of the peppered moth was rapid and visually dramatic. A relatively invariant species of white and black moth became almost exclusively completely black, over a period of just 50 years. Second, the change was seen to be associated with a specific environmental factor, the darkening of tree bark due to air pollution. Third, a single and easily understood selective factor appeared to be sufficient to explain the evolutionary change in the peppered moth. Finally, the change had occurred recently, within the living memory of some of those that unravelled the causes of the change. This was not an evolutionary event that had occurred in the long distant past, thousands, hundreds of thousands, even millions of years ago. It had occurred in the last hundred years, and in some parts of Britain, continental Europe, and North America, changes were still happening. Here was a case of evolution in action. As the renowned evolutionary geneticist, Professor Sewall Wright, wrote in 1978 of the peppered moth story, it is: 'the clearest case in which a conspicuous evolutionary process has been actually observed'. During the second half of the twentieth century, the story of the rise of the melanic form of the peppered moth in Britain has become one of the most often and widely quoted examples of evolution occurring as a result of natural selection, the central mechanism of Darwin's theory of evolution.

Everyone knows the basic peppered moth story, because it is in all the textbooks. However, in textbooks space is at a premium, with the result that the story is usually stripped of all but the bare essentials. Detail is lacking, both in respect of the ecology and behaviour of the moth or its predators, and in the way that the selective factors involved were uncovered. For example, although the first dramatic rise in melanic frequency occurred in the second half of the nineteenth century, it was not until the second half of the twentieth century that evidence was obtained to show why the melanic form had spread through many populations.

In this chapter and the next, I shall relate the history of melanism in the peppered moth. First, I will give the basic story as outlined in most textbooks. I will then describe the way in which this story was uncovered, including the alternative theories that were proposed during the first half of the twentieth century. In Chapter 6, I will consider the validity of the basic story, by considering the ecology, behaviour, and genetics of the peppered moth in detail.

The 'textbook' story of the peppered moth

The typical form of the peppered moth is white or off-white in colour and is liberally speckled with black scales (Plate 3a). As most texts relate, the first black form of the peppered moth was caught in Manchester in the middle of the nineteenth century. This form, called *carbonaria* (Plate 3a), increased in frequency rapidly in highly polluted industrial regions, but not in rural areas. The typical and melanic forms are under genetic control, the melanic form being produced

---

Box 5.1   Components of the 'textbook' peppered moth story

(a)   The peppered moth has two distinct forms, one, the typical form, being white with black speckling, the other, f. *carbonaria*, being almost completely black.

(b)   The two forms of the peppered moth are genetically controlled by a single gene, the *carbonaria* allele being completely dominant to the *typica* allele.

(c)   Peppered moths fly at night and rest on tree trunks by day.

(d)   Birds find peppered moths on tree trunks and eat them.

(e)   The ease with which birds find peppered moths on tree trunks depends on how well the peppered moths are camouflaged.

(f)   Typical peppered moths are better camouflaged than melanics on lichen-covered tree trunks in unpolluted regions, while melanics are better camouflaged in industrial regions where tree trunks have been denuded of lichens and blackened by atmospheric pollution.

(g)   The frequencies of melanic and typical peppered moths in a particular area are a result of the relative levels of bird predation on the two forms in that area, and migration into the area of peppered moths from regions in which the form frequencies are different.

---

by a single dominant allele. The peppered moth is a night flier, resting by day on tree trunks, where it relies on camouflage to help it avoid detection by predatory birds. In rural areas, tree trunks are frequently encrusted with lichens, and on such trunks, the typical form is well camouflaged. However, in industrial regions, two pollutants, sulphur dioxide (SO_2) and soot, change the appearance of tree trunks. The SO_2 kills lichens and the soot then blackens the denuded trunks. The result is that in polluted areas, the light speckled typical moths are no longer well camouflaged and are easily seen by birds. They thus suffer much higher levels of bird predation than does *carbonaria*, which consequently spreads through populations in polluted parts of the country. Of course, in rural areas, the typical form retains its effective camouflage, and it is the black form which is poorly camouflaged against the lichened tree trunks. The melanic form thus has not spread into rural parts of the country to any significant extent. Those melanic moths that migrate into unpolluted areas are quickly eliminated by bird predation, just as are typical moths in industrialised areas. It is only where rural and industrial areas meet that both forms occur together at reasonably high frequencies.

This basic story of the peppered moth can thus be split into a series of component facts (Box 5.1). In Chapter 6 each of these components will be examined in turn with additional elements being added. Here I will consider, in more or less chronological order, the observations, experimentation, and speculation, that led to the composition of this story.

The early history of the black peppered moth and other industrial melanics

The first significant observation was made in 1848, in Manchester, although it was only reported some 16 years later. This was the first capture of a black peppered moth (Edleston 1864). From that time, melanic peppered moths began to be found in greater and greater numbers in Manchester, and this form began to be found in other parts of the industrial north-west, later turning up in the Midlands, the north-east, and eventually London. The spread was rapid and noticeable, being documented in a number of papers and several general books on moths in the latter half of the nineteenth century. Edleston (1864) indicates that by 1864, *carbonaria* was already the commonest form in his garden near Manchester. By 1882, W.F. Kirby wrote that:

> the 'pepper and salt moth' (an old name sometimes used for the species) has an almost black variety, named Doubledayaria Mill. (syn = *carbonaria*), which is not very uncommon in many parts of England.

By searching the entomological literature and examining museum collections Steward (1977*a*) compiled a list of the earliest known records of f. *carbonaria* from different parts of Britain (Fig. 5.1). Examination of this map suggests that the occurrence of *carbonaria* across Britain is a result of a single mutation event.

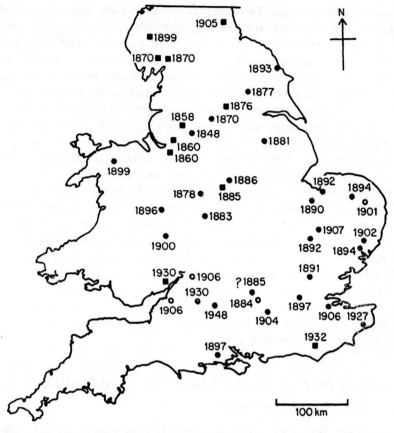

Fig. 5.1 Earliest records of *carbonaria* form of the peppered moth from localities in Britain, showing spread from Manchester. Dates by solid circles give first recorded date of *carbonaria*. Dates by solid squares give dates when *carbonaria* was already at appreciable frequency. Dates by open circles indicate latest date when *carbonaria* was known to be absent. (Adapted from Steward 1977a.)

The form is first recorded from industrial regions close to Manchester earlier than from more distant localities. The whole impression is of a radiation out from a Mancunian point of origin, leading to the deduction that *carbonaria* has reached new areas by migration. This does not prove that all British *carbonaria* originated from a single mutational event. Mutation is a recurrent process and it is certain that the same mutation must have occurred a number of times during the last century. Indeed, there are definite records of independent mutational events occurring on the Continent. For example, a form phenotypically identical to *carbonaria* was first recorded in Hanover, Germany, in 1884.

The *carbonaria* form did not just increase its geographic distribution. Once established in a population, it often increased in frequency in that population very rapidly. In Manchester, *carbonaria* had reached a reported frequency of 98 per cent by 1895. In less than 50 years since it was first noted, this black form had all but replaced the typical form.

Many scientists (e.g. Tutt 1891; Barrett 1902; Porritt 1907) commented on this dramatic change, offering a range of explanations. It is important to recall the state of both genetics and evolution at this time. Mendel's laws of inheritance, although published in 1862 in the *Journal of the Brno Natural History Society*, were virtually unknown to the scientific fraternity until the turn of the century. The resemblance between parents and offspring was accounted for by so-called 'blending inheritance', the characteristics of offspring being an average of those of their parents. In respect of evolution, many scientists had faith in Creation Theory (p. 77). However, a growing body of scientists believed in biological evolution. These scientists could be split into two schools: those that followed Lamarck and those that followed Darwin. Jean-Baptiste Lamarck had proposed that features which an organism developed during its life-time would be passed on to its offspring. So, for example, the giraffe's long neck could be explained by postulating that successive generations of ancestral giraffes had had to stretch up to graze on high branches, and in doing so had developed elongated and strong necks, each generation with a marginally longer neck than its predecessor. Similarly, traits that were little used would be underdeveloped, and in time would diminish and be lost. Lamarck's theory of inheritance and evolution is thus based on the use and disuse of traits and is often termed 'the inheritance of acquired characteristics'. The alternative view was that of Charles Darwin and Alfred Russell Wallace, who both came to the view independently, that the primary mechanism of evolutionary change was natural selection (p. 75).

The majority of explanations of the increase in melanic moths, proposed by entomologists in the late nineteenth century, were Lamarckian in character, or suggested that background colours in the environment influenced the colour of moths. For example, William Tugwell (1877) speculated that the cause of melanism was:

> the powerful impression of surrounding objects on the female during the all important period of life, viz., that of propagation, coupled with an instinctive provision for the protection of its future progeny.

Some entomologists noted that in experiments with butterflies, rearing at low temperature and high humidity produced melanic adults, and consequently they argued that melanic moths might be produced as a direct result of environmental factors. However, Tutt favoured a Darwinian explanation. Tutt (1896) was the first to suggest that the typical form of the peppered moth was well camouflaged when at rest on surfaces covered by foliose lichens, and that the *carbonaria* form had increased in industrial areas because the nature of the surfaces that the moths rested upon had changed as a result of pollution. On the dark and relatively homogeneous soot blackened trunks, Tutt suggested that *carbonaria* would be better camouflaged from avian predators than would the typical form. Tutt's view was challenged primarily because entomologists and ornithologists did not regard birds as major predators of cryptic day-resting moths.

The importance of industrial melanism to views on evolution at the turn of the century, may be gauged by the fact that the Evolution Committee of the Royal Society of London highlighted the phenomenon for special attention in 1900. This initiative did not bear much fruit in terms of the reasons for the increase in melanic forms, largely because of a lack of precision in published material, which rarely gave raw data. Doncaster (1906a,b) summarising the early results of this initiative notes the evidence collected up to 1906 was: 'somewhat meagre in amount'. In particular, there was a lack of data from the borders of the chief melanic areas, so that the progress of the spread of melanics was not well documented. However, advances were made in the understanding of the inheritance of melanic forms. Doncaster (1906a) showed that in some species (e.g. waved umber, peppered moth) melanism is inherited, but does not provide sufficient detail to show the exact mode of inheritance. Bowater (1914), on the other hand, reviewed data showing the mode of inheritance of the melanic forms of 12 species of moth (from a list, compiled by Bowater, of 211 species of British Lepidoptera with melanic forms). In general these melanics were found to be controlled by single dominant genes. Bowater (1914) argued that the case of the peppered moth was of special interest because:

> the melanic variety, *doubledayaria*, in the 65 years since it first appeared has multiplied and spread all over England and is now far commoner than the type, and is often quoted as a good example of a distinct alteration in a whole species occurring in our own times and apparently not due, directly at any rate, to man's influence.

Perhaps the most significant contribution to the peppered moth story during the first half of the twentieth century, was Haldane's (1924) calculation of the fitness advantage necessary to account for the spread of *carbonaria*, at the expense of *typica*, in Manchester, during the second half of the nineteenth century. He found that the melanic would have to have been one and a half times as fit as the typical form. This finding had a profound effect on evolutionary thinking. Darwin and the early selectionists believed that evolution was a very gradual process, and that differences in fitnesses between varieties would be small: of the order of up to 1 per cent. This was still the general consensus in the 1920s, except in the case of highly deleterious mutations. However, Haldane's calculation showed that very large fitness differences could occur in natural populations, without the presence of lethal or semi-lethal mutations. However, Haldane did not identify the causes of the fitness differences between the forms.

It is perhaps worth noting that the accuracy of Haldane's fitness estimates during the spread of the *carbonaria* form, depend on the initial and final frequencies of *carbonaria*, and the time taken. It is possible, although not very likely, that the *carbonaria* form was already at a frequency higher than 1 per cent in 1848. If so, Haldane's calculation may overestimate the fitness advantage to *carbonaria*. It is more likely, however, that Haldane's calculation underestimates the fitness advantage to *carbonaria*. This would be the case if *carbonaria* reached high frequency (close to 98 per cent) long before 1895. Edleston's (1864) initial note suggests that this may indeed have been the case, for he writes:

> I placed some of the virgin females in my garden (*in Bowdon, nr. Manchester*), in order to attract the males, and was not a little surprised to find that most of the visitors were the 'negro' aberration: if this goes on for a few years the original type of *A. betularia* will be extinct in this locality.

After the First World War, J.W. Heslop Harrison (1920) reviewed previous work on melanism. He specifically rejected Tutt's (1896) explanation based on differential bird predation, favouring instead the view that melanic forms arise as a consequence of metallic salts in pollution fallout on to leaves, which are eaten by larvae. Arguing that pollutants ingested in this way may affect the 'soma' and subsequently the 'germplasm', he predicted that it should be possible to induce the production of heritable melanic forms experimentally. In 1926, Harrison published the first of a series of papers on the effects that feeding larvae of some species of moth on leaves impregnated with lead nitrate and manganous sulphate had on the phenotypes of adults produced (J.W.H. Harrison 1926a,b, 1927a–d, 1928a,b; J.W.H. Harrison and Garrett 1926). In brief, these appeared to show that such treatment could lead to the production of melanic forms after a few generations.

Porritt (1926) criticises Harrison's experiments at some length, commenting:

> . . . but there are so many contradictory anomalies in connection with the subject that I think I shall be able to show that lepidopterists generally will require a good deal of further explanation before they can admit that the authors have advanced (if at all) our knowledge of the real causes of melanism in nature by more than a very small amount.

Porritt's main criticism was that the species in which Harrison obtained a response frequently did not show melanism in the wild, even in industrial regions, where Porritt maintained that their larvae must be feeding on leaves polluted with the salts Harrison used.

Hasebroëk (1925, 1929) conducted experiments of a similar type, but using pollutant gases, such as methane, ammonia, pyredin, chloroform, and sulphuretted hydrogen vapour, that might be absorbed through the tracheal system of pupae. His results from work on a dozen species of moths and butterflies showed that some individuals did show a darkening of the wings. However, Hasebroëk's experimental approach and interpretation of results were severely criticised by Cockayne (1926). These criticisms included the fact that Hasebroëk did not use appropriate species, for example, that industrial melanism is not found in butterflies, that no attempt was made to use stocks from which melanic forms were likely to be absent, and that Hasebroëk did not demonstrate that the darker forms produced were heritable.

While Harrison did not work on the peppered moth itself, his work affected much that was written on the subject of melanism for the next quarter of a century. In particular, Ford, who wrote extensively on the phenomenon of industrial melanism, is particularly critical of the suggestion that mutation might be directed by environmental influences such as pollutants. In numerous papers written between 1937 and 1964, Ford explains that the spread of industrial melanics is a consequence of heterozygote advantage (p. 89) (e.g. see Ford 1937, 1940a,b, 1945b). Because the majority of industrial melanics are controlled by single dominant genes, the mean viability of melanics will be greater than that of the non-melanics, if heterozygotes are generally fitter than homozygotes. Ford's adherence to this view is not surprising. For many years he worked closely with Sir Ronald Fisher, and was a strong supporter of Fisher's theory of the evolution of dominance, a theory that leads to the expectation of heterozygous advantage (p. 90). Ford proposes that the greater viability of melanics over non-melanics is emphasised in adverse conditions, such as those that are likely to prevail in highly polluted industrial regions. The fitness advantage over non-melanics would be reduced in the less harsh conditions of rural Britain, where the melanics would be conspicuous and eliminated by bird predation.

Later, in his extraordinary book, *Ecological Genetics* (1964), Ford admitted that he had previously probably overemphasised the greater hardiness of melanic forms in adverse conditions. However, in his earlier writings on melanism it is also clear that Ford supported Tutt's view that birds did eat day-resting cryptic Lepidoptera, and that the efficiency of a moth's camouflage would influence its fitness. Indeed, Ford had a considerable influence on Bernard Kettlewell, in part being responsible for Kettlewell's decision to leave general medical practice to become a full-time research biologist. He gained research funding for Kettlewell from the Nuffield Foundation, and through discussions at Oxford, influenced the design of Kettlewell's now classical experiments on the peppered moth in the 1950s.

Kettlewell's experiments on the predation of peppered moths by birds

The studies of the peppered moth by Bernard Kettlewell have been described and reviewed many times (e.g. Ford 1975; Lees 1981; Lambert *et al.* 1986; Majerus 1989b; Berry 1990). These investigations were painstaking in their detail and diverse in their approach. At this point, it is only pertinent to describe the main experiments that established beyond reasonable doubt that differential predation of the two forms by birds was the principal cause of the rapid spread of the *carbonaria* form.

Kettlewell initially considered his chosen subject material from the point of view of a predator. Believing, as the literature asserted, that bird vision is similar in nature to human vision, Kettlewell formulated an index of conspicuousness for the different forms using students and researchers as moth detectors. The index was made up of six grades, from −3 (most conspicuous), through −2, −1, +1, +2 to +3 (most cryptic). Having devised this index, he ran a pilot experiment in an aviary, at a research station in Madingley, near Cambridge, using peppered moths and other species of moth, released on to various types of background, and a pair of great tits, *Parus major*, as predators. The number and order in which moths with different cryptic indices were taken, was recorded. In general, the more conspicuous moths, as judged by the human eye, were taken before those considered to be cryptic. Kettlewell puts down the fact that some of the *carbonaria* with high cryptic indices were found and eaten to their proximity to conspicuous individuals. Once a conspicuous insect was found, it immediately put other insects close by at risk, because the tits tended to search intensively the area where they had had a success. The full results of these cage experiments supported both the view that on polluted oak trunks *carbonaria* is less heavily predated than *typica*, and that there is a positive correlation between cryptic index and survival. The results were sufficient to convince Kettlewell that birds do find and eat many species of moth, including both typical and melanic peppered moths, and that the accuracy of a moth's camouflage plays a large part in its likelihood of being found and eaten.

In 1953, Kettlewell conducted the first of his tests on the predation of peppered moths by wild birds in the field. His chosen site was the Christopher Cadbury Bird Reserve in Birmingham. This is a mixed deciduous woodland. The wood was heavily polluted by fallout from surrounding industries, and the trees supported virtually no lichens. However, despite this heavy pollution, the wood was inhabited by a large number of birds of a diverse array of species. Kettlewell found the peppered moth to be common in the wood in the summer of 1953, and was able to collect a sample of over 600, the majority of these being *carbonaria* (Table 5.1). Kettlewell released large numbers of peppered moths on to trees of different types and scored them using his index of conspicuousness. Only 2.2 per cent of the *carbonaria* released in this way were judged to be conspicuous ($n = 366$), while 88.96 per cent of the *typica* moths were scored as such ($n = 154$). To

Table 5.1 The percentage frequencies of the forms of the peppered moth, at Kettlewell's experimental sites, 1955. (For description of f. *insulaira*, see Chapter 6.)

	typica	insularia	carbonaria
Birmingham	10.1	4.8	85.0
Dorset	94.6	5.4	0

Kettlewell at least, there was no doubt that *carbonaria* was on average much more cryptic in the wood than was f. *typica*.

Two main experiments were carried out in this wood. The first involved releasing live moths on to tree trunks in the early morning, and checking to see which moths were still at their release points late in the day. Some of the moths released were kept under continuous observation throughout the day. Observations of moths seen being eaten by birds were recorded. As in the aviary experiments, the more conspicuous moths were taken more often than those that were more cryptic (Table 5.2). This means that, because of the bias in the conspicuousness scores for the two forms in this wood, the *typica* form was taken more often than

Table 5.2 Direct observations of the predation of peppered moth by two species of birds in a polluted woodland. (From Kettlewell 1973.)

Robin	Oak	4 July 1953 order of take	Robin	Oak	2 July 1953 order of take	Hedge sparrow	Oak	1 July 1953 order of take
f. *typica*	−3	1	f. *typica*	−3	1	f. *typica*	−3	1
f. *carbonaria*	+1	2	f. *typica*	−3	2	f. *typica*	−3	2
f. *typica*	−3	3	f. *carbonaria*	+3	3	f. *carbonaria*	+3	3
f. *typica*	−2	4	f. *typica*	−3	4	f. *typica*	−1	4
f. *carbonaria*	+3	Not taken by	f. *carbonaria*	+3	Not taken by	f. *carbonaria*	+3	Not taken by
f. *carbonaria*	+3	7 p.m.	f. *carbonaria*	+2	7 p.m.	f. *carbonaria*	+2	7 p.m.

Fig. 5.2 A mercury vapour moth trap with an assembly trap in Sir Cyril Clarke's garden, in Caldy. The traps are not normally set adjacent. (Courtesy of Sir Cyril Clarke.)

carbonaria, suggesting that the latter was at a selective advantage over the former. The same conclusion is reached if the number of moths which disappeared during the day is considered. Some 54.21 per cent ($n = 107$) of the typical moths released disappeared, while only 37.43 per cent ($n = 366$) of *carbonaria* could not be found.

The second experiment involved releasing large numbers of moths of both types which had been marked on the underside with a small dot of paint. Using mercury vapour moth traps (Fig. 5.2) and assembling traps (Fig. 5.3), large samples of moths were also captured. Each of these moths was checked to see whether it had been marked. The recapture data are given in Table 5.3. They show that the proportion of f. *carbonaria* that was recaptured was more than twice that of f. *typica*.

The results of the mark–release–recapture experiment could be explained in several ways. The melanics might be attracted to traps more readily than the typicals. Alternatively, the typical moths might be more dispersive than the melanics, and so be more likely to move out of the capture area. Another possibility is that

Fig. 5.3 A pheromone trap, of the type used by Clarke on the Wirral. Virgin females are placed in the central core to attract males which become trapped in the outer holding area. (Courtesy of Sir Cyril Clarke.)

Table 5.3 Recovery of marked *typica* and *carbonaria* forms of the peppered moth released in an unpolluted (Dorset) and polluted (Birmingham) woodlands. The recapture rate for *typica* in unpolluted woodland is approximately double that of *carbonaria*. Precisely the reverse is true in the polluted woodland. (From Kettlewell 1955a, 1956.)

		typica	*carbonaria*	Total
Dorset, 1955	Released	496	473	969
	Recaptured	62	30	92
	% of releases recaught	12.5	6.3	
Birmingham, 1953	Released	137	447	584
	Recaptured	18	123	141
	% of releases recaught	13.1	27.5	
Birmingham, 1955	Released	64	154	218
	Recaptured	16	82	98
	% of releases recaught	25.0	52.3	

carbonaria males are generally more long-lived than *typica* males for physiological reasons. The final possibility is that the melanics were eaten less by birds than the non-melanics. Given his direct observations of birds taking moths from tree trunks, Kettlewell was inclined to support this latter explanation, and indeed there was considerable circumstantial evidence against each of the other possibilities (see Kettlewell 1973).

Kettlewell (1955a) published his results in *Heredity*. They were received with scepticism in some quarters. As Kettlewell (1973) himself notes, ornithologists doubted that a single pair of great tits could eat 16 peppered moths in half an hour, and lepidopterists doubted that birds ate cryptic day-resting moths, because they had never seen it. In addition, some evolutionary geneticists were still suspicious that the recorded selective advantage of *carbonaria* over *typica* was so great. Consequently, Kettlewell resolved to produce a visual record of the experiments by filming them, and persuaded Professor Niko Tinbergen to collaborate with him.

In 1955, the experiments were repeated in Birmingham, with a second set of similar experiments being conducted, in parallel, in an unpolluted woodland, in Dorset. The basic procedures in Birmingham were similar to those employed in 1953, with the exception that a proper hide was built, to enable predation of moths from tree trunks to be filmed. The filming was a considerable success, Tinbergen succeeded in recording, on 16 mm film, the feeding activities of redstarts, *Phoenicurus phoenicurus*, over a 2-day period. Over this period equal numbers of moths of the two forms were released on to tree trunks in full view of the hide, so that the number of moths eaten could be recorded accurately. Three moths of each form were put out at a time, and a new six were put out as soon as all of one form had been eaten, any remaining of the other form being retrieved. Over

Table 5.4 Direct observation on predation of peppered moths by wild birds. (a) Data from the Christopher Cadbury Bird Reserve, Birminham, 1955; (b) data from Deanend Wood, Dorset, 1955. (Observations by N. Tinbergen and H.B.D. Kettlewell.) (Adapted from Kettlewell 1956.)

(a) Predation by redstarts

	typica	carbonaria	Total
19 July a.m.	12	3	15
20 July a.m.	14	3	17
20 July p.m.	17	9	26
Total	43	15	58

(b) Predation by five species of wild bird

	carbonaria	typica	Totals
Spotted flycatcher	46	8	52
(*Muscicapa striata* L.)	35	1	36
Nuthatch	22	8	30
(*Sitta europaea* L.)	9	0 (first day)	9 (first day)
	9	3 (second day)	12 (second day)
Yellow hammer	8	0	8
(*Emberiza citrinella* L.)	12	0	12
Robin	12	2	14
(*Erithacus rubecula* L.)			
Thrush	11	4	15
(*Turdus ericetocum* L.)			
Total predation observed (for days when records where kept)	164	26	190

Note: On all occasions, these observations commenced with equal numbers of both phenotypes being presented on tree trunks. They were replaced when all of one phenotype had been taken.

the two days, Tinbergen recorded 58 moths that were eaten, almost three-quarters of these being *typica* (Table 5.4a).

The mark–release–recapture experiments in Birmingham in 1955 produced higher recapture rates for both forms than those of 1953 (Table 5.3), a feature that Kettlewell (1956) attributes to the smaller area of the wood being worked. However, again the proportion of *carbonaria* recaptured was approximately double that of *typica*, corroborating Kettlewell's findings of two years before.

Kettlewell's decision to conduct a similar set of experiments in an unpolluted woodland, in the same year as those undertaken in Birmingham, was inspired. He reasoned that should such experiments produce contrary results to those from the polluted woodland, his critics would have to accept his demonstration of Tutt's thesis that differential bird predation was the main cause of the increase in *carbonaria* in industrial regions of Britain.

The wood that Kettlewell chose was Deanend Wood in Dorset, part of a relict deciduous woodland that appeared little affected by industrial fallout. Lichens were abundant in the wood, festooning most tree trunks and branches. Here, Kettlewell already knew that *typica* was the predominant form, a sample of 20 peppered moths, caught on a single night in 1954, containing no *carbonaria* (see Table 5.1).

The work in Deanend Wood replicated that in Birmingham in all important respects. The conspicuousness of moths of both forms on trunks in the wood were scored according to Kettlewell's cryptic index. Direct observations were made of moths of both forms released on to trunks. A large mark–release–recapture experiment was conducted. Finally, Tinbergen again used hides to allow him to record the work on cine film.

The results of each type of experiment were more or less a mirror image of those obtained in Birmingham, with *typica* appearing to be at an advantage over *carbonaria*. The rurality and unpolluted nature of Deanend Wood is reflected in the fact that everyone of some 120 *typica* released were classed as inconspicuous, with 85 per cent being given a score of +3. Conversely, all the *carbonaria* released (*n* = 75) were classed as conspicuous, some 80 per cent scoring –3.

Observations of live moths of both types released on to trunks did produce one unexpected set-back. More or less as soon as the experiments commenced, Kettlewell became suspicious that the efficiency of the crypsis of the *typica* form was making it difficult to find again some of the moths released, with the result that when scoring which moths had been removed and which remained in place, some *typica* appeared to have disappeared, simply because he was unable to find them again against the lichened trunks. He carefully checked this interpretation by releasing a set of *typica* and *carbonaria* moths on to trees in a small region of the wood, and then remaining close to them, so that his presence during the day would deter any birds from entering the area. He then attempted to find all of the moths at the end of the day. While he was able to retrieve all the f. *carbonaria*, 30 per cent of the typical moths released could not be found and remained unaccounted for, despite the absence of birds. The result of this set-back was that these experiments were discontinued and Kettlewell had to rely on moths that were seen from Tinbergen's hide being predated for data on the predation rates of the two forms.

The methods involved in releasing moths on to tree trunks were the same as those used in Birmingham, six moths (three of each type) being released on to a trunk, and these being replaced only after all of one form had been taken. Five species of bird were observed to take moths, and all took the melanic moths more often than f. *typica* (Table 5.4b). Some 86.32 per cent (*n* = 190) of the moths eaten were of the melanic form. These results are the opposite of those from Birmingham, but again accord with the differential predation thesis.

Some of the initial mark–release–recapture experiments in Deanend Wood were considered by Kettlewell to be unsatisfactory, largely because, initially,

moths were released in too small an area, thus producing a high density of moths
which attracted bird predators. This would produce a bias in the results, such that
the predation of the less conspicuous form would be unnaturally high. This is
because birds that have found one prey item tend to search the immediate area
more intensely than they would otherwise. This behaviour, known as area
restricted searching, is well known among many predators. The consequence of
this type of behaviour is that the presence of a conspicuous moth on the same
trunk as an inconspicuous one would increase the probability of the more cryptic
moth being found, once the conspicuous moth had been eaten. To reduce this
possibility, releases were later made over a much wider area. The results of the
mark–release–recapture experiments are presented in Table 5.3. When all results
are considered, the recapture rate for *carbonaria* is roughly half that of *typica*.
When those results that Kettlewell argues contain bias are excluded, the recapture rate for *carbonaria* drops to about a third of that for *typica*.

It was the reciprocal nature of the results from the two woods, together with
the visual record on film, that had such impact on the scientific community and
finally convinced the sceptics. Here was a polymorphic moth in which the fitness
of the two forms varied from one environment to another because of obvious
differences in the nature of the environments. Differences in the characteristics
of the forms meant that one form, *carbonaria*, was at an advantage in polluted
woodland, while the other form, *typica* was favoured in unpolluted woodland.
The mechanism of selection, differential bird predation, had been identified and
demonstrated. Furthermore, this natural selection apparently did have an effect
on the evolution of the natural populations of peppered moths, causing f. *carbonaria* to reach a high frequency in the polluted woodland, but preventing it
from invading the population in Deanend Wood.

The frequency of f. *carbonaria* across Britain

The apparent correlation between the fitnesses of forms as a result of bird predation in a particular location, and the frequency of forms in that location, was
crucial because it demonstrated that natural selection produces evolutionary
change. On the basis of this connection, one should be able to predict, at least
roughly, the frequencies of melanic and non-melanic forms in different geographic regions, on the basis of pollution levels and their influence on habitats.
The *carbonaria* form should attain high frequencies in heavily polluted areas,
while f. *typica* should still predominate in unpolluted regions. Areas with intermediate pollution levels should have both forms present, at intermediate
frequencies.

Kettlewell, therefore, embarked on a survey of the frequencies of the forms of
the peppered moth throughout Britain, to see if they correlated with measured
pollution levels from different locations. He collected form frequency data in
three ways. First, prior to 1952, he collected data from published records, museum

and private collections, and from the notebooks of other lepidopterists. Second, he personally sampled peppered moths at many locations around Britain from 1952 onwards, using light and pheromone traps. Third, he co-opted the help of a group of about a hundred amateur lepidopterists across Britain, to collect form frequency data of the peppered moth and many other species with melanic forms, from 1952 onwards.

The data collected from these sources and later surveys were collated and published by Kettlewell in 1973 in his book on melanism. It is therefore unnecessary to repeat it here in full. However, three important deductions from these surveys must be described and explained.

The selective advantage of f. *carbonaria* over f. *typica*

Comparison of the frequency of *carbonaria* at specific locations over time allowed Kettlewell to estimate the selective advantage of *carbonaria* over *typica* during its spread. Unfortunately, most of the data from before 1952 contained rather sparse information, often without the numerical details necessary to allow formal statistical analysis. So, for example, the first records of the *carbonaria* form in many locations report nothing more than that the variety had been taken, with no detail of sample size being provided. Similarly, the survey initiated by the Evolution Committee of the Royal Society in 1900 provided some indication of the comparative frequencies of the forms, but in the main the data were given in rough percentages, without sample sizes. The result is that the selective advantages obtained by Kettlewell (1973) can only be viewed as rough approximations. Despite this, the data support the high fitness differences between the two forms that Kettlewell and Tinbergen observed in their predation experiments.

The correlation between melanic frequency and pollutants

The various surveys conducted during the 1960s and 1970s allowed frequency maps to be drawn and considered alongside maps showing pollutant concentrations (Figs 5.4 and 5.5). Such comparisons suggested a very strong correlation between melanic frequency and pollution levels. High melanic frequencies occurred primarily close to, or to the east of, industrial centres. Rural regions in the west of England and Wales, and the north of Scotland were characterised by an absence, or very low frequency of *carbonaria*. The high frequencies in apparently rural parts of eastern England were explained by Kettlewell as the result of the prevailing south-westerly winds blowing pollutants across from the industrial Midlands and London. This explanation seems convincing given the scarcity

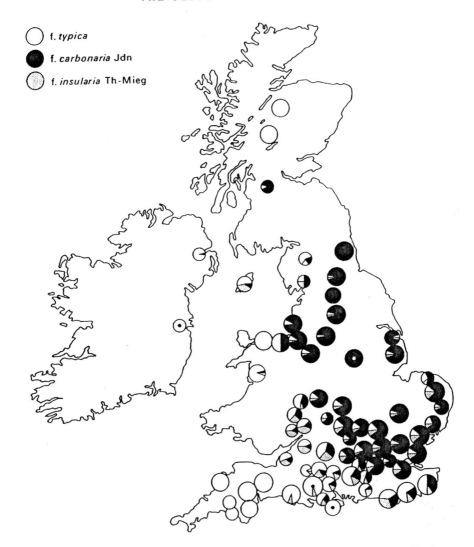

Fig. 5.4 Frequencies of melanic forms of the peppered moth in Britain. The black segments of each pie = the frequency of *carbonaria*; the shaded segments = the frequency of *insularia*; the white segments = *typica*. (Data from Kettlewell 1973; Lees and Creed 1975; Steward 1977*a*; Bishop *et al.* 1978*a*; Lees 1981.) (After Lees 1981.)

of lichens on tree trunks in the eastern counties. Indeed, in 1954, when Kettlewell was seeking an unpolluted woodland as a site for predation experiments, he states that he could find no lichened woods in the eastern half of England south of Yorkshire.

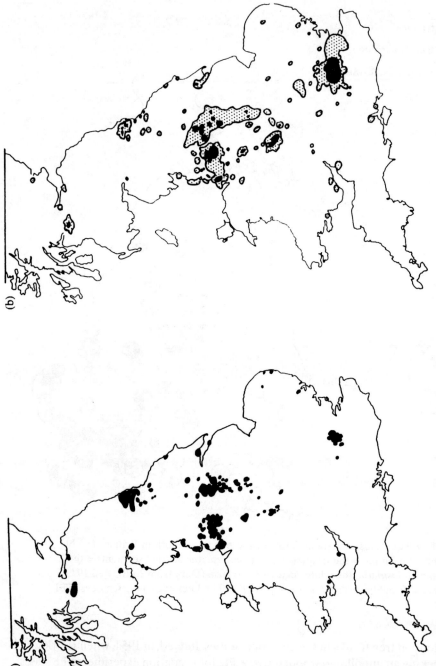

Fig. 5.5 Distribution of pollutants in Great Britain in 1975–76. (a) Smoke pollution: black areas = mean winter concentrations greater than 45 mg/m³. (b) Sulphur dioxide: grey areas = mean winter concentration of 50–100 mg/m³; black areas = mean winter concentrations greater than 100 mg/m³. (Data from the Warren Spring Laboratory.) (After Lees 1981.)

The decline in *carbonaria* as a consequence of anti-pollution legislation

Some later samples of the peppered moth (those taken from 1962 onwards) show an increase in the frequency of the typical form in areas where *carbonaria* was the predominant form. The general explanation of this evolutionary reverse is that anti-pollution legislation, brought in during the 1950s, had begun to lead to a reduction in both particulate soot fallout and SO_2 emissions, and that as the environment became cleaner, the advantage of being melanic in industrial regions was reduced, so that the frequency of *typica* began to rise. Many other frequency surveys of peppered moths have been conducted over the last 25 years, and many of these support the view that in recent times it is the typical form that is increasing, while *carbonaria* declines in frequency. These data, dealing with the fall in frequency of melanic peppered moths, will be considered in detail in Chapter 6.

Bernard Kettlewell's observational and experimental work on the peppered moth had a considerable influence on the subject of evolution, for it moved the primary perceived mechanisms of biological evolution from theory to fact. As MacArthur and Connell (1964) put it: 'It used to be argued that natural selection was only a conjecture, because it had not been actually witnessed'. But Kettlewell, along with Tinbergen and other co-workers did witness it, and furthermore, they filmed it, so that others could see natural selection in operation.

The importance of this case for advocates of biological evolution is one of two reasons why interest in it has continued over the four decades since Kettlewell's first predation experiments in Birmingham and Dorset. The second reason is that the Clean Air Acts and the ensuing rise in the typical form of the peppered moth, at the expense of *carbonaria*, offered an opportunity to fill a major gap in scientific knowledge. Although it is often said of the melanic peppered moth that we witnessed its rise, in fact very little precise data of the increase in the frequencies of *carbonaria* exist. By the time scientists began looking at the case in detail, most of the increases had happened. Ford's (1975) sadness at the failure of biologists to follow the rise of *carbonaria* in the nineteenth century reflected this. But the decline in *carbonaria* following the Clean Air Acts has raised the potential to make up for this failure.

Expansion of the work over the last four decades, and in particular during the last two, has concentrated on the fine detail, using both theoretical and experimental approaches. In the main, this work has confirmed the thrust of Kettlewell's findings, while at the same time revealing an understanding that many of the subtleties and nuances of the system depend on the fine detail of the ecology and behaviour of the moth. It is with this expansion of the work, and the complexity that is often obscured in texts giving the story in a page or two, that the next chapter will deal.

6 The peppered moth story dissected

Introductory remarks

The importance of industrial melanism in the peppered moth as one of the first, and still most cited examples of evolution in action, places emphasis on the need to be sure that the story is right. In the 40 years since Kettlewell's pioneering work, many evolutionary biologists, particularly in Britain, but also in other parts of Europe, the United States, and Japan, have studied melanism in this species. The findings of these scientists show that the prècised description of the basic peppered moth story is wrong, inaccurate, or incomplete, with respect to most of the story's component parts. When details of the genetics, behaviour, and ecology of this moth are taken into account, the resulting story is one of greater complexity, and in many ways greater interest, than the simple story that is usually related. In this chapter I consider each of the components of the basic story in turn, and then discuss a number of further elements that are rarely included in the basic story.

Before beginning the dissection of the story, two points should be stressed. First, it is important to emphasise that, in my view, the huge wealth of additional data obtained since Kettlewell's initial predation papers (Kettlewell 1955a, 1956), does not undermine the basic qualitative deductions from that work. Differential bird predation of the *typica* and *carbonaria* forms, in habitats affected by industrial pollution to different degrees, is the primary influence on the evolution of melanism in the peppered moth.

Second, during the course of the more recent work, it has become obvious that much of both the experimental and theoretical work suffers from artificiality. In many experiments, moths are placed into artificial situations, which may affect their own behaviour or that of their predators. For example, in most predation experiments peppered moths have been positioned on vertical tree trunks, despite the fact that they rarely chose such surfaces to rest upon in the wild.

In theoretical assessments of the evolution of melanism in the peppered moth, simplifying assumptions are often made to keep the complexity of the mathematical models used within reason. While this process is normal practice, sadly, in the case of the peppered moth, the assumptions made have later often been shown to be unjustified. To give just one example at this stage, one model (Cook and Mani 1980) of the effect of migration on the frequencies of the forms over geographic distance, makes the assumption that:'the males are first selected, then they migrate and mate with females who have also encountered selection'. The painstaking work of Tony Liebert and Professor Paul Brakefield has shown that this assumption is incorrect in two respects. First, males fly each night of their lives in search of mates, so that selection and migration cannot be sequential (Brakefield and Liebert 1990). Second, the model implies that females do not migrate at all, yet Liebert and Brakefield (1987) have shown that females do have a dispersal flight on the first night following eclosion. In females then, it would be reasonable to treat migration and selection sequentially, but only if migration is assumed to precede selection, rather than following it.

Simplifications of this kind, made for logistical reasons or because relevant empirical data are not available, are excusable, as long as they are recognised. Unfortunately, this has not always been the case, with the result that additional factors have been included to improve fits when models and observation do not agree. I will point to some of these as I consider the components of the basic story because some of them are rather persistent, despite having no good foundation in experiment or observation.

The components of the peppered moth story examined

The peppered moth has two distinct forms

If a large number of peppered moths, collected across the British Isles, are examined, it is obvious that the majority of individuals are either f. *typica* or f. *carbonaria*. However, a significant proportion of individuals do not fit the descriptions of either of these forms. The variation in patterning extends from white, with just a very fine dusting of dark scales, to almost completely black, with all grades of intermediates between the two (Bowater 1914). The intermediates between *typica* and *carbonaria* comprise a range of forms collectively known as f. *insularia* (Plate 3e, Fig. 6.1). This name embraces all intermediates between *typica* and *carbonaria*. As there is variation in the speckling on *typica* peppered moths, and occasionally individuals bearing the *carbonaria* allele have a very light dusting of pale scales, it is sometimes difficult to allocate individuals to a particular form with certainty. The palest 'genetic' *insularia* are paler than some *typica* and the darkest are indistinguishable from some *carbonaria*. As the *insularia* complex of forms embraces a continuous range of forms between *typica* and *carbonaria*, large collections contain an apparently continuous range of variants from the least to the most melanic.

On the other hand, if random samples are taken from around the country, the

(a)

(b)

Fig. 6.1 Dark and pale *insularia* forms of the peppered moth.

proportion of the wing that is black is measured and plotted as a frequency distribution, the distribution is obviously bimodal. One mode corresponds to f.
typica, the other to f. *carbonaria*. The frequency of forms with percentages of
black between these two modes is always low, partly because in most regions

Phenotypes in decreasing order of darkness	Allele	Genotypes
carbonaria	C	CC, CI^3, CI^2, CI^1, CT
		↑ ↓↑
insularia 3	I^3	I^3I^3, I^3I^2, I^3I^1, I^3T
insularia 2	I^2	I^2I^2, I^2I^1, I^2T
		↑
insularia 1	I^1	I^1I^1, I^1T
typica	T	TT

Vertical arrows indicate that the genotype may have the phenotype to which the arrow points.

Fig. 6.2 The inheritance of the *typica*, *insularia*, and *carbonaria* forms of the peppered moth. The inheritance involves five alleles of a single gene locus: dominance is not entirely complete. (Adapted from Lees 1981: based upon Lees and Creed 1977; Steward 1977*b*; Clarke 1979; with additional data from Majerus (unpublished).)

insularia is scarce, and partly because there is so much variation in the amount of black on different *insularia* individuals.

The genetics of melanism in the peppered moth

The basic forms of the peppered moth are indeed controlled by two alleles at a single genetic locus, and the *carbonaria* allele is dominant to the *typica* allele. However, the situation is complicated by the existence of the *insularia* phenotypes. The *insularia* complex of forms is controlled by at least three additional alleles. All are more or less recessive to *carbonaria* and dominant to *typica* (Clarke and Sheppard 1964; Lees 1968; Lees and Creed 1975, 1977; Steward 1977*b*). In general, dominance increases with increasing darkness, although there are minor exceptions (Lees and Creed 1977; Steward 1977*b*; Clarke 1979). The dominance relationships between these five alleles are shown in Fig. 6.2.

There is evidence that modifier genes have an influence on the progression of dominance. Indeed, Kettlewell (1973) notes that in the nineteenth century, the dominance of *carbonaria* over *typica* was not complete. Part of Kettlewell's evidence is based on the finding of '*carbonaria*' individuals in Victorian collections, with a dusting of white scales centrally on the forewings and streaks on the hindwings (Kettlewell 1958*a*). In an elegant series of crosses between *carbonaria* and either Cornish *typica* or the Canadian subspecies *Biston betularia cognataria* (Plate 3f), Kettlewell (1965) showed that the dominance of the *carbonaria* form may be broken down. For example, a cross between a heterozygous *carbonaria* from Birmingham and a Canadian *typica*, gave a normal 1:1 segregation of *carbonaria*:*typica*. However, when these were inbred, within three generations, the clear segregation was completely broken down. The dominance of the *carbonaria* allele over the *typica* allele had been destroyed in this line. Kettlewell then took moths of the next generation and crossed these to *typica* individuals of industrial origin, in order to begin to reconstitute dominance. Despite Kettlewell's own belief that it would take several generations to build up the dominance again by increasing the proportion of the genetic background that was of industrial British origin, a clear-cut segregation of *carbonaria* to *typica* was obtained in just one generation. Kettlewell (1973) argues that this breakdown of dominance is the product of the segregation out of dominance modifier genes, which are present only in populations which have been exposed to selection favouring the full expression of the *carbonaria* allele, and which would thus have been absent from the Cornish and Canadian populations, which lack melanic forms. The difference in the number of generations required to break down and build up dominance is explained if the dominance modifiers are genetically dominant to their allelomorphs.

While this evidence on the evolution of dominance is fully in accord with theory (p. 90), it is not uncontested. Several authors (Lees 1971, 1981; Bishop *et al.* 1978*a*) have pointed out that individuals fitting the description of Kettlewell's Victorian *carbonaria* are regularly found in modern samples from the Liverpool and Birmingham areas. Furthermore, Mikkola (1984) obtained no breakdown of dominance in crosses between Liverpudlian *carbonaria* and Finnish *typica*, while West (1977) also failed to obtain dominance breakdown when crossing *carbonaria* with Appalachian *B. b. cognataria*, although here the situation is complicated by the presence, at low frequency, of the American melanic, f. *swettaria*, which is absent from Canada.

Given that the *carbonaria* form is now declining rapidly in many British populations (p. 115), it is interesting to consider what may happen to the dominance of the *carbonaria* allele over the *typica* allele in the future. Fisher's theory of the evolution of dominance predicts that, over time, modifying genes which increase the expression of favourable alleles in heterozygotes will be selected. It is probable that in populations where the melanic form of the peppered moth has never been favoured, the modifiers that usually make *carbonaria* dominant over *typica* do not exist, or at least have never been fixed. These populations will now act as a reservoir of modifiers which would cause the dominance of *carbonaria* to break down, without the need for new modifiers to be produced by mutation. Migration from such populations into regions where *carbonaria* has occurred, will begin to spread these breakdown of dominance modifiers into regions where they will be favoured, because they reduce the expression of the now deleterious *carbonaria* allele. The question is, will this actually happen? The evolution of dominance, through selection acting on modifiers, is a relatively slow process, and will depend in this case on whether *carbonaria* heterozygotes are phenotypically variable enough for selection to favour those in which the *typica* allele is not fully dominated by the *carbonaria* allele. As will be seen, obtaining accurate measures of the relative fitnesses of even quite distinct forms is fraught with difficulties. Thus it does not seem practical to attempt to measure the fitnesses of *carbonaria* heterozygotes in which dominance is complete, compared with those in which dominance is breaking down. Consequently, the only test of this possible natural breakdown of dominance will be to score phenotypes in the field, seeking *carbonaria* individuals that have small patches of white scales on the central portions of the wings, as occurred on Kettlewell's Victorian '*carbonaria*'. If any individuals meeting this description are found, they should be subjected to genetic analysis by performing controlled crosses to typical moths of both industrial and non-industrial origin.

Modifier genes also influence the fine pattern of the typical form of the peppered moth. This is shown by the response, in true breeding lines of this form, to artificial selection: crosses between paler individuals tend to produce pale progeny, and crosses between dark f. *typica* tend to produce darker moths (Majerus 1989*b*). The *insularia* form would probably respond in a similar way, although the presence of three alleles controlling this complex, and their dominance over the *typica* allele, makes the production of homozygous stocks of *insularia* difficult.

The resting behaviour of peppered moths

When Kettlewell released peppered moths in Birmingham and Dorset to study bird predation, he did so by placing live moths on to tree trunks where they could be observed (Kettlewell 1955*a*, 1956). Most subsequent experiments to determine predation rates on peppered moth morphs have used dead moths, glued in 'natural postures' to tree trunks. The estimates of the relative fitness of the forms that have been obtained in this way have generally been, at least qualitatively, consistent with the thesis that, compared with *typica*, melanic moths are more heavily preyed upon in unpolluted woods, and less heavily preyed upon in polluted woods. However, a number of workers have questioned the quantitative accuracy of these estimates, because peppered moths do not naturally rest in exposed positions on tree trunks (Mikkola 1979, 1984; Howlett and Majerus 1987; Liebert and Brakefield 1987).

Data on the natural resting sites of the peppered moth are pitifully scarce, and this in itself suggests that peppered moths do not habitually rest in exposed positions on tree trunks. Many nocturnal moths do rest by day on tree trunks and searching trunks has long been a recognised method of collecting employed by lepidopterists. However, the number of published records of peppered moths being found on tree trunks is negligible. This is emphasised in an admission by Sir Cyril Clarke (Clarke *et al.* 1985): 'all we have observed is where the moths do not spend the day. In 25 years we have only found two *betularia* on the tree trunks or walls adjacent to our traps (one on an appropriate background and one not), and none elsewhere'.

The largest data set of the resting positions of wild pepper moths found in truly natural positions (i.e. not near moth traps or other light sources which may have attracted the moths) is of just 47 moths found over a period of 34 years (Howlett and Majerus 1987). Analysis of these results (Table 6.1) and an additional data set of the resting positions of moths found close to moth traps or street lights (Table 6.2), led Howlett and Majerus to conclude that peppered moths generally rest in unexposed positions, using three main types of site: (a) tree trunks, a few inches below a branch/trunk join so that the moth is in shadow; (b) the underside of branches (Plate 3a–c); and (c) foliate twigs (Plate 3d and Fig. 6.3). They also note that their data are bound to be biased towards the lower parts of trees, as these are most easily searched. These findings corroborate those from experiments to investigate the resting positions of captive male (Mikkola 1979, 1984) and female (Liebert and Brakefield 1987) peppered moths.

Mikkola watched male moths taking up resting positions in large experimental cages containing sections of trees, and concluded that in nature this species probably rests on the underside of horizontal branches in the canopy, where it may be less prone to bird predation, or may be exposed to different predators from those which habitually search tree trunks.

Liebert and Brakefield (1987) conducted similar experiments using females and obtained similar results. Most of the moths rested under, or on the side of, horizontal branchlets in the tree canopy, rather fewer resting on non-horizontal branches, main branches, or trunks.

Anecdotal support for the proposition that peppered moths tend to inhabit woodland canopies, high above the ground, comes from the finding that moth

Fig. 6.3 A typical peppered moth at rest in hazel foliage.

traps set on the roof of Juniper Hall in Surrey, caught more than four times the number of peppered moths than similar traps set at ground level (Clare Dornan and Bryony Green personal communication).

It is worth noting that the view that peppered moths do not always, or even usually, rest in exposed positions on tree trunks is not original. Kettlewell (1958b) himself was aware that tree trunks were a less commonly used resting site than under branches, for he wrote:

> whilst undertaking large-scale released of both forms (f. *typica* and f. *carbonaria*) in the wild at early dawn, I have on many occasions been able to watch this species taking up its normal resting position which is underneath the larger boughs of trees, less commonly on trunks.

If the relative fitness of the morphs of the peppered moth does depend on their crypsis, the resting position is crucially important to the estimation of fitness differences between the morphs. This is particularly the case in changing or intermediate habitats with respect to pollution, because in such habitats the distribution of lichens on trees is likely to be more heterogeneous than in very unpolluted or very polluted habitats. It is therefore valuable to consider, albeit briefly, on which parts of trees lichens of different types tend to grow in different situations.

In parts of rural Britain which have been exposed to only low levels of atmospheric pollution, foliose lichens occur mainly on the less shaded and wetter sides

Table 6.1 The resting positions of peppered moths found in the wild between 1964 and 1996. (Adapted, with additional data, from Howlett and Majerus 1987.)

	typica	insularia	carbonaria
Exposed trunk	3	1	2
Unexposed trunk	2	1	3
Trunk/branch joint	10	4	6
Branches	5	3	7

Table 6.2 Resting positions of peppered moths found in the vicinity of mercury vapour moth traps at various locations, between 1965 and 1996. (Adapted, with additional data, from Howlett and Majerus 1987.)

	typica	insularia	carbonaria
Exposed trunk	25	7	16
Unexposed trunk	10	6	6
Trunk/branch joint	33	10	23
Branches	9	4	7
Foliage	9	5	8
Man-made surfaces	13	5	7

Fig. 6.4   A branch of a Scots pine in Rannoch Black Wood. Notably, the growth of foliose lichens is richer on the upper side of the branch.

of trunks, upper surfaces of main branches, and more generally within the canopy (Liebert and Brakefield 1987). Furthermore, these lichens tend to grow more on upper surfaces of lateral branches, rather than under them, even in regions unaffected by pollution (Fig. 6.4). Crustose lichens occur on these surfaces as well, and dominate on trunks and the lower surfaces of main branches. In areas affected by high pollution levels, lichens are almost non-existent on trees, usually only *Lecanora conizaeoides* being found, and even this species is absent from the most heavily polluted locales (Cook *et al.* 1990).

These then are the two extremes. In areas subjected to moderate air pollution, foliose lichens, which are more susceptible to pollutants than some crustose species, tend to be rare and restricted to small patches on the upper surfaces of some branches or branchlets (Liebert and Brakefield 1987). The pattern of recolonisation is also of interest, because of the reduction in pollution levels over the last three decades. Initial lichen recolonisation of older trees, made barren by past pollution, is likely to be on new growth of branches in the tree canopy, with low residual surface pollution and comparatively high light intensity (Brakefield 1987). The last areas of the trees to be recolonised will be the trunks, largely as a result of the down-wash of pollutants on to the trunk (Disney personal communication). The first species to return is usually *Leucanora conizaeoides*, followed by other crustose lichens, with foliose lichens lagging some way behind

(Cook *et al.* 1990). If lichens return to some parts of trees faster than to others, the rate of change in the relative fitnesses of the forms of the peppered moth, as pollution levels decrease, will undoubtedly depend upon the natural resting positions of the moths. A concerted effort to obtain a substantial data set showing where peppered moths normally rest in the wild is urgently needed.

Birds find and eat peppered moths

One of the main reasons why Tutt's (1896) suggestion that bird predation was influential in the evolution of industrial melanism was initially ignored, was that lepidopterists and ornithologists tended to agree that birds were not major predators of trunk-resting adult moths. However, following Kettlewell and Tinbergen's production of a filmed record of wild birds taking live peppered moths off tree trunks, the evidence that birds are major predators of adult moths has slowly increased. Many ornithologists and entomologists have now shown by experiment and observation that birds do eat significant numbers of moths, and it is now the accepted view that bird predation is one of the main causes of mortality in many species of adult moth. Although observations of peppered moths being taken from natural resting positions are still lacking and are urgently needed, it is highly probable that predation levels are significant.

The level of bird predation of peppered moths is dependent on the moths' crypsis

The evidence from Kettlewell's predation experiments and other similar experiments strongly suggests that the likelihood of a moth being eaten by a bird will increase in environments where it is less likely to find a resting situation where it is well camouflaged. Indeed, this suggestion is the basis of the accepted peppered moth story. Yet, surprisingly, experiments to show formally that the degree of crypsis of the different peppered moth forms does affect the level of predation inflicted upon them by birds have never been carried out.

The design of experiments to demonstrate the importance of crypsis could be very simple. All that would be needed is two large flat transportable surfaces to act as artificial backgrounds to place moths upon. One of these should be the same basic colour and pattern as f. *typica*, the other resembling f. *carbonaria*. Rory Howlett (1989) developed a method of producing just such backgrounds, using a stipple brush, for his work on resting site selection, where his need was for backgrounds with the same mean and variance reflectance characteristics as the wings of the two forms. Once feeding boards have been produced, these should be placed in a situation where they may be easily watched. Given the artificiality of the boards, it may be necessary to condition birds to come to feed from them by placing some conspicuous food items on the boards over a period of days (black, white, or grey items should be avoided). Then put one or two moths of each form

on each board and record how long it takes birds to find and eat them. This procedure should be repeated at intervals so that a sufficiently large data set can be
built up for statistical analysis.

One other point pertinent to this component of the basic story is worth making.
Many of the predation experiments have involved placing peppered moths out
on tree trunks in some specific way, e.g. on the lightest patch that could be found
on the trunk's surface, or in a position where it was most cryptic. Such methods
make the implicit assumption that the bird predators of peppered moths see in
the same way that humans do. Until the 1980s, this was assumed to be the case.
However, it is now thought that the vision of most, if not all birds is different
from ours in a number of ways. For example, while we have three types of colour
vision cells, called cones, which are sensitive to red, green, and blue light wavelengths, almost all birds have at least one additional cone that is sensitive to ultraviolet light (UV) (Bowmaker 1991; Goldsmith 1991; Jacobs 1992). Indeed,
analysis suggests that birds are more sensitive to UV than to light in the 'human-
visible' part of the electromagnetic spectrum (400-700 nm), with greatest sensitivity being about 370 nm (Chen *et al.* 1984; Chen and Goldsmith 1986; Burkhardt
and Maier 1989; Maier 1992). Not only do birds see well in the UV, but they can
also discriminate different wavelengths in the UV range; Emmerton and Delius
(1980) have already shown that pigeons can discriminate differences of 7–16 nm
in the 360–380 nm range. In addition, birds have oil droplets of varying sizes that
lie in front of their cones and may act as filters and give birds greater colour discrimination abilities than we have (Lythgoe 1979; Jane and Bowmaker 1988;
Partridge 1989; Goldsmith 1991; Bennett and Cuthill 1994). While many of the
details of bird vision, and that of many other organisms, still have to be illumi-
nated, it is already clear that birds have greater visual acuity than we do. This
means that many of the previous studies of the behaviour of peppered moths and
their predation, which have been designed by humans on the assumption that
birds see as we do, may need to be reappraised (see also p. 136).

Non-melanic peppered moths are at an advantage in unpolluted areas, while melanics have an advantage in polluted regions

Kettlewell's famous mark–release–recapture experiments in Birmingham and
Dorset provide strong evidence for the differences in fitness of the forms in polluted and unpolluted environments. Furthermore, his observation that live moths,
which had been released on to tree trunks, were differentially preyed upon by
birds provides evidence that the intensity of bird predation on the morphs varies
according to habitat. Similar differences in fitness of the morphs, in polluted and
unpolluted regions, have been reported by other workers (e.g. Clarke and Sheppard 1966; Bishop 1972; Lees and Creed 1975; Steward 1977c; Howlett and
Majerus 1987).

Jim Bishop's (1972) work in the industrial north-west and north Wales is par-

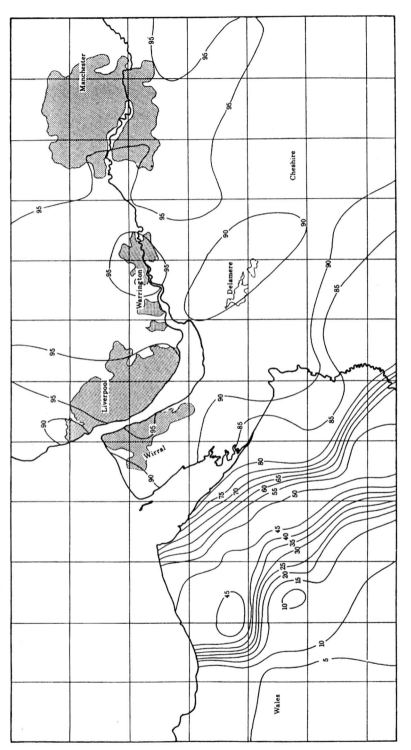

Fig. 6.5 Contour map of the frequencies of the *carbonaria* form of the peppered moth along the cline examined by Bishop (1972). (From Bishop *et al.* 1978*a*.)

ticularly notable. He worked along a transect running south-west from Merseyside into rural north Wales, surveying the frequencies of peppered moth forms
(Fig. 6.5). His aim was to gain as much knowledge as possible of the relative survival values of the different morphs, at different places along the transect, and
then, by use of a computer model incorporating all the information gained, test
whether the model fitted the observed frequencies of the forms along the transect. Estimates of the predation rates of the morphs were obtained by gluing dead
moths to tree trunks at seven sites along this transect. He collected data on the
number of eggs laid by captive females of each form, and the nights after eclosion when these were laid, to get measures of fecundity and oviposition rates.
Then he calculated the proportional contribution of each form to the next generation, by combining this information with the relative survival values.

Bishop also made assessments of the dispersal distances of the peppered moth
using a mark–release–recapture experiment, in which the release point was at the
centre of a grid of traps. From this he estimated that male peppered moths move
somewhere in the region of 2 km per year. Bishop then used a computer model
to simulate the rise, increase, and spread of *carbonaria* along a 54 km transect
from Liverpool, south-west into rural north Wales, assuming that at some time in
the early nineteenth century the frequency of *carbonaria* was close to the mutation rate. Because of his estimate of the peppered moth's migration rate, he split
the transect into 27 2 km blocks and allowed migration between adjacent blocks
each generation. The different genotypes were given relative fitnesses for different points along the transect on the basis of the predation experiments and the
oviposition data, and a variable level of heterozygote advantage was included,
largely because previous workers (e.g. Ford 1937, 1964; Haldane 1956; Clarke and
Sheppard 1966) assumed that the heterozygote had to be fitter than either
homozygote for the polymorphism to be maintained. Interestingly, Bishop also
included a weak frequency dependent selection component, on the basis that the
rare form towards either end of the cline would be less liable to be actively sought
by birds, because of its rarity. Although there was no evidence for this negative
frequency dependence in the peppered moth at the time, recent analysis of the
decline in *carbonaria* on the Wirral has shown that the selection on the rarer form
is indeed negatively frequency dependent (L.D. Hurst and Majerus, in preparation). Finally, Bishop cycled his model for 160 generations to bring the simulation from the early nineteenth century to 1970 (Fig. 6.6).

The frequencies of *carbonaria* along the transect produced by the model were
then compared with the frequencies of forms in wild samples caught at various
points along the cline. The fit was not good (Fig. 6.7), for although the model
predicted that the frequency of *carbonaria* would decline as one moved south-
west from Liverpool, the predicted position of this decline was much closer to
Liverpool than observed. Furthermore, the rate of decline in the observed data
was more rapid than expected from the model simulation. As Bishop (1980)
put it:

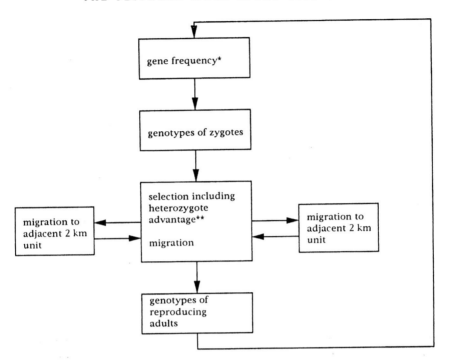

Fig. 6.6 Schematic flow diagram of Bishop's model to predict the 1972 *carbonaria* frequencies of the Liverpool/north Wales cline (Bishop 1972). The cline was split into 27 2 km units, and the model was cycled for 160 generations for each unit. *Bishop used an initial gene frequency for the *carbonaria* allele of 10^{-6}. **Bishop used a variable heterozygote advantage ranging from 5 to 15 per cent.

Additional selective forces need to be invoked to explain its (the cline's) position and the continued polymorphism in both Liverpool and north Wales. Some theories were discussed by Bishop *et al.*, (1978a) but interpretations were not supported by sound data.

In a series of elegant theoretical analyses, Mani (1980, 1982) has shown that the fit between simulated and observed frequencies for this transect can be greatly improved if strong non-visual advantages to *carbonaria* are introduced into the model. These treatments, and that of Cook and Mani (1980) also demonstrated that heterozygote advantage need not be invoked to account for the geographic variation in form frequencies. It has been suggested that analysis of deviations from expected ratios in segregating laboratory reared broods of the peppered moth provides some evidence in support of a non-visual selective advantage to *carbonaria* (Creed *et al.* 1980; Brakefield 1987). However, this contention has been challenged by Howlett and Majerus (1987). We suggested that the poorness of fit of Bishop's simulation to observed frequencies may result from quantitative errors in the relative fitnesses obtained, because moths were

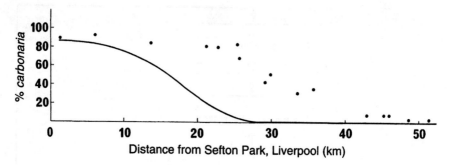

Fig. 6.7 Computer simulation of the decrease in *carbonaria* with distance from Liverpool (solid line). The points represent the actual data. (After Bishop 1972.)

put out for predation in unnatural positions, exposed and low down on tree trunks, rather than in their probable natural resting positions under branches or twigs in the canopy.

To support this view, Howlett and Majerus (1987) cite results from a predation experiment in which *typica* and *carbonaria* were positioned either exposed on tree trunks, or on the trunk 5 cm below the joint between a major branch and the trunk. The experiment was carried out at two oak woodland sites, one a polluted wood on the edge of Stoke-on-Trent, the other a relatively unpolluted wood on the western edge of the New Forest. At both sites 50 *typica* and 50 *carbonaria* were glued in pseudo-natural resting postures in each of the two situations, and the number that had been eaten recorded 72 hours later. The results are given in Table 6.3. The predation level for *typica* is significantly greater than that for *carbonaria* in the polluted wood, the reverse being true in the unpolluted wood, for both types of resting situation. These experiments thus support previous qualitative differences in survival values of the forms, *carbonaria* gaining an advantage in polluted areas, and *typica* an advantage in unpolluted regions. However, a comparison of the predation of moths placed in the two situations produces two further interesting points. First, the level of predation is significantly greater for moths placed in exposed positions than for those placed below trunk/branch joints. This is true for both forms in both woods. Second, the reduction in predation that accrues from being in an unexposed situation is greater for *carbonaria* than for *typica*, with the greatest reduction being for *carbonaria* in the unpolluted wood. The lower predation level for moths below branch trunk joints compared with those in exposed positions on trunks was endorsed by Carys Jones (1993), who conducted experiments placing dead moths of two species (peppered moth and brindled beauty) on trunks, or under horizontal branches, in Madingley Wood, Cambridge, at low density (two per acre). She found that those moths under lateral branches were about half as likely to be preyed upon than those on the trunks.

The results of these experiments seem to be intuitively sensible, for it seems

Table 6.3 Results of predation experiments to determine the effect of resting position on predation of the peppered moth in a polluted woodland (Stoke-on-Trent) and an unpolluted woodland (New Forest). (From Howlett and Majerus 1987.)

	typica		carbonaria	
	Eaten	Not eaten	Eaten	Not eaten
Stoke-on-Trent				
Exposed trunk	29	21	20	30
Trunk/branch joint	25	25	14	36
New Froest				
Exposed trunk	16	34	31	19
Trunk/branch joint	13	37	20	30

probable that a moth on a trunk is likely to be more obvious than one in shadow under a horizontal branch. Furthermore, it is not difficult to envisage that a uniformly dark moth will gain more of an advantage from resting in a dark shaded situation than a paler speckly moth. Nor does it seem unreasonable that the dark moth gains greatest advantage from resting in shadow in a wood where predation on it is most intense. The data sets from Howlett and Majerus' (1987) study are small and the difference in the reduction in predation of unexposed compared with exposed moths is not statistically significant in respect of either form or wood. However, qualitatively they appear to follow a reasonable course. Repetition of this type of experiment on a larger scale, and preferably with moths being placed under or on the side of lateral branches, or twigs in the canopy would be most valuable.

The results from the experiment in Stoke and the New Forest should be treated with scepticism until verified or refuted by a more extensive study. However, Howlett and Majerus (1987) state that this experiment may explain part of the lack of fit between Bishop's simulated frequencies and the observed frequencies for the Liverpool to north Wales transect. The greater reduction in the level of predation of *carbonaria* compared with *typica*, that results from resting in shaded rather than exposed positions, will have the effect of increasing the relative fitness of *carbonaria* all along the transect, compared with Bishop's (1972) values. Furthermore, because the change in predation is greatest for *carbonaria* in the unpolluted woodland, the increase in the relative fitness of *carbonaria* will be greatest at the south-western end of the transect. These changes in the fitness values will increase the predicted frequencies of *carbonaria*, particularly towards the southwestern end of the transect, thereby pushing the position of Bishop's predicted decline in *carbonaria* frequency appreciably away from Liverpool.

One other point, first stressed by David Lees (1981), is worth raising here. In Kettlewell's and some of the other predation experiments carried out on the forms in different parts of Britain, there is a good correlation between morph frequency and the species taken least (e.g. Kettlewell 1955*a*, 1956; Clarke and Sheppard 1966; Howlett and Majerus 1987). However, this is not the case for all the experiments. For example, in Bishop's (1972) experiments, *carbonaria* was taken more often than *typica* at some sites where its frequency was well over 50 per cent. The same was true in tests in East Anglia, where over 60 per cent of the population was *carbonaria* (Lees and Creed 1975). Conversely, at Tongwynlais in south Wales, where *carbonaria* represents only 18 per cent of the population, with *typica* and *insularia* both having much higher frequency, although *insularia* was least heavily preyed upon, *carbonaria* was taken less than *typica* (Steward 1977*c*). Steward (1977*a*) investigated the relationship between predation and morph frequency further by an indirect method. He assessed the 'relative crypsis' of the forms at 52 sites in southern England and compared the values obtained with the frequencies of the forms. Although he found that there was a correlation, it was weak, only accounting for 18 per cent of the variation in the frequencies in *carbonaria* between different sites, and 3 per cent of the variation for *insularia*. He contrasts this weak correlation with that for two other species, the green brindled crescent, *Allophyes oxyacanthae* (Linnaeus), and *Diurnea fagella* (Schiffermüller), which gave much stronger correlates, with over 60 per cent of the variance in melanic frequencies being accounted for by differences in crypsis. He concluded that for southern Britain, selective predation is only of secondary importance with regard to melanic frequency of the peppered moth, some other factor being the main determinant of melanic frequency.

For two reasons Steward's (1977*a*) conclusion must be deemed unsound. First, unfortunately, Steward assessed relative crypsis of moths by comparing them with tree trunks. While this is reasonable for the green brindled crescent and *Diurnea fagella* which both naturally rest on tree trunks, it is not reasonable for the peppered moth, which rests elsewhere. This will be particularly the case as the work was conducted in the mid-1970s, when the effects of anti-pollution legislation were beginning to be felt. The regrowth of lichens on horizontal branches, generally precedes that on vertical trunks (Brakefield 1987; Disney personal communication). Consequently, the relative crypsis of the forms of the peppered moth may have been quite different on lateral branches, where peppered moths probably do rest (p. 121), from on trunks, where they do not. Second, Steward's assessment of relative crypsis suffers from the criticism that it was made with a human's eye, not a bird's eye. As recent work by Jim Stalker has shown, different lichen species have very different UV reflectance characteristics from one another (Stalker personal communication). Thus the relative crypsis of the morphs of peppered moths to birds will depend on the lichen species present.

The frequencies of the forms of the peppered moth are determined primarily by selective predation and migration

If one form of the peppered moth is at an advantage in a particular habitat because it is more cryptic than other forms, the frequency of the fitter form will gradually increase, and, if the population is isolated, will eventually reach fixation, other forms being reduced to the level of rare recurring mutations. That the frequency of *carbonaria* has never reached 100 per cent in any population is attributed to two main factors. First, because *typica* is recessive to *carbonaria*, when the *typica* form becomes rare, declining to say just 1 per cent, the frequency of the *typica* allele will still be 10 per cent, but the majority of these alleles (90 per cent) will be 'hidden' in *carbonaria–typica* heterozygotes, where they will not be exposed to higher predation. In crosses between two heterozygotes, 25 per cent of progeny will be *typica*. Second, industrial regions, where *carbonaria* has reached high frequencies, are surrounded closely, or distantly, by more rural areas where *typica* has retained high frequencies. Continual migration from these rural areas into urban areas will ensure that *typica* retains some presence even in highly polluted areas. Similarly, migration of *carbonaria* out from industrial areas will cause this form to spread into surrounding rural regions. Here it may be at a selective disadvantage, so that a balance between selective elimination and immigration results.

Circumstantial evidence, supporting the importance of migration to the peppered moth case, comes from a comparison between the frequency distributions of melanic forms of this and two other species, the scalloped hazel (Fig. 6.8), and the pale brindled beauty (Plate 4a and b), in industrial north-west England and rural north Wales (Lees 1981). The frequency of *carbonaria* in the late 1960s and early 1970s between the river Dee and Manchester, was consistently over 85 per cent. In contrast, the frequency of the melanic form of the scalloped hazel, f. *nigra*, showed considerable variation, ranging from 80 per cent down to 10 per cent (Askew *et al.* 1971; Bishop *et al.* 1975; Bishop *et al.* 1978a,b). Moreover, the frequency of *nigra* drops dramatically to the west of Liverpool and is absent from rural north Wales. Ecologically, these two species differ in a number of respects. The most pertinent to this comparison is that while the peppered moth is a fairly mobile species, at least in respect of males, the scalloped hazel is extremely sedentary (Bishop *et al.* 1978b). In the absence of significant migration the frequencies of the forms in populations of the scalloped hazel will be a product of the local fitnesses of the forms.

The migration rate of the pale brindled beauty is thought to be intermediate between those of the peppered moth and the scalloped hazel (Conroy and Bishop 1980). If migration is important in determining form frequencies, the melanic forms of the pale brindled beauty should spread further into rural north Wales than melanic scalloped hazels, but not as far as melanic peppered moths. Consideration of a survey of a melanic forms of these species from Liverpool into rural north Wales confirms this expectation (Bishop and Cook 1975).

Although circumstantial, this evidence indicates that migration has an important influence on the frequencies of melanic forms in different regions. It should be noted, however, that although differences in the relative rates of migration in these species are consistent with their clines in melanic frequencies, this does not

Fig. 6.8 Typical (above) and melanic, f. *nigra* (below) forms of the scalloped hazel.

mean that the migration rates used to model morph frequency clines are accurate. Three points are worth noting. First, most assessments of migration rates
have involved only adult male moths (Bishop 1972; Bishop *et al.* 1978*b*), on the
assumption that females are sedentary. This assumption is probably safe, for
although Brakefield and Liebert (1990) have shown that females do have a dispersal flight on their first night after eclosion, this flight is generally short.
However, Brakefield (1987) has suggested that basing migration rates purely on
adult flight may have led to serious underestimates. Brakefield asserts that larvae
may disperse over long distances. After hatching, first instar larvae suspend themselves from silk threads, and may then be dispersed by air movements as part of

the aerial plankton. In mid-summer, strong thermals may be produced above tree canopies, taking minute larvae to high altitudes from where they may be passively moved considerable distances. Brakefield suggests that this type of migration may contribute to the unexpectedly high frequencies of melanics in some rural areas, such as East Anglia, which may receive many larvae carrying the *carbonaria* allele on the prevailing winds from the industrial Midlands and London.

Second, as the success of assembly traps in obtaining large samples of male peppered moths has shown, male moths are strongly attracted to the pheromones produced by virgin females. Males on the scent of females are thus likely to fly up a pheromone gradient, i.e. against the wind. In general this means that they will have a bias to fly in a south-westerly direction, against the prevailing wind. In Bishop's (1972) model simulation of the Liverpool/north Wales cline, random migration direction along the transect was assumed. If instead, the peppered moths showed a bias in their movement along this cline from Liverpool into north Wales, this would have the effect of increasing the frequency of *carbonaria* in north Wales, thereby improving the fit to the observed data.

Third, migration rates are not constant over time. Although a figure of 2 km per generation is generally quoted for the peppered moth, on the basis of Bishop's work, there is no doubt that in some years the dispersal distances may be much further. Indeed, Bishop himself recorded moths flying up to 5.8 km in a single night, and notes that moths may have flown further than this, but no moth traps were set at greater distances. My own view is that, in exceptional circumstances, adult male peppered moths may move very considerable distances. I have only two pieces of circumstantial evidence to support this view, but to my mind they are suggestive. In July 1990, there was a period of 4 days with a constant and strong, hot easterly wind. At the time I was trapping at Llanbedyr near the west coast of north Wales. Over a period of four nights, I was surprised to capture five *carbonaria* in a total of 37 peppered moths. Similarly, on the same dates, four *carbonaria* were taken in a trap on Exmoor, these being the first *carbonaria* that had been taken in 9 years trapping at the site. Had there been only one case, I might have expected that someone had bred and released melanics locally. However, the two instances together, I think, speak of long-range wind-driven migration. At both sites it is likely that these *carbonaria,* and any melanic descendants they had, would have been quickly eliminated by selective predation. But this type of occasional long distance dispersal may have been influential in the nineteenth century spread of the *carbonaria* allele around the country to polluted localities where it would not be eliminated by selection. Certainly, the speed of the movement of *carbonaria* in its radiation out from Manchester after 1848 (Fig. 5.1, p. 100) suggests an occasional migration rate greater than 2 km per generation. For example, in just 49 years *carbonaria* managed to move some 300 km almost due south to the New Forest.

Other factors pertinent to melanism in the peppered moth

The foregoing discussion of the basic peppered moth story should have shown that not only is melanism in this species very much more complex than suggested in most school texts, but also that there is some controversy over the quantitative accuracy of some of the estimates of the factors involved. A further complication to the story arises because some potentially important factors are not included in the basic story. I will here discuss three of these factors: the visual acuity of birds; resting site selection; and the effect of mating on selective predation.

The visual acuity of birds

A crucial element of Kettlewell's initial work on differential predation on the peppered moth involved the evaluation of the cryptic indices of *typica* and *carbonaria* forms of the peppered moth on tree trunks showing different levels of pollution. Initially, these indices were obtained by placing peppered moths of either form on oak trunks and assessing the distance from the trunks that observers found them indistinguishable from their background. Similarly, the relative crypsis of the forms in different locations obtained by Steward (1977a) used peppered moths viewed against trunks by humans. As I have already pointed out (p. 126), birds do not see as we do, having greater discriminatory prowess, and being highly sensitive to UV wavelengths, to which we are virtually blind. The exact way in which birds use their UV sensitive cones is not known. However, it seems more likely that they effectively have tetrachromatic vision, in which all four types of cone are stimulated simultaneously; thereby producing a mean wavelength from any point being viewed, rather than having the ability of using the UV sensitive cones independently from their other cones. If this is the case, it might be argued that as Kettlewell (1955a) and Steward (1977a) were mainly interested in the relative crypsis of the forms of the peppered moth in particular places, what birds actually see compared with humans does not actually matter as long as the differences are the same for both forms of peppered moth and for the backgrounds upon which they rest. This argument only holds if the UV reflectance characteristics of *typica* and *carbonaria*, of lichens of various species and of the bark of appropriate trees are the same. This is not the case.

Jim Stalker, has assessed the UV reflectance characteristics of all the common forms of the peppered moth, together with six lichen taxa taken from seven tree species. He found that while the black scales and hairs of all forms of the peppered moth absorb UV wavelengths, the white scales reflect UV strongly. Stalker's results in respect of lichens are of particular interest. A number of authors have stressed the correlation between the frequency of *typica* and foliose or vegetative lichens. Stalker thus considered foliose and crustose lichens separately. All parts of the foliose lichens analysed, including *Hypogymnia* and

Evernia spp., absorbed UV. Conversely, some parts of the crustose lichens examined, e.g. *Leucanora* spp., reflected UV, other parts absorbing it giving an overall speckled appearance in pure UV. This meant that, although *carbonaria* was very obviously more conspicuous than *typica* in the human visible spectrum when the two were set against bark covered in foliose lichens, the reverse was true in the UV spectrum. On the other hand, when set against bark covered with *Leucanora conizaeoides*, *typica* was less visible than *carbonaria* both in human visible and UV wavelengths (Stalker *et al.* in preparation).

Stalker's results led to two deductions. First, none of the assessments of the relative crypsis of moths as determined by humans should be applied to birds. Second, it seems likely that the patterning of the typical form of the peppered moth has evolved to be cryptic against crustose lichens, not foliose lichens. If this is so, it seems probable that the ancestral daytime resting site for the peppered moth is under horizontal branches, which in unpolluted regions are usually encrusted with crustose lichens. The moths are unlikely to rest on the upper surfaces of the branches, as these tend to support a rich foliose lichen flora (Liebert and Brakefield 1987).

Resting site selection

Much has been written about the possibility that the different forms of the peppered moth behave in different ways when choosing resting sites. Indeed, I have written at some length upon this subject (Howlett and Majerus 1987; Majerus 1989*a*). However, I was tempted in this account of the work on the peppered moth, to omit all of it, because it suffers from the fact that the UV dimension is lacking. I decided against this course because, while the tests of resting site preferences chosen by moths of different forms are certainly undermined by the fact that the UV characteristics of the backgrounds offered or chosen have not been assessed, the debate over whether the forms do, or even should, show differential resting site selection is still of interest. Throughout the following section, however, the fact that the experiments are designed on the basis of human and not bird perception should be borne in mind.

The environments in which peppered moths live, whether urban or rural, are highly heterogeneous. The idea that a mechanism will evolve which enables different forms of a moth to rest in different positions appropriate to their appearance is intuitively seductive. For example, melanic peppered moths may evolve to rest preferentially on relatively dark homogeneous backgrounds, while *typica* retains an ancestral behaviour of resting on paler and more speckly backgrounds.

Kettlewell (1955*b*) first conducted experiments to determine whether the different forms of the peppered moth actively choose resting positions which maximise their crypsis. He placed peppered moths in barrels lined with equal surfaces of black and white card. His results showed that *typica* rested preferentially on

Table 6.4 The resting preferences of the *typica* and *carbonaria*
forms of the peppered moth in Kettlewell's barrel experiments, in
which the moths were presented with a choice of black and white
surfaces of equal area. (After Kettlewell 1955*b*.)

	typica	*carbonaria*	Totals
Black background	20	38	58
White background	39	21	60
Totals	59	59	118

the white surfaces, *carbonaria* preferring black (Table 6.4). Other workers have
obtained similar results with the peppered moth (Boardman *et al.* 1974; Kettlewell and Conn 1977) and other species (Steward 1985). However, others who
have tried to replicate the experiments with the peppered moth have failed to
obtain evidence of resting site selection (Mikkola 1984), or have obtained contradictory evidence (Howlett and Majerus 1987; C. Jones 1993).

Howlett and Majerus (1987) analysed the light reflected from the wings of peppered moths. They showed that due to the partial translucence of the typical
moth's wings, the wings are more similar to a plain black surface than to a plain
white one when place against most natural surfaces (Fig. 6.9). So, when presented
with a choice of black or white surfaces, both forms should prefer the darker
surface. Such tests were made and the expectation confirmed (Table 6.5). Offering a choice of white against black resting alternatives is obviously a very artificial situation, and other factors, apart from the reflectance of resting substrate,
will be involved in the choice of resting site. So, for example, several workers
(Sargent 1969; Sargent and Keiper 1969; Kettlewell 1973; Lees 1975; Howlett
1989) have shown that texture is an important component in the selection of
resting positions for melanic moths. Work on other species of moth with melanic
forms has given highly variable results, some providing evidence that morphs of
some species do have the ability to select appropriate backgrounds, others suggesting that they do not (e.g. Boardman *et al.* 1974; Steward 1976, 1977*d*, 1985;
Majerus 1982*b*; Sargent 1985; Grant and Howlett 1988).

Leaving aside the issue of whether the forms of peppered moth actually do
choose resting sites appropriate to their wing pattern, we can ask how such sites
might be chosen. Two hypotheses have been offered. Kettlewell (1955*b*) suggested that after landing on a surface but before finally clamping down, light
stimuli received from the resting substrate are compared by the moth with the
colour and pattern of the tufts of scales which surround the eye. If the two are
dissimilar, then a state of what Kettlewell calls 'contrast/conflict' occurs, and the
moth moves until this contrast/conflict is reduced. Sargent (1968) attempted to
test this contrast/conflict hypothesis by painting the tufts of scales around the
eyes of two North American species of moth, *Catocala actinympha* and *Campaea
perlata*. He was unable to provide any evidence to support it. Rather he suggested

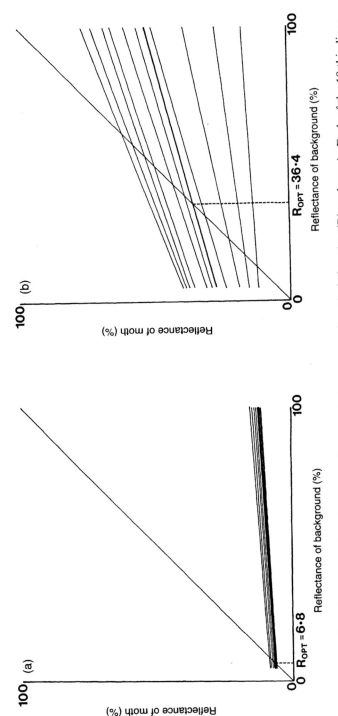

Fig. 6.9 The effect of background on forewing reflectance in the peppered moth: (A) *typica*; (B) *carbonaria*. Each of the 10 thin lines is a plot of the reflectance of a single, randomly selected point on the forewing of a moth against the reflectance of the background. For each point on the wing, measurements were taken against backgrounds with reflectances of 5, 25, 50, 75, and 100 per cent of a white, barium sulphate standard. The relationships are all exactly linear, all data points falling on the lines. The thick line shows the mean for the 10 points. R_OPT is the optimal background reflectance. (From Howlett and Majerus 1987.)

Table 6.5 The resting preferences of the *typica*, *insularia*, and *carbonaria* forms of the peppered moth, when presented with a choice of black and white surfaces of equal area in cylinders. (From Howlett and Majerus 1987.)

	typica	insularia	carbonaria	Totals
Black side	58	30	70	158
White side	20	7	14	41
Floor	21	5	36	62
Totals	99	42	120	261

that the ability to select appropriate resting sites is genetically determined. More recently, Grant and Howlett (1988) demonstrated that some individual peppered moths do appear to have preferences for backgrounds of particular colours, but that these preferences are not correlated to the moths' phenotypes. In addition, some individuals had no preferences.

The issue of whether and how the forms of peppered moth select appropriate resting positions is fraught with controversy and full of contradictory evidence. The latter may be partly due to the lack of UV assessment, but it is also worth noting that there are two theoretical difficulties with the initial, intuitively per-suasive argument, that morph specific resting site selection behaviour should evolve in industrial melanics. First, the existence of a sophisticated resting site selection mechanism would reduce the importance of differential visual selection in the maintenance of the polymorphism (Cook personal communication). The habitats in which peppered moths live are heterogeneous. Dark homogeneous and pale heterogeneous backgrounds occur in all such habitats, whether in rural or industrial areas. Peppered moths rarely occur at very high population densi-ties. Given these two facts, if the moths are able to select positions to maximise their crypsis, it is argued that all moths, irrespective of their form or frequency, would be able to find some patches of background which they would match to a high degree, in any habitat. All could then have a similar level of crypsis and so have the same fitness in respect of visual predation.

Second, unless wing pattern to background matching does operate through Kettlewell's contrast/conflict state, it is difficult to envisage how such a system could have evolved in the peppered moth, in which melanism is relatively recent, unless rather limiting assumptions are made. This is because, if differential resting site selection is genetically controlled, as suggested by Sargent (1968) and Steward (1985), then unless the gene controlling the colour polymorphism is tightly linked to the gene (or genes) controlling the resting site preferences, recombination will lead to rest site preference alleles becoming associated with inappropriate phenotypes. Howlett (1989) attempted to model the evolution of resting site preferences in the absence of linkage to the colour pattern locus, basing his model on the peppered moth. He assumed that prior to the rise of *carbonaria* in the nineteenth century, peppered moths would have selected backgrounds appropriate to their pale speckly pattern. He then introduced a new mutant allele, which gave moths a preference to rest on dark homogeneous backgrounds, into theoretical populations in which *carbonaria* was increasing. He assumed, because the frequency of *carbonaria* was increasing that the four combinations of phenotype and behaviour followed a fitness hierarchy, with *carbonaria* with a preference to rest on dark surfaces being fitter than *carbonaria* on pale, which in turn was fitter than *typica* on pale, *typica* on dark backgrounds being the least fit combination. He found that when *carbonaria* was at low frequency, the dark preference allele was eliminated from the population, because, due to recombination, it was frequently associated with the *typica* phenotype and so had low fitness. Alternatively, if *carbonaria* was already at high frequency when the new mutant was introduced, it rapidly spread to fixation, because the original preference for resting on pale heterogeneous backgrounds would be unfavourable for the now common *carbonaria* morph. The threshold in the *carbonaria* frequency, which determines whether the dark preferences mutation becomes extinct or spreads to fixation, depends on the relative fitnesses of the four morph-behaviour combinations, and on the dominance relationship between the two behaviour alleles. However, the basic result is the same even if the fitnesses of *typica* on pale and *carbonaria* on pale are reversed.

These results suggest that the selection of appropriate resting sites will only evolve if there is linkage between the colour pattern locus and the behavioural locus. The likelihood of such a system evolving will thus be, at least in part, a function of the lengths of time that the various alleles for the different forms and the different resting preferences have existed in the population. Therefore, while it is possible to envisage such a system evolving in species in which melanic polymorphisms have existed for long periods of time, it is difficult to imagine it when melanism has evolved only recently following the industrial revolution, except as a one-off case in the unlikely event that the relevant behaviour and morph genes just happen to be tightly linked.

The difficulty inherent in modelling the evolution of genetically based rest site preferences could be taken to support, albeit by exclusion, Kettlewell's theory that the fact of being a particular colour, in itself confers upon the moth a preference to rest on an appropriate background. However, Howlett's findings offer another possible explanation of the contradictory evidence from background-choice experiments. The occurrence of different frequencies of *carbonaria* in populations of the peppered moth over the last 140 years, mean that if alleles which cause moths to rest preferentially on dark surfaces arise by mutation, but are not linked to the *carbonaria* locus, some of these will spread to fixation, in areas of high *carbonaria* frequency, while others are lost from populations, where *typica* predominates. Consequently, three types of population may be envisaged.

(a) Those in which a dark-preference allele has never arisen or has failed to spread, so the peppered moths in the population all tend to rest on pale heterogeneous backgrounds.

(b) Those in which a dark-preference allele has recently arisen and is in a state of transient polymorphism as it spreads towards fixation. In these populations some peppered moths will have a preference to rest on pale heterogeneous backgrounds, others will prefer dark homogeneous ones, but there will be little correlation between form and behaviour.

(c) Those in which a dark-preference allele has become fixed in the population so that all the moths show a tendency to rest on dark homogeneous surfaces.

Owing to the recent decline in pollution levels since the introduction of anti-pollution legislation, *carbonaria* is decreasing in frequency. Therefore, it is possible that a fourth type of population may exist in which a preference to rest on pale heterogeneous backgrounds is now spreading in populations in which the dark preference allele had previously become fixed.

Given that different populations may contain moths with quite different rest site preferences, it becomes imperative to know the origin of moths used in background choice experiments. For example, if all the moths used in Kettlewell's experiments came from the same population, the observed preference of *typica* for white and *carbonaria* for black resting sites could be explained. If the population were in a state of transient polymorphism and moths had already been exposed to selection in the wild, the *carbonaria* with a pale preference and the *typica* with a dark preference may have already been eliminated by predators. Thus moths used in experiments may have been only those that selected backgrounds appropriate to their colour pattern. Alternatively, if the moths used came from different populations, for example *typica* from rural areas and melanics from industrial regions, it may be that the *carbonaria* simply came from a population in which the dark allele had been fixed. If this was the case, Kettlewell's results would be explained. The same theory could explain other apparently contradictory results. So, for example, the results of Howlett and Majerus (1987) in which both *typica* and *carbonaria* moths showed a preference for black surfaces would be explained if the moths were from a population in which the dark allele had been fixed. It is perhaps relevant to note that the moths used by Howlett and Majerus came from the Cambridge area, at a time when a frequency of *carbonaria* was about 40 per cent, and was declining from a level of over 90 per cent in the late 1950s.

In trying to unravel the causes of the great variation in the results of background choice experiments in the peppered moth, I have made exhaustive but unrewarding enquiries to find out where Kettlewell obtained the moths he used in his experiments. It is possible that he only used moths caught in Oxford, in traps which he ran nightly at the time (Lees personal communication), but he also had extensive stocks of bred material and as he was working at that time both in Dorset and Birmingham, he may have used moths from any of these sources. The problem is therefore unresolved.

One way to resolve this matter would be to test the resting site preferences of moths from different parts of the country. Such work was undertaken by Carys Jones. Using a 'dawn box' originally designed by Rory Howlett (Fig. 6.10), Jones assessed the resting site preferences of moths from five different populations, some coming from areas (e.g. Cambridge, Surrey, Telford) where *carbonaria* had attained high frequency, others coming from regions where *carbonaria* had never been more than a rarity (Devon and County Kerry, Eire). The dawn box, which is essentially a cylinder with a top mounted lamp attached to a dimmer switch, enabled Jones to simulate four 'dawns' every 24 hours, thus obtaining a dozen or more resting choices for each moth. The results produced provided no evidence of morph specific resting site preferences. Moreover, the results from four of the populations were broadly in support of Majerus' (1989*b*) hypothesis that different populations would exhibit different levels of preference for black or white resting surfaces (C. Jones 1993). Moths from Cambridge, Surrey, and Telford, all showing a preference to rest on black, and those from County Kerry resting randomly. However, moths from Devon, which would be expected to show no preference for black, as *carbonaria* is rare in that county, did have such a preference. More data from a greater range of populations are needed before it will be possible to explain with any conviction the reason why different workers have obtained such different results. Furthermore, any future work should take into account UV, as well as the human visible spectrum, when preparing alternative surfaces to offer moths as potential resting sites.

It is imperative to gain full understanding of any resting site preferences that exist in the peppered moth, for they may be crucially important in understanding other factors that influence the polymorphism. For example, the level of selective elimination of *typica* in Birmingham and *carbonaria* in Dorset could be unrealistic if the live moths released on to trees took up positions based on resting site preferences from other populations (note that Kettlewell records the frequency of *carbonaria* to be zero at Deanend Wood, so that at least the *carbonaria* he used in predation experiments in Dorset must have come from elsewhere).

Of more interest perhaps is the possibility that these ideas on different types of population, in respect of resting site preferences, may explain the second inconsistency between the observed and simulated frequencies of *carbonaria* along the Liverpool/north Wales cline. The rate of decline in *carbonaria* frequency with distance from Liverpool, was less abrupt in the model prediction than in the observed data (see Fig. 6.7, p. 130). Suppose that an allele to rest on dark backgrounds had become fixed in the Liverpool population, where *carbonaria* is common, and that this allele had not become fixed in rural north Wales, where *carbonaria* is rare. The *typica* moths in Liverpool and the *carbonaria* moths in rural north Wales will have resting site preferences characteristic of their

(a)

(b)

Fig. 6.10 Howlett's 'dawn box' used for obtaining multiple resting choices per day: (a) the box ready to be covered; (b) inside the cylinder.

respective populations, so *typica* in Liverpool will preferentially rest on dark surfaces and *carbonaria* in rural north Wales on pale ones. The rarer form at each end of the transect will thus be exposed to an exaggerated level of selective predation, because they will preferentially rest on inappropriate backgrounds. This effect has not yet been modelled, and to some extent it is difficult to see how it should be modelled, for the time that a dark-preference allele entered and became fixed in the Liverpool population would be an important factor in determining the rate of gene flow of such an allele along the transect. Yet this time is unknown. However, it is intuitively obvious that different resting site preferences would affect the levels of differential predation along the cline. In his selective predation experiments, Bishop (1972) glued moths out on tree trunks in pseudonatural positions, where they matched their background to some extent. However, this procedure will underestimate the selective elimination of *typica* at the north-eastern end of the transect, and of *carbonaria* at the south-western end, if the resting site preferences are as suggested above. Greater differential predation of the rarer morphs at each end of the cline must have the effect of making the decline in frequencies more abrupt than that predicted by Bishop's model. Such a scenario may therefore explain the lack of fit between model prediction and observed frequencies in this respect.

One final possibility is worth considering. C. Jones (1993), trying to explain the preponderance of dark site resting preferences in her work and that of Howlett (1989) and Grant and Howlett (1988), tried to imagine resting strategies from the moth's point of view. She pointed out that tree trunks are an uncommon natural resting site for peppered moths, the moths usually choosing darker shaded positions, under branches or branchlets in the canopy. This, she assumes, was a specialised resting behaviour, which had evolved long before the advent of industrialisation and the subsequent pollution of the environment. She proposes that throughout most of Britain, in pre-industrial times, as the trunks and branches of tree saplings grew, lichen communities would have formed a more or less continuous cover over the bark. As the crown of the tree formed, the lower parts would have become shaded, and the micro-environment radically altered. The lichen species would have been selected along the trunk in response to individual needs, resulting in a floristic composition at the base that differed from that in the canopy. If the peppered moth, being a lichen mimic, had a tendency to rest preferentially on or near lichen species, i.e. pale heterogeneous backgrounds, then the trunks would have provided this growth, as well as the crown of the trees. However, if a peppered moth actively sought dark sites in which to rest, they would select a position in the canopy, probably under a branch, or below a trunk/branch joint. These sites have abundant lichen growth. Indeed the internodes of twigs are the first site of lichen colonisation of adult trees, and possibly the first site of recolonisation after pollution levels decrease. Such positions have the advantage over trunks as resting sites in that they are in the shade. A moth resting in such a position would be more difficult to see than its trunk resting counterpart, a statement reinforced by the finding that peppered moths glued on to trunks are taken significantly more from trunks than those put out under branches (Table 6.3, p. 131).

Consequently, C. Jones (1993) argues that, a moth resting on lichen on a trunk is more likely to be eaten by a predator than one resting upon similar lichen under a branch. Any moth that actively sought out these shaded sites would still have any cryptic advantage afforded by resting upon lichen, but would also gain the selective advantage of being difficult to see due to the shadow. Thus, if the peppered moth is to express a resting site preference in artificial cylinder experiments, it might be expected to select a dark surface, reflecting its natural resting position on trees. This preference would be independent of morph as the crypsis of all forms of the moth would be maximised by resting in shaded areas.

The effect of mating on selective predation

The adult life span of the peppered moth is short. Estimates of average longevity vary from 2 to 8 days, females possibly living slightly longer than males (Majerus, unpublished data). Following emergence from subterranean pupae, usually shortly before dusk, females fly to trees where they 'call' males by releasing pheromones, and copulation takes place shortly after dusk. Pairs remain together for approximately 20 hours, and males will not depart from the vicinity of the female until after dusk. In consequence, the pair of moths remain in close proximity throughout a whole day which, because the life span is short, represents a considerable proportion of the time that an adult peppered moth is exposed to predation. The question of whether mating moths are exposed to different levels of predation is therefore an important one. Considering just the *typica* and *carbonaria* forms, there are effectively three mating combinations. It has been suggested that a mating pair of typicals resembles a patch of foliose lichen more closely than does a single typical (Mikkola 1984; Brakefield 1987; Liebert and Brakefield 1987). It could also be argued that a mating pair of *carbonaria* would be less moth-like in shape than a single moth. Consequently, both similar mating combinations may be associated with reduced predation. However, the increased size of the target, and the fact that, if detected, two moths will be eaten, must be weighed against this suggestion. Of more importance is the argument that in any habitat, one or other of the moths in the third mating combination, *typica* × *carbonaria*, must be poorly camouflaged. As the discovery of one individual increases the likelihood of any others close by being found (Kettlewell 1955a), it is probable that both partners of such a pair will be at risk. Overall, therefore, *typica* × *carbonaria* pairs are likely to have lower fitness than either of the other two mating combinations.

Work to test the effect of mating with a similar or different morph on levels of selective predation is urgently needed. Not only will such work help to increase the accuracy of estimates of morph fitness, but a finding that morphs do tend to mate assortatively (like tending to mate with like) would be of great significance

PLATES

(a)

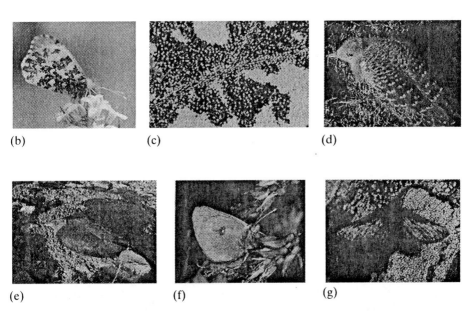

(b) (c) (d)

(e) (f) (g)

Plate 1   The diversity of melanism. (a) A 'black' flowering cherry in Pitlocherie, Scotland. The black covering on trunks and branches is due to growth of a melanised sooty mould (inset). (b) The underside of the orange-tip butterfly gives the impression of being dappled green and white. (c) The underside hindwing of an orange-tip butterfly has no green scales, the dappled green impression being produced by a mixture of yellow, black, and white scales. (Courtesy Mr Gerald Burgess.) (d) A non-melanic chick of the Arctic skua. (Courtesy Dr Peter O'Donald.) (e) A melanic chick of the Arctic skua. (Courtesy Dr Peter O'Donald.) (f) The dark scales on the underside of the northern clouded yellow, aid it to heat up when at rest. (g) Black mountain moth sunning itself.

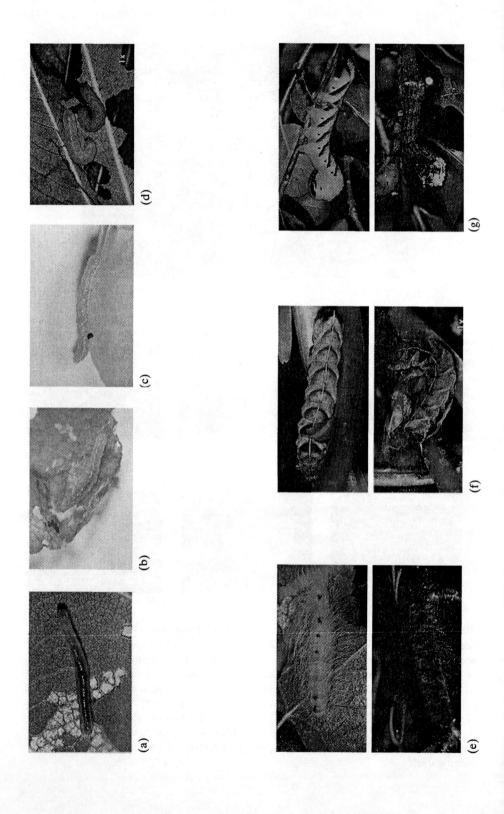

(a)

(b)

(c)

(d)

(e)

(f)

(g)

Plate 2 Larval melanism. (a–d) In their early instars, larvae of the angleshades moth take on the colour of their food plant, the gut contents showing through an almost transparent body wall. Only in later instars is the colour genetically controlled: (a–c) first instar larva feeding of dock leaf, pink rose petal and yellow broom petal, respectively; (d) green and brown (melanic) fourth instar larvae. (e) Melanic and non-melanic larvae of the miller moth *Acronicta leporina* Linnaeus. (f) Melanic and non-melanic larvae of the dot moth *Melanchra persicariae* Linnaeus. (g) Typical and melanic larva of the death's-head hawk moth.

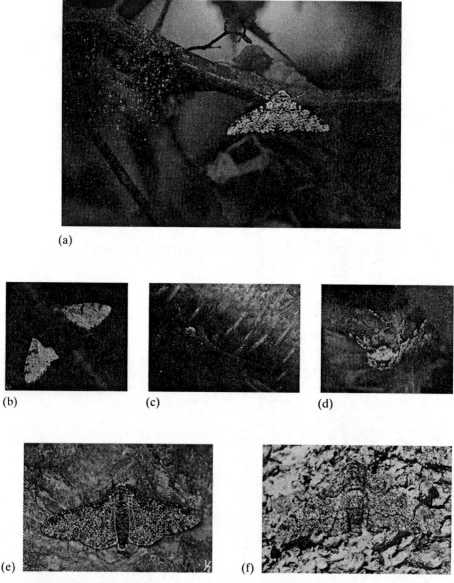

(a)

(b) (c) (d)

(e) (f)

Plate 3 The peppered moth. (a) *Typica* and *carbonaria* forms of the peppered moth on an horizontal birch branch. (b) A pair of peppered moths on a twig at dawn. The *carbonaria* male is much less conspicuous than the *typica* female. (c) A *carbonaria* peppered moth in shadow under a horizontal branch, showing how this positioning may reduce the likelihood of detection. (d) Typical form of the peppered moth at rest during the day in hazel foliage. (e) An intermediate, *insularia* form, of the peppered moth. (f) The non-melanic form of the peppered moth from North America, *Biston betularia cognataria* (courtesy of Professor Bruce Grant).

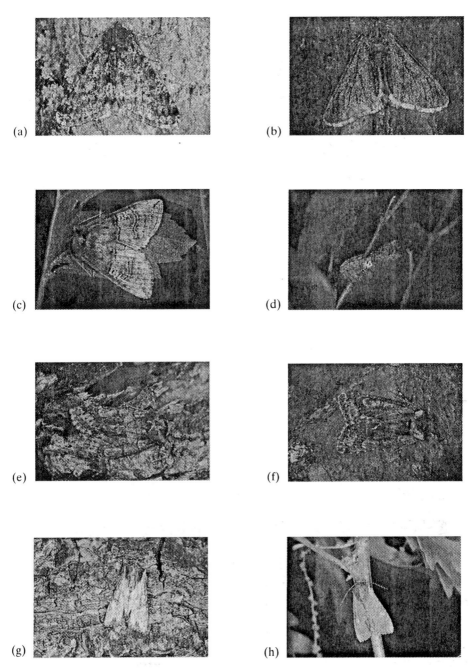

Plate 4 Industrial and partial industrial melanism. (a) Typical and (b) melanic forms of the pale brindled beauty. (c) Typical and (d) melanic forms of the figure of eighty moth. (e) Typical and (f) melanic forms of the green-brindled crescent. (g) Typical and (h) melanic forms of the clouded-bordered brindle.

(a)

(b)

(c)

(d)

Plate 5 Non-industrial melanism. (a) The melanic female form, f. *valesina*, of the silver-washed fritillary, gains a thermal advantage over the paler typical female, due to her darkness, but is less likely to mate: top, male; middle, typical female; bottom, f *valesina* female. (b) Melanism in the mottled umber may be an example of anti-search image melanism. (c) and (d) Northern melanism in the Hebrew character: (c) f.

(e)

(f)

(g)

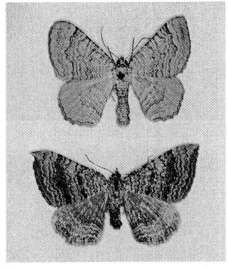

(h)

gothicina, Rannoch Black Wood, Scotland, (d) typical, Cambridge. (e) and (f) Northern melanism in the water carpet *Lampropteryx suffumata* Denis and Schiffermüller: (e) f. *piceata*, Rannoch Black Wood, Scotland, (f) typical, New Forest. (g) Northern melanism in the northern spinach: above, typical, Dyfed, Wales; below, melanic, Shetland. (h) Western coastline melanism in the yellow shell: above, typical, Surrey; below, melanic, f. *hibernica*, western Ireland.

Plate 6 Melanism and habitat selection. (a) and (b) Pictures taken with the same exposure settings outside (a) and under (b) the yew canopy in Juniper Bottom, Box Hill, Surrey, showing the relative darkness under the canopy. (c) Under the yew canopy in Juniper Bottom, Box Hill, Surrey. (d) Melanic and non-melanic forms of the tawny-barred angle. Strong morph specific habitat preference was first demonstrated in this species. (e) The melanic form, f. *rebeli*, of the willow beauty, which exhibits a strong morph specific habitat preference for flying in dark woodlands. (f) Unexpectedly, the full melanic form of the lunar underwing preferentially flew within a mature apple orchard rather than in the open. Paler forms of this species showed no such preference.

(a)

(d)

(b)

(e)

(c)

(f)

Plate 7 Melanism in conspicuous Lepidoptera. (a) Forms of the scarlet tiger moth: top left, f. *dominula*; top right, f. *medionigra*; bottom left, f. *bimacula*; bottom right, f. *rossica*. (b) Melanism in the ruby tiger moth appears to make the moths less visible to avian predators when the moths fly during the day: above, dark form, f. *borealis*, Perthshire; below, typical, Cambridge. (c) In the ghost moth, male wing colour has evolved in response to both mating and defensive pressures: top left, male typical, Cambridge; top right, female typical, Cambridge; bottom left, male f. *thulensis*, Shetland; bottom right, female, Shetland. (d) The white ermine moth shows melanism towards the north of its range, possibly as a consequence of thermal factors: above, typical, Cambridge; below, melanic, Shetland. (e) Location dependent sexual dimorphism in the muslin moth: top, female, Cambridge; bottom left, male, Cambridge; bottom right, male, Ireland. (f) The 5 spot burnet: above typical; below, the rare melanic f. *obscura*.

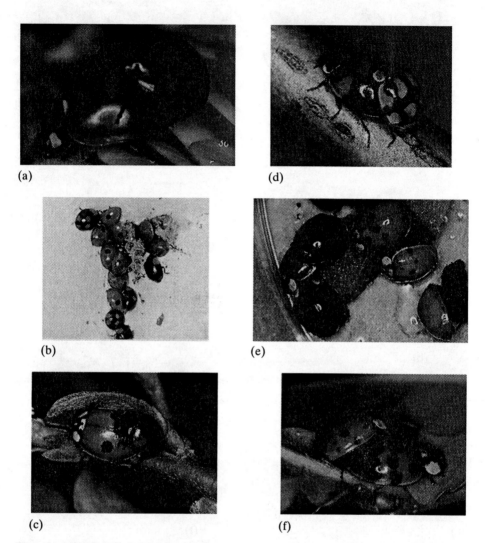

Plate 8   Melanism in ladybirds. (a) A rare melanic male 7 spot ladybird mating with a typical female. (b) Overwintering group of 2 spot ladybirds, showing a range of non-melanic and melanic forms. (c) A male melanic, f. *quadrimaculata* 2 spot ladybird, mating with a non-melanic female. (d) An f. *decempustulata* male 10 spot ladybird mating with an f. *bimaculata* female. (e) Various forms of *Harmonia axyridis* from Sapporo, Japan. (f) Light and heavily spotted *Harmonia axyridis* mating.

in respect of general evolutionary theory. If it were shown that a preference to mate with a similar rather than with a different phenotype had evolved because the mating between different forms was at a selective disadvantage, this would perhaps represent the first stage in the process of speciation through disruptive selection. Furthermore, such evidence would suggest that prezygotic reproductive isolation might evolve in the absence of geographic separation or postzygotic reproductive isolation mechanisms.

Melanism in the peppered moth outside Britain

Melanic polymorphism in the peppered moth is not confined to Britain. On the Continent, f. *carbonaria* was first recorded in the Netherlands at Breda in Noord-Brabant in 1867 (Haylaerts 1870), when a pair of this form were found mating. This suggests that *carbonaria* may have already been at a significant frequency at that time (Brakefield 1990). By the turn of the century, *carbonaria* was well established across most of the Netherlands, the full range of *insularia* forms also being present. Professor J.B. Lempke, the foremost authority on melanic Lepidoptera in the Netherlands, surveyed form frequencies at 81 sites across the country from 1969 to 1973. Although many locations produced only small samples, the data were sufficient to allow Brakefield (1990) to produce a map of *carbonaria* frequency, based on the Dutch Provinces, which could be compared with Barkman's (1969) map of epiphytic vegetation across the Netherlands (Fig. 6.11). This comparison shows a similar correlation, to industrialisation and the lack of epiphytes, with that seen in Britain. However, there are two differences between the Dutch and British data. First, the frequency of *carbonaria* in Holland does not reach the highest levels recorded in Britain, not exceeding 75 per cent in any of Lempke's samples. On the other hand the frequencies of the *insularia* forms across the Netherlands tend to be higher than they are in Britain (Table 6.6).

Industrial melanism in the peppered moth has been reported in many other European countries. In Luxembourg, during the 1970s, the frequency of *carbonaria* was over 70 per cent in the southern industrial districts, but only about 30 per cent in Clervaux, in the Ardennes. By the late 1980s and early 1990s, the frequency of *carbonaria* had declined significantly in both regions (Majerus unpublished data). Both *carbonaria* and *insularia* are common in industrial regions of Belgium and northern France. In Germany, *carbonaria* was already common in industrial regions along the Rhine by the end of the nineteenth century. High melanic frequencies have also been reported in industrial regions of Denmark (Hoffmeyer 1948, 1949); Sweden (Kettlewell 1973); Finland (Kettlewell 1973); and the Czech Republic (Novak and Spitzer 1986; Kula and Kralicek 1995). In the latter, Novak and Spitzer (1986) showed a significant positive correlation between *carbonaria* frequency and levels of air pollution throughout Czechoslovakia. The lowest frequencies were from southern Bohemia, with the highest being from central and north-west Bohemia and

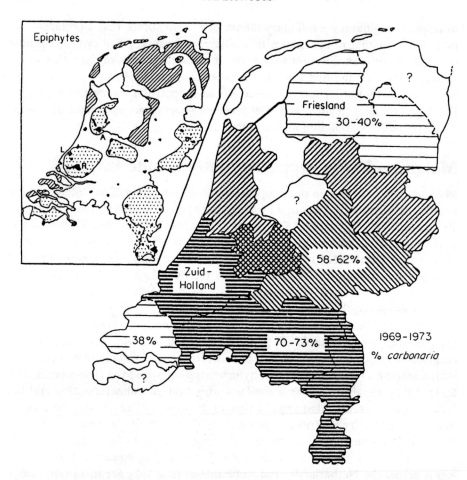

Fig. 6.11 Frequencies of forms of the peppered moth in regions of Holland in samples taken by Lempke, 1969-73. (From Brakefield 1990.) Inset map gives levels of epiphytic vegetation in the 1960s: dotted = none, clear = poor, hatched = normal. (After Barkman 1969.)

southern and central Moravia. They report *insularia* to be relatively rare in the west but its frequency increases eastwards. The frequency of *insularia* appears to be independent of levels of atmospheric pollution. This being so, populations of the peppered moth in the Czech Republic would well repay further attention to try to determine the factors that do influence the frequency of *insularia*. In Finland, according to Professor Esko Suomalainen (reported in Kettlewell 1973), industrial melanism comprises the occurrence of f. *insularia* in industrial areas. Here too, further work might be valuable in shedding light on the selective factors involved in the evolution and maintenance of this form.

Table 6.6 Percentage frequencies of the five major non-melanic and melanic phenotypes of the peppered moth, in pooled samples obtained in 10 Dutch Provinces by B.J. Lempke, from 1969 to 1973. Total sample sizes are also given. (From Brakefield 1990.)

| Province | Percentage frequency | | | | | Total sample size |
	typica	insul 1	insul 2	insul 3	carbonaria	
Friesland	19.1	16.3	16.3	14.7	33.6	194
Drenthe	16.2	20.9	17.4	4.7	40.7	86
Overijssel	13.9	9.6	9.3	8.4	58.8	330
Noord-Holland	12.5	10.1	8.8	6.4	62.3	377
Gelderland	7.0	8.5	17.4	7.9	59.1	328
Utrecht	9.0	10.2	13.3	9.8	57.8	256
Zuid-Holland	4.3	7.8	9.9	7.3	70.7	232
Noord-Brabant	3.7	3.0	10.4	10.8	72.1	297
Zeeland	12.1	18.2	22.7	9.1	37.9	66
Limburg	2.3	4.5	10.8	9.6	72.9	1706

Further afield, industrial melanism is also recorded in the American subspecies of the peppered moth, *Biston betularia cognataria*. The typical form of this species has a darker, more smoky appearance than European *typica* (Plate 3f and Fig. 6.12). However, the full melanic form, known as f. *swettaria* is indistinguishable from f. *carbonaria*, and is also inherited as a dominant allele at a single locus (West 1977). The first f. *swettaria* was recorded in 1906 near Philadelphia (Owen 1961, 1962). In parts of Michigan, particularly the south-east, the frequency of *swettaria* rose to over 90 per cent by 1960. Indeed, Owen reports that in Livingstone and Washtenaw counties *swettaria* increased from 1 to 80 per cent in less than 30 years. One of the reasons for this very rapid rise may be that *B. b. cognataria* has two generations per year, rather than the one of *B. b. betularia* in Britain.

Elsewhere in North America, f. *swettaria* has been recorded from Illinois, Massachusetts, and Delaware in the United States, and from Montreal, Toronto, and southern Ontario in Canada (Kettlewell 1958c; Owen 1961, 1962; Sargent 1974; West 1977).

A third subspecies of the peppered moth, *Biston betularia parva*, occurs in Japan, where it is known as Oo-shimofuri-eda-shaku, meaning frosted branch-measuring moth. This subspecies is intermediate in appearance between the typical forms of *B. b. betularia* and *B. b. cognataria*. Melanics of this subspecies are not known despite a recent attempt to find them. Takahiro Asami and Bruce Grant (1995) report that few records of this species were made prior to their studies of 1988, 1992, and 1993. During these years they trapped 307 individuals. None were melanic. All the moths taken were caught in the central mountainous region or further north. Traps set in coastal industrial regions produced no *B. b. parva*. Asami and Grant (1995) conclude that in Japan, this species does not occur in regions affected by industrial pollution to any significant degree, and for this reason, melanism has not evolved in Japanese *Biston*.

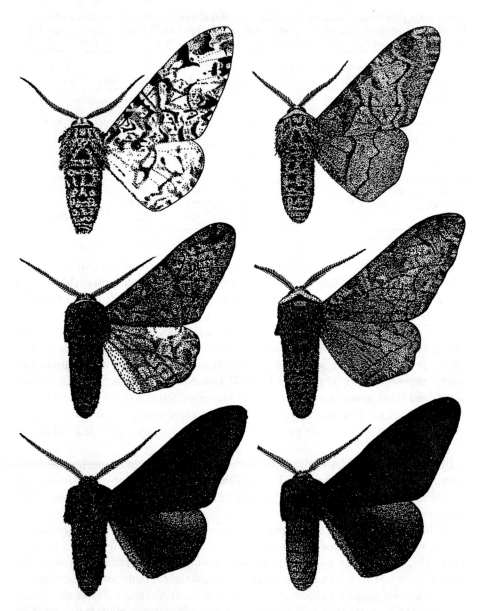

Fig. 6.12 The forms of the peppered moth from Britain (left-hand side) and America (right-hand side). The typical form from Britain is considerably paler than that from America. However, the melanic forms, *carbonaria* and *swettaria*, are indistinguishable. (Illustrations by Derek Whiteley; courtesy of Professor Bruce Grant.)

The decline in melanism in the peppered moth

I have already alluded on a number of occasions to the fact that melanism in the peppered moth is currently declining. This decline was instigated in Britain by the enactment of the Clean Air Acts in the 1950s, and other anti-pollution legislation subsequently. Unlike the original rise in *carbonaria*, this decline has been monitored. Much of the credit must go to Sir Cyril Clarke and his co-workers who have trapped peppered moths at his home on Caldy Common, West Wirral, every year since 1959 (Grant *et al.* 1996). Although other workers have reported on declines in the frequency of *carbonaria* over the last 40 years (Cook *et al.* 1970; Lees and Creed 1975; Whittle *et al.* 1976; Bishop *et al.* 1978a; Howlett and Majerus 1987; Brakefield 1987; Mani and Majerus 1993), no other data set from a single location spans the whole period. The full data set of form frequencies between 1959 and 1995 are shown in Fig. 6.13. The pattern of decline follows expectation well, in that there is a lag between the reduction in pollution levels and the decline in *carbonaria*. This is to be expected if the fitness of *carbonaria* relative to *typica* depends on its crypsis. Following a reduction in pollution levels, it would take some years before the surfaces of trees that peppered moths rest upon will

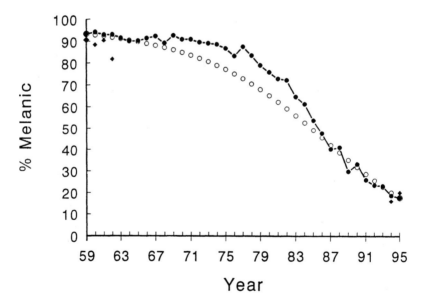

Fig. 6.13 Comparison of the forms of the peppered moth from Europe and America. Solid circles = frequency of melanic (*carbonaria*) at Caldy Common, north-west England. Solid diamonds = frequency of melanic (*swettaria*) at the George Reserve, Michigan, USA. The open circles give the theoretical expectations based on a constant selection coefficient of 0.153 against the melanic allele. (Courtesy of Professor Bruce Grant.)

lighten as a result of soot fall-out washing off, and the production of new unpolluted growth. There would also be a lag before lichens and other epiphytes began to recolonise and grow on the cleaner tree surfaces.

Although most of the surveys have been in the north-west of England, declines of the same type have been observed in several other parts of Britain (Mani and Majerus 1993; West 1994) (Table 6.7).

A similar decline has been observed in the Netherlands, although here f. *carbonaria* is being replaced not only by *typica*, but also by the darkest of the *insularia* forms (Fig. 6.14) (Brakefield 1990). These data suggest that the *insularia* form in the Netherlands and in Britain are either in some way inherently different, or are acted upon by different selective factors. Collaborative work on both sides of the North Sea would be valuable to address these differences.

One other decline in the frequency of a melanic form of the peppered moth has been reported. This is the decline in the frequency of f. *swettaria* in Michigan. Bruce Grant, Denis Owen, and Sir Cyril Clarke (1995, 1996) have recently published the results of parallel declines in the frequencies of melanics on the Wirral, Merseyside, and at the George Reserve, 30 miles west of Detroit, between 1959 and 1995. In both cases the decline in the full melanic form has been from over 90 per cent to less than 20 per cent (Fig. 6.13). Unfortunately, the data from the

Table 6.7 Percentage frequencies of the forms of the peppered moth, in Cambridge, 1952–96.

Year	Recorder	*typica*	*insularia*	*carbonaria*	*n*
1952–56	HBDK/BOCG	4.5	3.4	93.0	88
1957–64	BOCG	4.4	0.9	94.8	115
1970	DRL	15.0	10.0	75.0	?
1975	DRL/ERC	29.4	5.9	64.7	34
1981	MENM	34.1	20.0	45.9	85
1982	MENM	31.6	18.4	50.0	38
1983	MENM	40.3	16.8	42.9	119
1984	MENM	36.8	21.1	42.1	38
1985	RJH	48.8	11.6	39.5	43
1986	RJH	64.5	6.6	28.9	76
1987	RJH	53.3	10.0	36.7	30
1988	MENM	35.0	32.5	32.5	60
1989	MENM	64.0	14.6	21.3	89
1990	MENM/CWJ	55.6	22.2	22.2	36
1991	MENM/CWJ	61.5	13.1	25.4	122
1992	MENM	59.5	17.4	23.1	121
1993	MENM	61.7	14.9	23.4	94
1994	MENM	63.2	11.8	25.0	68
1995	MENM	67.1	13.7	19.2	73
1996	MENM	67.4	14.7	17.8	129

Data collected by Bernard Kettlewell (HBDK), Brian Gardiner (BOCG), David Lees (DRL), Robert Creed (ERC), Michael Majerus (MENM), Rory Howlett (RJH) and Carys Jones (CWJ).

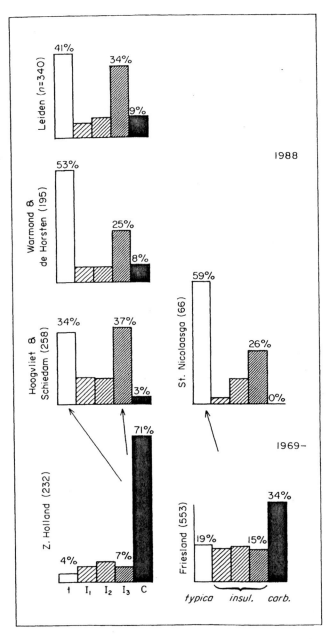

Fig. 6.14 The decline in *carbonaria* and increases in both *insularia* and *typica*, in the Netherlands, between 1969 and 1988. (From Brakefield 1990.)

George Reserve are very intermittent and it is not definite that the course of the decline in Michigan has exactly paralleled that on the Wirral. However, the similarity in the start and end frequencies is striking.

Grant and his colleagues report that, in both regions, declines in air pollution levels have followed anti-pollution legislation (Fig. 6.15), and that the drop in melanic frequencies on both sides of the Atlantic correlate well with the decline in sulphur dioxide and suspended particulates. They comment that in neither locality does there appear to have been any appreciable increase in the lichen flora over the period. However, given that the assessments of lichens were anecdotal, rather than systematic, and that the woodland canopies may not have been monitored, this comment should be viewed tentatively. The more so, because other authors have recorded increases in lichens following anti-pollution

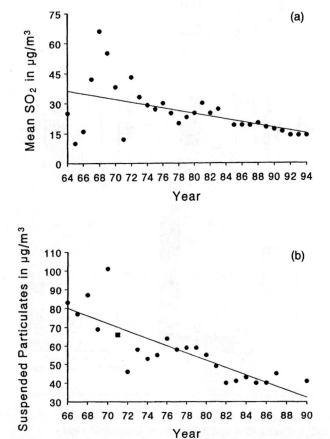

Fig. 6.15 Changes in air quality in south-eastern Michigan: (a) sulphur dioxide; (b) particulate soot pollution. (Courtesy of Professor Bruce Grant.)

legislation (Cook *et al.* 1990). But the importance of this case is not in the minutiae. As Dr Douglas J. Futuyma put it: 'When you have parallel changes like this it's like having different replicates in an experiment. The more you have, the more confident you are that you're getting a consistent result'.

The value of Sir Cyril Clarke's Caldy Common data set should not be underestimated. This is the only data set representing the whole of the current demise in *carbonaria*. It is to be hoped that Sir Cyril will continue the data set for many years to come to monitor the latter stages of *carbonaria's* decline. This is very necessary given the possibility that as it becomes rarer selection against it is diminishing, possibly because there is a frequency dependent aspect to the differential predation, as the result of search image formation (L.D. Hurst and Majerus in preparation). To this end, discussions will be held in June 1997, between Clarke, Grant, and myself, aimed at ensuring that monitoring will continue at Caldy Common into the twenty-first century.

Conclusions

The case of the peppered moth is undoubtedly more complex and fascinating than most biology textbooks have space to relate. Even here, I have had to be selective about which data sets to include, given the huge wealth of material on this case. I have tended to concentrate on work carried out in the industrial northwest of England, because this is the most extensive work, and East Anglia where I have worked myself. My aim in this and the preceding chapter has been to provide sufficient information for the reader to assess for themselves which parts of the story are complete, and where further work is needed (see Chapter 10). My view of the rise and fall of the melanic peppered moth is that differential bird predation in more or less polluted regions, together with migration, are primarily responsible, almost to the exclusion of other factors. I do not support the view that non-visual selection plays a very significant part in Britain. Lack of correlations and concordance between simulations and observed frequencies are, to my mind, more likely to be due to lack of understanding of the resting behaviour of the moths and their ecological interactions with their predators, than selective factors whose nature has yet to be identified.

My view of the story, is, I know, not held by all other entomologists or evolutionary biologists. However, in reviewing well over 100 papers written about this case, and having collected peppered moths in the wild for over 30 years, I am unconvinced that there must be other important selective factors involved in the evolution of melanism in the peppered moth in Britain. My view may well be proved to be incorrect. However, if it is to be refuted, research into the case must continue and be amplified. Recently (*New York Times*, 12 November 1996), the eminent evolutionary biologist Dr Douglas J. Futuyma said of the peppered moth story: 'We don't know the whole story yet, but it should be possible to find out what it is'.

  If recent predictions of the decline of the peppered moth are borne out,
Dr Futuyma's appraisal may be overoptimistic, for we probably only have
about two decades until *carbonaria* has gone from the country in which it is best
known.

7 Melanism in the Lepidoptera

A classification of melanism in the Lepidoptera

The prominence given in biological texts to industrial melanism, and the case of the peppered moth in particular, has tended to eclipse the fact that the majority of British Lepidoptera that now have melanic forms, already had melanic forms prior to the industrial revolution.

Kettlewell (1973) defines industrial melanism as follows:

> This must include dark forms which are distributed in and around industrial areas and far beyond in the direction of the prevailing smoke-drift. In some widely distributed species this frequently leads to clines. It is found in cryptic night-flying Heterocera only. Inheritance of melanic forms is usually dominant, less frequently multifactorial or polygenic; more rarely the character has no dominance. In Britain it is recessive in a few instances only.

This definition is inadequate for two reasons. First, it places too much emphasis on polymorphism, thereby excluding the gradual darkening of ground colours seen in many species in industrial areas over the last century. Second, it is too vague about the causes of the presence of melanic forms in regions affected by industrial pollution, effectively including all species which have dark forms in such regions, whether there is a connection between industrialisation and melanism or not. A more precise definition is needed.

Industrial melanism is the proportional increase of melanic pigments in members of a population, whether this be as a result of an increase in the frequencies of distinct melanic forms, or as a result of a general darkening of some or all forms within a population, where this increase is associated with the effects of pollutants from industrialisation. This definition may be subdivided into three classes.

(a) Fully industrial melanic polymorphism: in which distinct melanic forms have arisen since the industrial revolution and have increased as a direct consequence of the effects of industrialisation on the environment.

(b) Partial industrial melanic polymorphism: in which melanic polymorphisms occurred prior to the industrial revolution, but have increased in frequency since, as a direct consequence of the effects of industrialisation on the environment.

(c) Polygenic industrial melanism: in which the average ground colour of some populations has darkened gradually as a consequence of the effects of industrialisation on the environment.

Of course, not all melanism is the result of the effects of pollution. To these classes of industrial melanism must be added two other main classes: non-industrial melanism in cryptic species and melanism in warningly coloured species. These classes also require some subdivision. Non-industrial melanism in cryptic species may be divided into eight categories, according to the ecological reasons for the evolution and maintenance of melanism:

(a) thermal melanism;

(b) rural or background choice melanism;

(c) northern latitude melanism;

(d) western coastline melanism;

(e) pluvial melanism;

(f) melanism associated with fire-resistant trees;

(g) ancient conifer melanism;

(h) anti-search image melanism.

In these categories, I have retained Kettlewell's subclasses for consistency, but have added one new class, anti-search image melanism, which might also be called apostatic melanism (Clarke 1962), to encompass the existence of melanism in a number of common species. In some common species, the existence of melanic forms may have less to do with being black or dark as to being different. The benefit gained by polymorphism in such cases is that the variation across a species with many different forms makes it less likely that birds will form strong search images for a particular form (p. 92).

Among warningly coloured species of Lepidoptera, three melanic subdivisions may be recognised:

(a) recessive melanism;

(b) mimetic melanism;

(c) sexual melanism.

A summary of this classification is given in Box 7.1. I will consider the rather special cases of melanism in warningly coloured species in Chapter 8. In this chapter I will discuss each of the subdivisions in the industrial and non-industrial melanism classes. I will deal with non-industrial melanism first so that industrial melanism can be put into chronological context more easily, and because many cases of partial industrial melanic polymorphisms have their origins in non-industrial melanism.

Box 7.1   Classes of melanism in the Lepidoptera

Industrial melanism

(a) Fully industrial melanic polymorphism: melanic polymorphism only since the industrial revolution.

(b) Partial industrial melanic polymorphism: melanic polymorphisms known prior to the industrial revolution, but have increased since.

(c) Polygenic industrial melanism: average ground colour has darkened gradually in industrial areas.

Non-industrial melanism

(a) Thermal melanism.

(b) Rural or background choice melanism.

(c) Northern latitude melanism.

(d) Western coastline melanism.

(e) Pluvial melanism.

(f) Melanism associated with fire-resistant trees.

(g) Ancient conifer melanism.

(h) Anti-search image melanism.

Melanism in warningly coloured species

(a) Recessive melanism.

(b) Mimetic melanism.

(c) Sexual melanism.

Non-industrial melanism

Thermal melanism

Some Lepidoptera are melanic, or have melanic forms not as a defence against predation, but because of the thermal properties of dark compared with light surfaces. Dark surfaces both absorb and radiate heat more rapidly than pale surfaces. Consequently, melanic Lepidoptera will warm up more quickly than their light counterparts given a heat source such as sunlight. The reverse is of course also true, for melanics will also cool down more rapidly after dusk. In cool or temperate climes, melanics may be at a significant advantage if they can warm up and become active when non-melanics cannot.

Examples of thermal melanism in Lepidoptera, as in many other taxonomic groups, are most common at high latitudes and high altitudes, because of the low temperatures frequently encountered. Intra-generic comparative analysis of *Erebia* and *Colias* butterflies, shows that species living at high latitude or altitude are generally darker than those living in the lowlands or further south. In arctic species of clouded yellow butterflies (Plate 1f), extensive research by Ward Watt and his collaborators has shown that increased melanic pigmentation, particularly in the basal third of the wings, causes the butterflies to heat up more rapidly (Watt 1968, 1969, 1974). Temperature is important for many species of butterfly, and most arctic species only fly in the sun. In the Abisko region of Swedish Lapland, where the pale arctic clouded yellow, *Colias nastes* Zetterstedt, and the northern clouded yellow, *Colias hecla* Aurivillius, fly at altitudes from 400 m to over 1000 m, completely clear days are rare, even in high summer. The butterflies, which rest low down in the dwarf scrub, must therefore make the most of the brief interludes when the sun is out. It is remarkable how quickly these butterflies then appear. I have stood many times at the top of an appropriate slope (Fig. 7.1), waiting for these butterflies to appear, or more pertinently watching for a brief break in the clouds. Once the sun is out, the butterflies appear from their

Fig. 7.1 Habitat for pale arctic clouded yellow and northern clouded yellow, in Swedish Lapland.

hiding places within a minute or two, scudding at great speed over the tundra looking for mates, food, or oviposition sites.

Melanism is also seen in many arctic or alpine moth species. The black mountain moth (Plate 1g), and the netted mountain moth, are both day flying and have grey to black wings. Early on dry days, both species climb out of their night-time retreats among the low scrub, to find a vantage point, usually on a rock, where they sun themselves, orienting their wings to catch the maximum heat.

Several species of butterfly have darker wings in arctic and alpine conditions than in lowlands further south. For example, in the Hebrides, the dark-green fritillary, *Argynnis aglaja* (Linnaeus), is represented by a form, f. *scottica*, in which the tawny brown colour is suffused by black. Similarly, the extent of black venation in the green-veined white butterfly, *Pieris napi* (Linnaeus), increases both in northern Scotland and Scandinavia, and a similar form f. *bryoniae* occurs in the Swiss Alps. In the USA, the same sort of phenomenon is seen, with the amount of melanic pigmentation on the wings of the checkered white, *Pieris protodice* (Boisduval and Leconte) ssp. *occidentalis* Reakirt, varying with season, the spring flying generation being more melanised than that flying in the summer (Kingsolver and Wiernasz 1987). Kingsolver and Wiernasz (1991), demonstrated that the variation was produced environmentally, being a consequence of the day length and temperature to which larvae were exposed. They conclude from their studies, and a review of similar work on *Colias* butterflies, that in these genera, degree of melanisation is adaptive, as a result of its role in thermoregulation. Similar work on the small Apollo, *Parnassius phoebus* Fabricius, shows that degree of melanisation in this species also affects activity; however, in this case, greater melanisation in the basal third of the wings led to an increase in flight time rather than lower temperature flight initiation (Guppy 1986a,b). Guppy (1986b) argues that this is probably a consequence of greater absorption of solar radiation by darker butterflies while in flight. The adaptive involvement of thermodynamics in the occurrence of melanic forms has been suggested or demonstrated in many other butterflies. Indeed, in one species, the monarch, *Danaus plexippus* (Linnaeus), thermal factors have been shown to be involved in both adult basking behaviour, and larval melanism (James 1986).

In England, one butterfly shows melanic polymorphism which has a thermal component. The silver-washed fritillary has a sex-limited form, f. *valesina* (Plate 5a), which occurs at appreciable frequencies in females in parts of Hampshire and Dorset. Magnus (1958) has shown that this form, in which the orange-brown ground colour of the females is replaced by a dark purple-olive hue, is less attractive to males than the typical form. This disadvantage is balanced by the thermal advantage gained by the darker females which fly in cooler conditions, gaining more food and having longer to find the violets on which they lay their eggs.

Thermal melanism is not confined to the northern hemisphere. In New Zealand, Hudson (1928) notes that 'all members of the genus *Oracrambus* which replaces the related genus *Crambus* at high altitudes, are very dark, enabling them to heat up rapidly' and thus take advantage of the fitful periods of hot sun-shine characteristic of alpine climates.

Kettlewell (1973) writes that among the Lepidoptera, thermal melanism is con-fined to day flying species. While this may be true, it is possible that some thermal advantage accrues to night flying moths that have specific habitat preferences. For example, recent research has shown that the melanic forms of some night flying British moths have a strong habitat preference for dense canopied wood-lands (p. 184). In such woodlands, temperature drops more slowly after dusk than in the open. It is probable that the morph specific habitat preferences exhibited by these moths has evolved primarily for predator avoidance. However, the dis-advantage of being melanic, in terms of heat loss for a night active species, may be less for a species which confines itself to such woodlands.

Rural or background choice melanism

Kettlewell's (1973) subclass of rural or background choice melanism is itself sub-divided, there being four different ecological causes for the melanisms that he includes therein. The first involves the effects of the angle of incidence and bright-ness of the sun at high latitudes on disruptively patterned moths. This cause of melanism is, I feel, more a feature of northern melanism, and will thus be included in that section (p. 167). Under the heading 'predator–prey relationship', Kettlewell discusses the likelihood that melanic forms of common species may exist to increase the number of forms that exist, thus reducing the ease with which predators may form search images. While I have no doubt that Kettlewell is correct in attributing melanism in some species to this cause, I feel that the impor-tance of this mechanism, which will itself maintain polymorphism, because the selection is necessarily frequency dependent, demands that it be considered in a separate class.

This leaves us with two situations in which melanism may have evolved in response to the heterogeneity in the colour of backgrounds that moths may use as resting sites. In one situation, the heterogeneity in backgrounds is within the same habitat and may lead to polymorphism, possibly with morph specific behav-iours also evolving. The second situation occurs where backgrounds differ on a larger scale, as for example when one moves from peat soils to chalky soils. In this case, there is the possibility for whole populations to adapt to local condi-tions, with geographic or local races resulting.

Sadly, we have rather little good data on the natural daytime resting sites of many of our nocturnally active moths. Nor is it easy to see how good data could be gained, except by exhaustive and time consuming field observations. Even these observations are problematic for they are open to the criticisms that it is impossible to search all parts of most habitats, that we only see those individu-als that happen to be in suboptimal positions, and that the way we perceive moths at rest is not the way that the predators would perceive them. Recent work has improved the situation for a small number of species (p. 188), but for most species we are ignorant of where moths spend the day.

If the information we have on where particular species of moth rest is poor, that on whether different forms of a species select different substrates to rest upon is pitiful. As far as I can discover, only four studies have addressed this question in the field. In the most recent of these, experiments were designed primarily to investigate whether different morphs show differential habitat selection, rather than differential resting site selection (p. 187). However, the results do indicate that the melanic, banded, and non-melanics of the mottled beauty may show weak resting site preferences. Of the other three studies, in two, involving *Diurnea fagella* (Stewart 1977*d*) and *Cyclophora pendulinaria* (Sargent 1985), no morph specific behavioural differences were detected. The fourth study involved the pine beauty, and in this species, morph specific resting site preferences were found (Majerus 1982*b*).

The pine beauty is polymorphic in most British populations. The commonest form is typically orange-red, sometimes more or less suffused with ochreous, while the hindwings are pinkish brown. Two melanic forms occur. In f. *griseovariegata* Goeze the forewings are flushed with bluish-grey or greenish-grey, although some red is still apparent, and the hindwings are grey-brown. In f. *grisea* Tutt, the red colour is almost completely absent, being replaced by bluish-grey, the hindwings being dark grey-brown. The inheritance of these forms is due to a single biallelic gene, the typical and *grisea* forms being homozygotes, and the *griseovariegata* form being heterozygous (Majerus 1982*b*). Data obtained by collecting moths at rest on pines in southern England suggested that two of the forms, typical and *griseovariegata*, showed differences in their resting site preferences, typical moths being found most often on trunks and *griseovariegata* most on foliate twigs (Table 7.1). Only one f. *grisea* was found in these collections, so an experiment was set up using captive bred material to test the resting preferences of all three forms in the field. One hundred and fifty moths (75 male and 75 female) of each phenotype were released at 7.00 a.m. in a small isolated stand of seven mature Scots pines on Picket Hill Heath, near Ringwood in Hampshire. The pines were examined between 2.00 and 6.00 p.m. later the same day, and the position of all the moths found was recorded. Positions were split into four categories: trunks, branches, twigs, and needles. The results (Table 7.2) show significant differences in the behaviour of the forms, particularly among the females. Taking account of the bias produced by the ease with which the different sites could be searched, it appears that in females, typicals prefer to rest on the trunks, *griseovariegata* prefers branches and foliate twigs, and *grisea* prefers foliate twigs and needles. The same trends are seen in the data for males, but to a lesser extent. The correlation between the colours of the forms and their resting site preferences, suggest that selection by predators that use colour vision to detect prey has been influential in the evolution of both the morphological and behavioural polymorphism in this species.

Table 7.1 Positions of wild pine beauty imagines, of three phenotypes, *typica* (*typ.*), *griseovariegata* (*gvt.*), and *grisea* (*gris.*), taken at rest on Scots pine, in England. Only one *grisea* was found. (From data in Majerus 1982*b*.)

	Male *typ.*	Fem *typ.*	Male *gvt.*	Fem *gvt.*	Total
Trunks	16	21	7	2	46
Branches	11	8	3	4	26
Foliate twigs	12	6	6	11	35
Totals	39	35	16	17	107

Table 7.2 The positions of released, pine beauty moths, of different phenotypes, recollected from different parts of Scots pine stand. Abbreviations as in Table 7.1. (From Majerus 1982*b*.)

Site type	Male *typ.*	Male *gvt.*	Male *gris.*	Female *typ.*	Female *gvt.*	Female *gris.*	Totals
Trunks	15	11	9	16	4	3	5
Branches	7	7	5	6	7	5	3
Twigs	9	7	10	4	9	5	4
Needles	1	2	4	2	5	12	2
Totals	32	27	28	28	25	25	16

Interestingly, the frequencies of the three forms in four populations (Table 7.3) suggest that the relative availability of different types of resting sites may affect the frequencies of the forms. Scots pine, *Pinus silvestris*, shows two main growth habits, depending on the situation. Typically, these trees are dome shaped, with the leader shoot showing little dominance over side shoots. However, when grown close together in plantations, the leader shoot becomes conspicuously dominant, leading to tall spire-like trees, on which foliate side branches are confined to the top third of the trees (Watson 1981). Consequently, in plantation trees, the number of needled branches and twigs, compared with non-foliate sites, is reduced. In the sampling areas at Northwood and Keele, the Scots pines were mature solitary trees and the frequency of the *grisea* allele was higher than at Picket Hill and Englefield Green, where the sampling sites were in Scots pine plantations (Majerus 1982*b*).

Although the demonstrated morph specific resting site preferences in the pine beauty is a unique case among adult Lepidoptera, it is likely that the phenomenon occurs in many other species. Experiments on other species would be most valuable. The species perhaps most suitable for study are those which are most habitat or food plant specific. This is because in these there is more opportunity for precise adaptations to particular situations to evolve. For example, several species of moth, which live mainly in reed-beds, show melanic polymorphism. The moths rest on old reed stems and leaves, which differ in character, with age, and

Table 7.3 Details of sexes and phenotypes of samples of pine beauty, recorded at four sites in England, with frequency estimates for the *typical* and *grisea* alleles. (From Majerus 1982*b*.)

Location	Year	*typica* Male	*typica* Female	*griseovariegata* Male	*griseovariegata* Female	*grisea* Male	*grisea* Female	Total	Allelic frequencies *typica*	Allelic frequencies *grisea*
N	1968	12	4	5	2	0	0	23	0.848	0.152
N	1969	10	0	4	0	0	0	14	0.857	0.143
N	1970	6	1	0	0	0	0	7	1.0	0.000
N	1971	9	1	6	0	0	0	16	0.813	0.187
N	1972	21	5	5	0	0	0	31	0.919	0.081
N	1973	7	1	4	0	0	0	12	0.833	0.167
N	1974	19	6	8	4	1	0	38	0.816	0.184
N	1975	30	6	19	4	2	0	61	0.779	0.221
N	1976	35	11	23	9	1	1	80	0.775	0.225
Total		149	35	74	19	4	1	282	0.817	0.183
EG	1972	2	0	0	0	0	0	2	1.0	0.000
EG	1973	9	0	1	0	0	0	10	0.950	0.050
EG	1974	7	0	0	0	0	0	7	1.0	0.000
EG	1975	6	1	0	0	0	0	7	1.0	0.000
EG	1976	14	1	1	0	0	0	16	0.969	0.031
EG	1977	6	0	0	1	0	0	7	0.929	0.071
EG	1978	9	1	0	0	0	0	10	1.0	0.000
EG	1979	2	0	0	0	0	0	2	1.0	0.000
Total		55	3	2	1	0	0	61	0.975	0.025
PH	1975	11	2	0	1	0	0	14	0.964	0.036
PH	1976	18	2	3	0	1	0	24	0.896	0.104
PH	1977	14	3	2	1	1	0	21	0.881	0.119
PH	1978	4	0	0	1	0	0	5	0.900	0.100
PH	1979	15	2	4	1	0	0	22	0.886	0.114
PH	1980	3	1	1	0	0	0	5	0.900	0.100
Total		65	10	10	4	2	0	91	0.901	0.099
K	1979	7	2	2	1	1	0	13	0.808	0.192
K	1980	9	2	2	0	0	1	14	0.857	0.143
Total		16	4	4	1	1	1	27	0.833	0.167

N, Northwood: approximate grid reference TQ082906; EG, Englefield Green: SU996698; PH, Picket Hill: SU185065. K, Keele University: SJ823440.

the growth of black smutty mildews. Younger dead leaves and stems are beige in colour, while old ones are dark brown or sooty grey. Consequently, species such as the twin-spotted wainscot, *Archanara geminipunctata* (Haworth), the bulrush wainscot, *Nonagria typhae* (Thunberg) (Fig. 7.2), and the brown-veined wainscot, *Archanara dissoluta* (Treitschke), would certainly repay attention.

When crucial elements of the habitats differ in character between geographic regions, populations of a species may become locally adapted, thus producing

Fig. 7.2 Melanic (above) and non-melanic (below) forms of the bulrush wainscot.

local forms or races. Several species of British moth may be cited as examples. Perhaps the most striking is the annulet, *Gnophos obscuratus* (Denis and Schiffermüller), which occurs on limestone, chalk, and peat soils. On the limestone and chalk soils, the moths are pale, while on peat heathlands, the moths are very much darker (Fig. 7.3). In the white-line dart, *Euxoa tritici* (Linnaeus), specimens from moorland habitats are usually darker than those from coastal sandhills. In Archer's dart, *Agrotis vestigialis* (Hufnagel), the darkest forms are found on the acid heathlands of Surrey, Hampshire, and Dorset, while those from the East Anglian Brecklands and coastal sand-dune systems are lighter. The feathered rustic, *Agrotis cinerea* Denis and Schiffermüller, and the sand dart, *Agrotis ripae* (Hübner), both show considerable intra-population variation, but again,

Fig. 7.3 Variation in the annulet moth is correlated to soil colour: top left, chalk land; top right, limestone, bottom left, heathland, Cornwall; bottom right, heathland, Hampshire.

populations from areas with dark soils are on average darker. In each of these species it is probable that the variation in the colour patterns is controlled polygenically, with stabilising selection acting in each population to maintain an optimum determined by the colour of the local soils. However, this has yet to be demonstrated by relevant crosses between moths from different localities.

An interesting case is that of melanism in the latticed heath, *Semiothisa clathrata* (Clerck) (Fig. 7.4). The melanic form, f. *alboguttata*, is not uncommon in the south of England, and the melanic appears, somewhat unexpectedly, to be just as common on chalk downland as on peaty heathlands. The species flies both at day and at night, and would certainly repay detailed investigation.

Northern melanism

I have already discussed the role of the thermal qualities of dark versus light surfaces on the high incidence of melanism at high latitudes (p. 160). There are, however, two other features of high latitudes that are likely to promote melanism. First, the angle of incidence, and the strength of the sun are both relatively low, thus changing the quality of light, particularly in respect of the strength of shadow that is cast. Second, in high summer, in high latitudes, it is light around the clock,

Fig. 7.4 Typical (above) and melanic, f. *alboguttata*, (below) forms of the latticed heath.

with the result that moths that are nocturnal farther south do not have the cover of darkness as a protection for their activities.

Kettlewell (1973), describes a general rule for species showing disruptive colour patterns, stating that the angle of incidence of the sun causes sharper shadows near the equator, which tends to produce sharply defined patterns, as seen in striped mammals such as tigers and zebras. However, with increasing latitude, the strength of shadows decays, with the result that disruptive patterns also become more diffuse. As an example, Kettlewell cites the Hebrew character, *Orthosia gothica* (Linnaeus). In England and Wales, this species has a sharply defined black mark around the orbicular stigma of the forewing. However, in the highlands of Scotland and in Scandinavia, f. *gothicina* Herrich-Schäffer, in which this mark is pale or absent is common (Plate 5c and d). Other species which display similar phenomena are the scalloped oak (Fig. 7.5) and the treble lines, *Charanyca tri-*

Fig. 7.5 Northern melanism in the scalloped oak: above, typicals male and female, Cambridgeshire; below. f. *unicolor* male and female, Perthshire.

grammica (Hufnagel) (Fig. 7.6). In these species, disruptive lines or bars across the forewings, that are obvious in southern individuals, are obscured in many northern Scottish moths.

Most British moths emerge and fly in the summer months. Those that are nocturnal feed, find a mate, and oviposit, under the cover of darkness to avoid diurnal predators, such as most birds. At high latitude, from about 55°N up into the Arctic circle, this is not possible in mid-summer, for although it gets darker, even around midnight, there is always sufficient light for birds to forage. Thus, moths have to be active in the light. The timing of main activity at these latitudes is a balance between choosing periods of least light, and the need to be warm enough to become active. However, although many species fly mainly in the hours around midnight, these moths are still flying in twilight. Melanism is promoted in these conditions, not so much for protection when the moths are at rest, but for protection when they are flying. This is because dark-coloured moths are more difficult to see in dim light than are pale moths. Because melanism at high latitude has been most studied in the northern hemisphere, particularly on the Shetland Isles, Kettlewell called this phenomenon northern melanism, although the same considerations must be pertinent to high latitudes in the southern hemisphere.

In the Shetland Isles, Kettlewell notes that 21 of 62 resident species show melanism. These may be broadly split into: (a) those in which the melanics represent geographic races, paler forms of the species being found further south; (b) those in which two distinct forms occur as a balanced polymorphism; and (c)

Fig. 7.6 Northern melanism in the treble lines: above, typical, Suffolk; below, f. *bilinea*, Rannoch Black Wood.

those showing a continuous range of forms from light to dark, probably controlled polygenically.

Before discussing specific examples, it should be stressed that in most instances, we have little evidence to indicate whether melanism is due to thermal qualities, to reduction in disruptive patterns, or to the necessity of flying in the light. In many instances, it is possible that all three factors play a part. For example, in the ingrailed clay, *Diarsia mendica* (Fabricius), a geographic race, f. *thulei* Staudinger, occurs in the Shetlands (Fig. 7.7). Not only is this form dark, but the black disruptive marks on the forewings, that occur on many southern specimens, are always absent.

The ingrailed clay is one of several species that are monomorphic for melanic races at high latitude. In this species, there is a second race, f. *orkneyensis*, occur-

Fig. 7.7 The typical southern form of the ingrailed clay (above), and its northern race, f. *thulei*, from Shetland (below).

ring on the Orkney Islands, which is slightly paler than that from Shetland, but still much darker than is usual in the south of Britain. Other cases of melanic geographic races on the Shetlands include the marbled coronet, *Hadena confusa* (Hufnagel), and the netted pug, *Eupithecia venosata* (Fabricius), while in the blue-bordered carpet, *Plemyria rubiginata* (Denis and Sciffermüller), a northern melanic race, f. *semifumosa* (Fig. 7.8), is found more widely, across northern England and Scotland.

Melanic polymorphism on the Shetlands is seen in the square-spot rustic, *Xestia xanthographa* (Denis and Schiffermüller), in which a very dark brown form occurs; in the northern spinach, *Eulithis testata* (Linnaeus), where a dark brown

Fig. 7.8 The southern (above) and northern (below) races of the blue-bordered carpet.

form, f. *masanaria* Freyer, is found with the yellow barred forms found farther south (Plate 5g); and in the autumnal rustic *Paradiarsia glareosa* (Esper) (Fig. 7.9), in which a melanic form, f. *edda* Staudinger, occurs at frequencies varying from 2 per cent in the south to 97 per cent in Unst. This latter polymorphism has been the subject of considerable research in the Shetlands and Orkneys.

The melanic, f. *edda*, of the autumnal rustic, is confined to the Shetlands, the Orkneys, Fair Isle, and a few coastal areas on the mainland of Scotland and Denmark. This form is dark brown in colour, with the reniform and orbicular stigma outlined in pale brown. The ground colour of the typical form is light grey, often tinged with pink, and with small black marks at the base and either side of the orbicular stigma. It is probable that the inheritance of *edda* and *typica* is due to a single biallelic gene, with the *edda* allele being almost fully dominant in some

Fig. 7.9 On the Shetland Isles, the autumnal rustic shows melanic polymorphism: above, typical form; below, f. *edda*.

populations, such as on Unst, and incompletely dominant elsewhere (Kettlewell and Berry 1969).

Surveys of the frequency of f. *edda* throughout the Shetland Islands show that the frequency of this form increases greatly from south to north (Fig. 7.10) (Table 7.4), and there is temporal variation in the frequencies of forms, both within and between years (Kettlewell 1973). Elsewhere, smaller samples have been recorded on Orkney and Fair Isle, where the frequency of f. *edda* also varies both temporally and spatially (Table 7.5).

In this species, bird predation appears to be intense. Evidence to support this view comes from both direct observations of birds feeding on the moths, and examination of the gut contents of shot birds. In 1960 and 1962 Kettlewell shot seven common gulls, *Larus canus*, and six hooded crows, *Corvus cornix*. Autumnal rustics were found in the stomach contents of both species. Interestingly, the frequency of the typical form in these stomachs was significantly higher than in

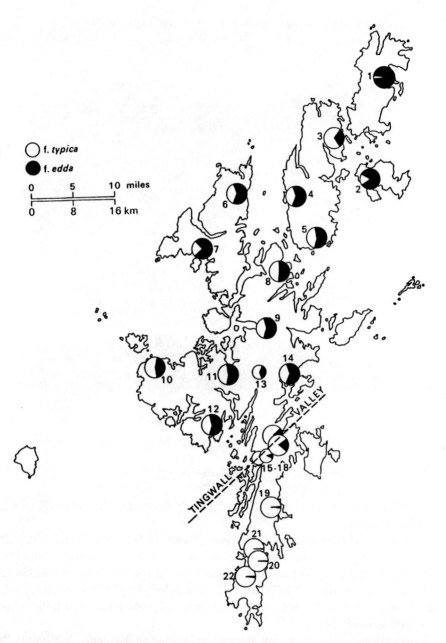

Fig. 7.10 The frequency distribution of the melanic and typical forms of the autumnal rustic on Shetland. (From Kettlewell 1973.)

Table 7.4 (a) Frequencies of the forms of autumnal rustic, at sites on Shetland: sites sampled 1959–62 (b) Sites sampled in 1969. (Adapted from Kettlewell 1973.)

(a)

Site	Year	f. *typica*	f. *edda*	Total	Overall % frequency of f. *edda*
Baltasound	1959	77	2462	2539	97.0
	1960	302	10356	10658	97.2
Hillswick	1959	4	19	23	82.6
	1960	21	66	78	75.9
	1961	222	683	905	75.5
	1962	195	634	829	76.5
Vatster	1961	116	69	185	37.3
	1962	88	48	136	35.3
Catwalls	1960	33	102	135	24.4
	1961	337	87	424	20.5
	1962	97	29	126	23.0
Scousburgh	1959	95	4	99	4.0
	1960	280	10	290	3.4

(b)

Area	Numbers caught f. *typica*	f. *edda*	Total	Overall % frequency of f. *edda*
1. Baltasound	16	513	529	96.9
2. Uyeasound	7	104	111	93.7
3. South Yell	29	32	61	52.5
4. North Roe	56	117	173	67.6
5. Hillswick	9	33	42	78.6
6. Voe	53	95	148	57.4
7. Vatster	79	65	144	45.1
8. Catwalls	155	46	201	22.0
9. Grimista	149	32	181	17.7
10. Quaff	187	41	218	14.2
11. Boddam	87	6	93	6.4

the areas where they were shot (Kettlewell 1973). Kettlewell interprets this as demonstrating that the typical form is at a significant selective disadvantage due to predation. While this is probably the right view, it is also possible that the birds had migrated from areas where the typical form was at higher frequency. In the north of Shetland, bird predation of moths is particularly high, as a result of the foraging of large populations of migrating birds, which arrive during the autumnal rustic's flight season. Here, the melanic form appears to gain advantage over the typical form for two reasons. First, against the dark peat soils, the dark form is very much more cryptic than the pale form. This is the reverse of the situation at the south of Shetland Mainland, where the autumnal rustic inhabits a vast

Table 7.5   Frequencies of the melanic form, f. *edda*, of the autumnal rustic on the Orkney Islands and Fair Isle. (From Kettlewell 1973.)

Site	Year	f. *typica*	f. *edda*	Total	% frequency of f. *edda*
Rousay (Orkney)	ffenell 1961	17	3	20	15.0
Orphit (Orkney Mainland)	Lorimer 1967	66	16	82	19.5
Orphit (Orkney Mainland)	Lorimer 1968	153	19	172	11.0
Binscarth (Orkney Mainland)	Lorimer 1968	37	22	59	37.3
Fair Isle	Hardy 1961	40	15	55	27.3

region of sand-dunes and the typical form is more cryptic. Second, the melanic form gains an advantage in being more difficult to see in flight than the paler typical form. Indeed, Kettlewell has demonstrated that there are behavioural differences between the two forms, the melanics eclosing earlier in the year, and taking flight less than the typicals.

The extensive work of Bernard Kettlewell and Professor Sam Berry on the autumnal rustic on Shetland has shown that populations may become adapted to local conditions, both in respect of their morphology and their behaviour. Furthermore, in many transfer experiments, reciprocally moving moths between the north and south of Shetland Mainland, they demonstrated that moths survive better on home-ground than elsewhere. The phenomenon of morphological and behavioural adaptation to specialised local conditions is also pertinent to cases of melanism in the next four categories.

Species showing northern melanism through a range of forms from light to dark include: the grey mountain carpet, *Entephria caesiata* (Denis and Schiffermüller) (Fig. 7.11); the garden carpet, *Xanthorhoe fluctuata* (Linnaeus), and the ghost moth, *Hepialus humuli* (Linnaeus) (Plate 7c). In the former two, dark forms are common in the Highlands and Islands of Scotland and in Scandinavia. In both, phenotypically similar melanic forms are found further south as industrial melanics (Kettlewell 1973; Skinner 1984). However, whether the northern and industrial melanics are genotypically similar has yet to be investigated.

The case of the ghost moth is particularly interesting. This species shows considerable sexual dimorphism. Males of this species have remarkable wings, which appear to shine white on their upper surface when the moths fly at dusk. These wings also reflect ultraviolet light strongly (Stalker personal communication). The females, on the other hand, have dull red markings on a yellow ochre ground colour. In the Shetland Isles and Faroes, most males have much darker forewings, due to brown patterning upon them, and the hindwings are almost black. Kettlewell (1973) cites previously unpublished form frequency data collected by Cadbury, in Shetland in 1964 (Table 7.6). These data show the darkest form of males to have higher frequency on the more northerly island of Unst, than on Mainland Shetland.

Fig. 7.11 Northern melanism in the grey mountain carpet, showing typical (above) and melanic (below) forms.

The bright wings of southern males (Fig. 7.12) are undoubtedly a sexual attractant to females. Courtship in this species is somewhat unusual. Males form leks, of up to several dozen individuals, at dusk, when they fly in specific small areas, hovering over the same spots, a few metres apart, for up to 3 minutes at a time. In this behaviour they are very visible. The males also produce an attractant pheromone from a brush-like scent gland on the tarsus of the hind legs. The much less conspicuous female flies directly at a male, knocking him to the ground, or simply alights in herbage below one of the hovering males. Once on the ground, the female begins calling, releasing a pheromone, which the male uses to locate her, before he mates with her.

It appears then that females use both vision and scent to locate males initially. In southern areas, the cost of the great visibility of hovering males at dusk is small, because there are few dusk-flying predators that find their prey by eye. In

Table 7.6 The phenotypic frequencies of male ghost moths, in Shetland, 1964. (Cadbury unpublished, cited in Kettlewell 1973.)

Locality (northernmost at the top)	Phenotypes				
	Ochreous — brown as in female	Cream with dark pattern	Immaculate cream	China-white	Total
Unst					
Burrafirth	7	0	1	0	8
Baltasound	28	6	1	0	35
	35	6	2	0	43
	81.5%	13.9%	4.6%		
South Mainland					
Kergord					
(light trap)	1	0	0	0	1
Tingwall	17	9	0	0	26
Easter Quarff	2	2	0	0	4
Cunningsburgh	2	0	0	0	2
Channerwick	2	1	1	0	4
	24	12	1	0	37
	64.9%	32.4%	2.7%		

Shetland, on the other hand, considerable numbers of common gulls and black-headed gulls, *Larus ridibundus*, quarter meadowlands searching for these moths in the long northern twilight. These birds must confer a considerable selective disadvantage on the more visible whiter males, compared with the darker individuals. However, dark males will be at a mating disadvantage, as they will be less visible to females searching for mates. The colour of male ghost moths on the Shetlands and Orkneys, seems then to be a compromise between avoidance of predation, and obtaining mates. Indeed, there is some anecdotal evidence to suggest that the pheromonal attraction of Shetland males is stronger than that of males from further south (Cadbury, unpublished, cited in Kettlewell 1973). If so, here we have a case where evolution of a melanic form to reduce the risk of predation has had a knock-on effect on the courtship of a species, shifting emphasis from visual to pheromonal recognition (Kettlewell 1973).

Western coastline melanism

Geographic clines in melanism occur not only with increasing latitude. There are a number of species which show east–west clines. In northern Europe, several species show melanism along western coastlines, with the darkness then diminishing inland to the east. In all these cases, melanism appears to be controlled polygenically, for in none does a clear-cut polymorphism exist.

The tawny shears, *Hadena perplexa* (Denis and Schiffermüller), is whitish or pale yellow in eastern and central England. In Devon, Cornwall, and most of Wales, mid-brown forms occur, while along the western coast of Ireland, the nom-

Fig. 7.12   Ghost moth males from southern England have bright white wings.

inate subspecies is completely replaced by *H. perplexa capsophila* Duponchel, an almost black form that is called the pod lover (Fig. 7.13).

The yellow shell *Camptogramma bilineata* Linnaeus exhibits the same sort of phenomenon (Plate 5h), although in this species, the dark forms f. *hibernica* Tutt and f. *isolata* Kane, are more strongly confined to the western coastline, for just 10 km inland, all moths are of the more usual yellow form. Dark forms inclining to black are also found on islands off the west coast of the Iberian peninsular, such as Tearach, 16 km off the coast of Valencia (Kane 1896).

Across central and eastern England, the marbled coronet, is easily recognised from the attractive black and white patterning of the forewing. Along western coastlines of England, Wales, the Isle of Man, Scotland, and Ireland, the pale markings become darker and less distinct (Fig. 7.14). Kettlewell (1973) suggests that in some related species, melanic forms found along western coasts have experienced full speciation. Indeed, several species of dark-coloured noctuid are confined to western coastal habitats, the grey, *Hadena caesia* (Denis and Schiffermüller) ssp. *mananii* Gregson, and the black-banded, *Polymixis xanthomista* (Hübner) ssp. *statices* Gregson, being examples.

In some of these species, notably the marbled coronet, the grey, and the yellow shell, the melanism along western coasts increases northwards, so that some species undoubtedly show both western coastline and northern melanism. For example, in the marbled coronet, although specimens from western coasts are

Fig. 7.13 The ground colour of the tawny shears darkens towards the west: top, Kent; middle, Devon; bottom, west Cornwall.

Fig. 7.14 Western coastline melanism in the marbled coronet: typical, Cambridge (above); melanic, Cornwall (below).

considerably darker and less patterned than those from elsewhere in England, the darkest individuals, of the form *obliterae* Robson, are found in Shetland.

The evolutionary reasons underlying western coastline melanism are unknown, but it is probable that the same phenomenon occurs in many Palaearctic regions. Several possible explanations have been suggested. The darkening of moths along western coasts may be an anti-predator device, for when these moths rest on rocks, their crypsis may be increased, as the rocks will often be darkened by sea spray or by the high rainfall in such regions. Kettlewell (1973) has also questioned whether dark forms may have greater tolerance to salt then paler forms. Two other alternatives seem possible. First, as cloud cover is high in these coastal regions, there may be a thermal advantage to being dark in such locations. Alternatively, the protective qualities of melanic pigments against abrasion (p. 30), on coasts exposed to the prevailing winds, may play a part. Until careful studies of this phenomenon are undertaken, it is impossible to distinguish these possibilities.

Pluvial melanism

An allied class of melanism may be that seen commonly in forested regions having very high rainfall. Such habitats are found in South Island, New Zealand, the Himalayas, and in the Olympic Mountains in north-west USA, where rainfall may exceed 500 cm per annum. In such situations light intensity is low and the backgrounds are dark, because of repeated wetting from the high rainfall and humidity. In such places, melanism is common (Hudson 1928). Again, as with western coastline melanism, virtually no experimental work has been conducted on this phenomenon, and the evolutionary causes are unknown, although again it seems likely that increased crypsis and possibly thermal melanism will be important.

Melanism associated with fire-resistant trees

In many parts of the world, fire from lightning or volcanoes has for aeons regularly set scrub and savannah aflame. More recently, human activities, whether intentional or accidental, have frequently led to a scorching of large tracts of vegetation. Some types of vegetation are extremely resistant to flash fires. Tropical grasslands frequently recover from fires very rapidly, for fast moving fires do no more than burn off the upper foliage, leaving the roots and growing points undamaged. The same is true of certain types of tree. For example, many species of eucalyptus, pine, and some oaks, can withstand burning, reshooting thereafter from charred stumps or trunks. In forests composed of fire-resistant trees, melanic polymorphism is not uncommon. In the eastern pine forests of the USA, several species of 'underwing' moths of the genus *Catocala* show melanic polymorphisms (e.g. *Catocala cerogama* Guenée, with its melanic form, f. *ruperti* Franc). Similarly, Kettlewell found the moth, *Cleora tulbaghata* Felder, to have a melanic polymorphism in South Africa, where *Acacia cyclops*, a fire-resistant tree introduced from Australia grows. Undoubtedly, other examples will come to light if suitable habitats in regions regularly exposed to fire are carefully investigated.

It is perhaps surprising that one instance of a correlation between an increase in melanic frequency and incidence of fire has been recorded in Britain. The moth in question is the horse chestnut. This small geometrid is found locally on heaths in southern England. Typically, the colour of the forewings is mid-grey, the hindwing being pale silvery grey. In Hampshire and east Dorset, a melanic form, f. *nigrescens* Lempke, in which all the wings are dark grey, is found as a polymorphism. Samples of this moth were collected on Picket Hill Heath, in the New Forest, twice each year between 1975 and 1980, to coincide with the emergence of each of the two broods per year. Other samples were also collected from other localities in Surrey between 1975 and 1978, and in Hampshire and Dorset in 1980 (Majerus 1981). The melanic form was shown to be controlled by a single recessive allele.

Table 7.7 The frequencies of the typical and *nigrescens* phenotypes of the horse chestnut moth, in samples taken between April 1975 and September 1980, on Picket Hill, New Forest, and the frequencies of the alleles controlling these forms.

Approximate dates when samples were taken	No. of each form			Phenotypic % frequencies		Allelic frequencies	
	Typical	*nigrescens*	Total	Typical	*nigrescens*	Typical	*nigrescens*
April/May 1975	68	10	78	87.2	12.8	0.613	0.358
Aug./Sept. 1975	50	9	59	84.7	15.3	0.609	0.391
April/May 1976	89	18	107	83.2	16.8	0.590	0.410
Aug./Sept. 1976	127	26	153	83.0	17.0	0.588	0.412
April/May 1977	63	11	74	85.1	14.9	0.614	0.386
Aug./Sept. 1977	61	8	69	88.7	11.3	0.659	0.341
April/May 1978	46	7	53	86.8	13.2	0.637	0.363
Aug./Sept. 1978	33	4	37	89.2	10.8	0.671	0.329
April/May 1979	38	5	43	88.4	11.6	0.659	0.341
Aug./Sept. 1979	52	7	59	88.1	11.9	0.656	0.344
April/May 1980	167	21	188	88.8	11.2	0.665	0.335
Aug./Sept. 1980	39	5	44	88.6	11.4	0.662	0.338
Total	833	131	964	86.2	13.6	0.629	0.369

The frequencies of the forms in the Picket Hill samples are given in Table 7.7. The frequency of f. *nigrescens* is as great, or greater, on Picket Hill Heath than elsewhere. More importantly, on Picket Hill Heath the frequency increased in late 1975 and 1976, but thereafter declined. Significantly, both 1975 and 1976 had long hot summers, with the result that large tracts of heathland in the New Forest were scoured by accidental fires, started by the many tourists that visited the region. Certainly, a large portion of Pickett Hill Heath was burnt in 1975, with a further fire charring additional areas of heather in 1976. Although the role of visual predation in the evolution and maintenance of melanic polymorphism in the horse chestnut moth has not been demonstrated, it seems likely that the additional crypsis of the melanic form, against the backgrounds available on burnt heathlands, was a contributory factor to its short-term increase.

Ancient conifer melanism (relict melanism)

Melanism is often found in regions of indigenous coniferous forest such as the relict Caledonian pine forests. It is likely that from about 10000 years ago, after the last glaciation, and before the subsequent colonisation of deciduous trees, Britain was covered with conifer species. These forests are now represented only by small pockets, such as Rannoch Black Wood, Perthshire. Here several species of moth, such as the mottled beauty, have melanic forms that appear, to the human eye, to be less well camouflaged when at rest than their non-melanic forms. However, whereas most moths which rely on crypsis for protection rarely fly by day, trunk resting species in these woods are frequently disturbed into flight by large wood ants that forage on the trunks. On the wing during daylight hours, the moths are at risk of predation by birds (Kettlewell 1973). However, in the dim light conditions under the canopy, melanic forms are less easy to follow on the wing than the non-melanics. It is for this reason that Kettlewell (1973) believes that melanic polymorphism occurs in a relatively high proportion of species that inhabit these ancient and unpolluted woodlands.

Morph specific habitat preferences in melanic Lepidoptera

In the foregoing discussion of the subdivisions of non-industrial melanism, the associations between melanism and the habitats where melanics occur have been obvious. A series of data sets, collected over the last decade, have added to this element of the melanism story.

The research on morph specific habitat selection began by accident, as a result of a single night's trapping in Dyfed. Two of my colleagues, John Spencer and Peter Kearns were in north-west Wales to collect samples of a copper tolerant grass, *Agrostis tenuis*. I had accordingly asked them to put out 'Heath' moth traps for me. On the final night, they set up the traps somewhat hurriedly prior to visiting a local hostelry, and set the traps a short distance apart, either side of a track between two types of woodland, a Douglas fir, *Pseudotsuga menziesii*, plantation and an open deciduous woodland. Stoppering the traps in the morning, they transported them back to Cambridge, where I scored the catches. I was extremely surprised to find that the proportions of melanic forms of two species, the mottled beauty and the tawny-barred angle, *Semiothisa liturata* (Clerck) (Plate 6d), were quite different between the two traps, although the traps had been set up less than 20 m apart (Table 7.8).

Three explanations of these results seem plausible. First, there might be very strong differential predation of the forms in the two habitats. Given the short dis-

Table 7.8 Details of the numbers of different forms of mottled beauty and tawny-barred angle, taken from Heath moth traps in different habitats, on 12 July 1984, at Ynys-Hir, Dyfed. Trap 1 was set up in mixed deciduous woodland, trap 2 in a Douglas fir plantation. The traps were set 20 yards apart. (From data in Kearns and Majerus 1987.)

	Trap 1	Trap 2	Totals
Mottled beauty			
typica	41	45	86
conversaria	0	11	11
nigricata/nigra	0	2	2
Totals	41	58	99
Tawny-barred angle			
typica	8	16	24
nigrofulvata	1	20	21
Totals	9	36	45

tance between the two traps, this seems unlikely. Second, the melanics and non-melanics may have been choosing different habitats in which to fly. Third, the result might be a freak of sampling.

In an attempt to test whether this was a freak result or whether results of this type might be found elsewhere, pairs of traps were set up a short distance either side of boundaries between other sharply contrasting habitats. The most protracted series of samples were collected in Juniper Bottom, below Box Hill in Surrey. This site is remarkable for its mature yew woodland which runs along the sides of this east–west running valley. The yew canopy is extremely dense, the under canopy being extremely dark, with virtually no ground vegetation (Plate 6b and c). The bottom of the valley is characterised by chalk grassland, close cropped by rabbits, and with a few standard deciduous trees and patches of mixed deciduous shrub (Plate 6a). The boundary between the yew slopes and the open bottom of the valley is sharply defined, consisting of what amounts to a mixed deciduous hedge. One moth trap was placed approximately 10 m inside the yew wood, and the other in the open, 10 m from the yew boundary. Traps were run using Honda EM650 petrol driven portable generators.

Microclimatic data, including wind speed and temperature readings inside and outside the yew canopy, were collected, and light levels were also recorded in the two habitats, using a Gossen master-six light meter, with profilux attachment. The results of these measurements indicated that in June, July, and August, light levels were roughly 35 times less under the yew canopy than outside it, that wind speed was generally less under the yew canopy than outside, and that the temperature under the canopy dropped less during the night.

Other sites where paired traps were run included inside and outside deciduous woodland in the New Forest, inside and outside pine woodland in the Forest of Dean, and under and outside an old apple orchard near Cambridge. The results of these tests showed that the data from Dyfed were not unique, several other species showing similar differences in the frequencies of forms either side of habitat boundaries. These results promoted a very much more detailed research programme funded by the National Environment Research Council.

The aims of this programme were clearly defined:

(a) to seek evidence of correlations between the frequencies of 13 species of polymorphic Lepidoptera and habitat type, concentrating primarily on open and closed canopy habitats;

(b) to investigate the natural resting behaviour of the target species by field observations and laboratory experiments;

(c) to estimate dispersal rates across habitat boundaries in the field, using multiple mark–release–recapture techniques;

(d) to assess relative rates of predation of the morphs of target species in the field, using formal predation experiments and mark–release–recapture techniques.

An additional aim was to collect data on other species of moth, during moth-trapping programmes, so that the habitat specificity of monomorphic and other polymorphic species of Lepidoptera could be investigated.

The results of this programme have or are being published in full elsewhere. Here, it is sufficient simply to summarise the data and their interpretation.

(a)   Using paired mercury vapour moth traps, again set 10 m either side of an abrupt habitat boundary, evidence of correlations between morph frequencies and habitat types (details of habitat types are given in Appendix: see Table A1) were obtained for 14 moth species. These included eight of the target species, namely: mottled beauty; willow beauty; tawny-barred angle; pale brindled beauty; dotted border, *Agriopis marginaria* (Fabricius); common pug, *Eupithecia vulgata* (Haworth); engrailed, *Ectropis bistortata* (Goeze); small engrailed, *Ectropis crepuscularia* (Denis and Schiffermüller); and six additional species: riband wave, *Idaea aversata* (Linnaeus); common marbled carpet, *Chloroclysta truncata* (Hufnagel); grey pine carpet, *Thera obeliscata* (Hübner); pine carpet, *Thera firmata* Hübner; large yellow underwing, *Noctua pronuba* (Linnaeus); and lunar underwing, *Omphaloscelis lunosa* (Haworth), (Jones *et al.* 1993; Aldridge *et al.* 1993; Fraiers *et al.* 1994; Majerus *et al.* 1994, Majerus in preparation *a*). The results for the mottled beauty, willow beauty and tawny-barred angle were in accord with previous findings (Kearns and Majerus 1987; Majerus 1989*b*). In three target species, the peppered moth, mottled umber, *Erannis defolaria* (Clerck) and figure of eighty, *Tethea ocularis* (Hübner), no correlations between morph frequencies and habitat boundaries were revealed, despite melanic and non-melanic forms being taken in reasonably large numbers. In the other two target species, either overall sample sizes were too small to reveal such correlations (black arches), or the frequency of one form was too low to allow meaningful analysis (scalloped hazel). In addition, in two non-target species, taken in large numbers (dark arches, *Apamea monoglypha* (Hufnagel), and July highflyer, *Hydriomena furcata* (Thunberg)) no correlations were found (Aldridge *et al.* 1993).

In each case in which significant correlations between morph frequency and habitat type were demonstrated, melanic and/or banded forms were taken at higher frequencies in close canopy coniferous habitats than in open habitats, such as open deciduous woodland or chalk grassland (Appendix, Table A2) (Majerus in preparation *a*).

In addition to investigations of morph specific habitat correlations, data, collected from the operation of paired traps in different habitats, were analysed to assess general levels of habitat specificity of all species of macro-Lepidoptera taken in appreciable numbers. Findings were considered in relation to factors which might affect habitat specialisation, including species morphology (robustness), larval food plants, roosting sites, and microclimatic differences between trap sites (Majerus *et al.* 1994; Majerus in preparation *a*)

(b)   Information on the natural resting sites of target and some non-target species was collected in three ways: by observations of the resting positions of moths found in the wild, with relevant data recorded prior to this study also being taken into account; by observation of the resting sites taken up by bred and light trap caught moths released into the wild at, or soon after, dawn; and, for a small number of species, by conducting formal 'barrel-type' choice experiments (p. 137). Results were considered in relation to both the general resting behaviour of each species, and any morph specific differences in behaviour within a species (Majerus in preparation *b*). A summary of the results obtained for each technique are presented in the Appendix (see Tables A3–A6). The basic conclusions reached on the basis of these data are given in Table 7.9.

(c) Estimates of the relative dispersal rates of different morphs of seven species across habitat boundaries were obtained using multiple mark–release–recapture techniques, with a grid of 12 mercury vapour moth traps. Two procedures were employed: either moths were released at fixed points between traps, or equal numbers of each morph were released at fixed points in each habitat (see Majerus in preparation *c*, for full details). A summary of the results of each procedure are given in the Appendix (Table A7). Data were analysed, both against a null hypothesis of equal likelihood of being recaptured in either habitat, and in the light of general habitat preferences of the species concerned, by incorporating a species specific, but not morph specific, habitat preference factor (based on results obtained in the same location and habitat in previous years) into the calculation of expected values. In five species, the mottled beauty, willow beauty, engrailed, tawny-barred angle, and riband wave, there was evidence of morph specific differences in the rate of dispersal across the boundaries between habitats. In each case exhibiting a significant difference, the rate of dispersal of melanic forms from open habitat into close canopy habitat was greater than that for the non-melanic form. The reverse was the case for movement in the opposite direction for four of these species, with the fifth species, the riband wave showing no significant difference in the rate of dispersal from close canopy habitat to open habitat, although the rate for the non-melanic form was higher.

(d) Predation rates on different forms of eight species in different habitats were obtained using formal predation experiments, with dead material placed in appropriate 'resting' positions in the light of the findings from the natural resting site observations. A summary of the results is given in the Appendix (Table A8). Observations of moths during these experiments indicated that the main predators were birds, but ground beetles and ants also took some of the moths. In the majority of species there was no evidence of any significant difference in the level of predation on different morphs of a species in a particular habitat.

In addition, multiple mark–release–recapture experiments were used to estimate the relative fitnesses of those species showing morph frequency differences between habitats, but not found to have differences in dispersal rates across habitat boundaries. Species showing morph specific dispersal differences across habitat boundaries were not used for these tests, because of the confounding effects that differences in dispersal behaviour would introduce into result

Table 7.9 Deductions from field records, observed releases, and laboratory experiments on natural resting positions and rest site preferences of some British Lepidoptera

Species	Probable common resting positions	Species specific rest site selection	Morph specific rest site selection
Mottled beauty	On tree trunks and under branches	√ prefers black	Banded rests more often across black/white border (compared with other forms). F. *nigra* and f. *nigricata* versus non-melanic: no difference
Willow beauty	On tree trunks and under branches	√ prefers black	√ melanic stronger preference for black than non-melanics
Tawny-barred angle	In tree foliage and on tree trunks: less commonly under branches	√ prefers black	Difference between Dyfed and Staffs populations. In the former melanics prefer black more than non-melanics; in the latter the reverse is the case
Pale brindled beauty	On tree trunks and under branches		
Dotted border	Under tree branches, but also on other parts of trees, and in leaf litter		
Common pug	In tree and other foliage, less commonly on tree bark		
Engrailed	Anywhere on trees out of the sun		
Small engrailed	Anywhere on trees out of the sun		
Riband wave	In foliage, less commonly on tree trunks	√ prefers white	none
Common marbled carpt	On tree trunks, under branches and in tree and other foliage		
Grey pine carpet	Anywhere on conifer trees out of direct sun		
Pine carpet	Anywhere on conifer trees out of direct sun		
Large yellow underwing	In ground herbage layer, in any low hidden situation		
Lunar underwing	Low in grass layer		

analysis. Releases of laboratory bred marked material were made in large areas of similar habitat, either open deciduous woodland or close canopy conifer woodland, with release points being not less than 300 m from the nearest habitat boundary. Grids of 12 mercury vapour moth traps, in a double diamond configuration, were used (for full details, see Majerus in preparation *d*). Estimates of the relative fitnesses and average longevity of the different morphs of each species were obtained. A summary of the results is given in the Appendix (Table A9).

The results of these detailed investigations demonstrate that the morph frequencies of many species of Lepidoptera depend crucially upon habitat type. In 14 of the 19 species in which two or more morphs were taken in sufficient numbers to allow meaningful analysis, significant differences in the frequencies of morphs were recorded over a distance of 20 m, when an abrupt change in habitat type occurred in this distance. For 12 of the 14 species showing habitat-related morph frequencies, records were obtained from two, or more, locations. In all these species, results from different locations, with respect to habitat-related morph frequency differences, were broadly consistent. Melanic, and/or banded forms were taken at higher frequencies in habitats with a continuous dense canopy, than in those with a highly discontinuous canopy, or no canopy.

The higher frequencies of banded forms of the mottled beauty and riband wave under conifer canopies may be protective. Banded forms may gain an advantage due to the disruptive effect of their phenotype on their characteristic moth outline, which is one of the cues utilised by birds when searching for predators. Such forms may thus gain an advantage under close canopies due to the increased shade offered and also because the light that does penetrate on sunny days does so in a dappled form.

Most of the close canopy sites comprised mainly conifers. At such sites, the differences in light levels, 1 m above the ground, between close canopy and open habitats, were extreme. However, at two sites (Springfield and Madingley Wood), while the light level differences were still appreciable, they were an order of magnitude smaller than those involving conifer woodland. Yet habitat-related morph frequency differences were recorded at these sites. Of particular note are the results for the lunar underwing. This species flies in the autumn as the canopy at this site (apple orchard) begins to open up. The contrast between the two habitats at this site is not so obvious as at other sites, and habitat differences of similar magnitude must occur commonly between adjacent habitats throughout Britain and elsewhere. It thus seems probable that correlations between morph frequencies and habitat type will be found to be extremely common and widespread.

Various hypotheses have been proposed to explain habitat-related morph frequencies in melanic Lepidoptera, since the initial observations of Kearns and Majerus (1987). First, they could be the result of differences in the comparative rates of predation of morphs in different habitats (Majerus 1989*b*; Jones *et al.* 1993). Second, they could reflect morph specific habitat selection that has evolved as a consequence of relatively weak selective predation in the past (Kearns and Majerus 1987; Majerus 1989b; Jones *et al.* 1993). Third, they may be the result of morph specific differences in habitat preferences for reasons not associated with crypsis or protection from predators. For example, night temperature is generally higher under dense canopies than in more open habitats. As the degree of irradiation from a surface is dependent on its chroma and tone, dark surfaces radiating heat faster than pale ones, it is possible that melanic forms are at a disadvantage if they fly in cooler more open situations (Aldridge *et al.* 1993).

The predation experiments were conducted to determine whether strong differential bird predation, occurring each generation, could account for the habitat-related morph frequency differences. These predation experiments were conducted in the light of the results of extensive observations and experiments on the natural resting sites of the species concerned, to avoid the criticisms aimed at predation experiments involving the peppered moth, when moths have been offered in positions where they would not naturally rest (p. 123). Broadly, three general conclusions were drawn from the predation experiments. First, in virtually every case, the level of predation was higher in the more open of two contrasting habitats. Second, in a small number of tests, involving the mottled beauty in Dyfed and Surrey; willow beauty in Hampshire; tawny-barred angle in Dyfed; dotted border in Hampshire and Suffolk and the small engrailed in Suffolk, the level of predation of melanic and banded morphs was lower than that of typical morphs in close canopy habitats, the reverse being the case in the more open habitats. However, in most instances the differences were not significant, and there was no strong consistency, either within or between species. Third, in the riband wave, the level of predation of the banded form was higher than that of the typical form in the closed canopy habitat, the reverse being the case for the open habitat.

The results of multiple mark–release–recapture experiments, conducted to obtain longevity and fitness estimates of different forms of some species in specific habitats, also show inconsistency between species. In the small engrailed, melanics had significantly higher fitnesses than non-melanics in the close canopy habitat, the reverse being the case in the open habitat. In the other three species investigated in this way, no significant differences were detected.

These results suggest that different factors are likely to be responsible for the observed correlations between habitat types and morph frequencies in different species. In the small engrailed, at least, the level of differences in differential predation of morphs in different habitats is sufficient to account for the differences in the frequencies of forms in adjacent habitats. In other species, in which habitat-related differences in levels of differential predation were observed (but were not significant), it is doubtful that they could, on their own, be responsible for differences in morph frequencies of the magnitude observed. In addition, in several species, no morph-related differences in predation or fitness levels were detected.

The data relating to dispersal rates across habitat boundaries indicated that in five species (mottled beauty, willow beauty, engrailed, tawny-barred angle, riband wave), melanics and banded forms showed a significant preference for close canopy habitats compared with the non-melanics. In the dotted border, no morph-related dispersal differences were detected. In one experiment using the common pug, the recaptured rate for the typical form was higher than that for the melanic form in the close canopy habitat, the reverse being the case for the open habitat, although the differences were not significant. In the other experiment with this species, dispersal was random with respect to morph.

Taken together, the results of the dispersal and predation experiments lead to the conclusion that in most, but not all, species, habitat-related differences in morph frequencies over short distances can be explained on the grounds of strong morph specific habitat preferences, or strong differential bird predation.

The question of why some species did not show correlations between habitat type and morph frequency remains to be addressed. In the peppered moth, melanic polymorphism has probably evolved too recently for a morph specific behaviour to have evolved (Howlett and Majerus 1987; Majerus 1989b; Jones et al. 1993). The same explanation might be put forward for the industrial melanic of the figure of eighty.

In the cases of the dark arches and July highflyer, Fraiers et al. (1994) have speculated that the genetics of melanism may be an important determinant in whether morph specific habitat preferences can evolve. In both these species it is probable that the polymorphisms are controlled by several loci. For a specific behaviour, such as actively preferring a particular habitat, to become associated with one phenotype by genetic linkage, such linkage may only be possible if melanism is controlled by one or two gene loci.

In the case of the remaining species, the mottled umber, it is feasible that the occurrence of melanic forms, among the great range of forms in this species, are examples of anti-search image melanism (p. 195). The principal selective pressure in the evolution of melanism in this species, may thus simply have been to be different (Majerus 1989b). The conclusions from this study may thus be summarised.

(a) Correlations between morph frequencies and habitat type are common in the Lepidoptera and may produce large and very abrupt changes in morph frequency over a few metres across habitat boundaries.

(b) In species showing morph habitat correlations, melanic and banded forms occur at higher frequencies in dark close canopy habitats, compared with lighter more open habitats.

(c) In some species, differences in differential bird predation between habitats, occurring each generation, could be sufficient to explain the morph frequencies observed.

(d) Some species exhibit morph specific habitat preferences of a magnitude sufficient to explain the morph frequencies observed.

(e) For some species, no causative explanations of the abrupt changes in morph frequencies across habitat boundaries have yet been revealed.

One basic question remains: how did these morph specific habitat preferences evolve. The first point of note is that in all the species showing these morph specific behaviours, melanism is probably fairly ancient. Undoubtedly, in some species there has been an increase in the frequency of melanic forms in industrial areas over the last century or so, but there is evidence, either from old collections, or because melanics currently occur far from industrial areas, that melanics of each of these species occur independently of the effects of industrialisation. It is also notable that several of the species showing morph specific habitat preferences are among the species that Kettlewell found to have melanic forms in Rannoch Black Wood. We thus have a different situation to that faced in trying to visualise the evolution of morph specific resting site selection in the peppered moth (p. 141), for here, time is on our side. In the peppered moth, as melanism has evolved so recently, Howlett (1989) has argued that there has simply not been sufficient time for specific behavioural and morph alleles to be brought together, into tight genetic linkage, so that they are inherited together. As it is possible that melanic forms may have evolved in conifer woodlands at least 10000 years ago, or, according to Kettlewell (1973), possibly much longer ago than that, the possibility of translocations occurring to bring appropriate loci together in these species, will have been two orders of magnitude higher, or more.

On the basis of these features, I have put forward a speculative evolutionary scenario that may explain the existence of at least some of these morph specific behaviours.

(a) In the long past, typical forms of relevant species had patterns which maximised their crypsis in their normal habitats.

(b) Melanic (or dark banded) morphs evolved in specific ecological circumstances, such as in the great conifer forests that periodically covered large areas of Europe over the last hundred thousand years and more.

(c) Melanic (or dark banded) forms were at a selective disadvantage if they moved from areas in which the specific ecological circumstances persisted.

(d) Consequently, these forms evolved behaviours which restricted them to habitats where they were not at a disadvantage. Indeed, as habitat types moved up and down across the map of Europe, with successive changes in climate during the cycles of ice ages and interstadials, the moths would have moved, albeit passively, with suitable habitats.

(e) Recent changes in forestry, land usage, and increases in pollution in more recent times, have provided new habitats (e.g. conifer plantations and woodland areas with high pollution levels and lacking lichens) in which the ecological conditions favoured melanic forms. This does not imply that the typical forms were, or are, necessarily less fit in such habitats than these forms were, or are, in other habitats.

The data suggest that habitat selection may be of crucial importance to the understanding of melanism. Therefore, species may need to be considered on a much smaller geographical scale, giving detailed consideration to habitat type and other ecological factors, in future work. Furthermore, the validity of previous work on many species may have to be reappraised.

Obviously, much further work into this question is needed. First, more detailed investigations are needed of those species in which no explanation of the correlations between habitat type and morph frequency have been found. In addition, knowledge of the mode of inheritance of melanism, in those species in which it is not known, would be of great value. Indeed, one investigation that is central to my hypothesis of the evolution of these habitat preferences, would be to test whether the same or different alleles or loci control melanic forms in different populations. Were different genes found to be responsible for melanic forms in industrial areas compared with those found in rural melanism, the thesis that the habitat preference shown by melanic forms of species such as the mottled beauty are of ancient origin, would have to be discarded. It was to this end that I began to investigate the genetics of melanism in the mottled beauty in 1992, with the aid of a most generous grant from the Trustees of the Cockayne Research Fund.

The basis of these investigations was simply to determine whether the melanic forms of the mottled beauty, from several different regions, were the result of the same allele, were the result of different alleles of the same gene, or were controlled by quite different genes.

While these aims appear simple, the work required to differentiate these different possibilities has proved quite exhaustive, over 150 broods of moths from five locations, and some 12 000 adult progeny being involved. Again, full results are to be published elsewhere (Majerus in preparation *e*). However, here it is sufficient to give the results of just some of the families bred from moths originating in Rannoch Black Wood and from Box Hill, to show the crux of the findings.

It should be noted from the outset that neither the melanic nor the non-melanic forms of the mottled beauty from Rannoch Black Wood and from Box Hill are identical. The full melanic form from Rannoch is attributed the name f. *nigricata* Fuchs, and is not quite as uniformly dark as the full melanic from Box Hill, f. *nigra* Tutt. Similarly, the non-melanic form from Rannoch is both lighter and more finely patterned than its Surrey counterpart. In addition, both populations contain the banded form f. *conversaria* Hübner, in which again Rannoch specimens have a much more contrasting pattern of light and dark markings than those from Box Hill. Thus, when scoring progeny from crosses between moths from these two sites, my initial intent was to score not only the basic form, melanic, banded, or typical, but also whether moths were more 'Rannoch-like' or more 'Box Hill-like'. In practice, this intent had to be abandoned almost immediately because progeny in the first F_1 families, while being easily attributed to the main classes, were more or less intermediate in the extent of patterning, and F_2 progeny showed a full and continuous spread from Rannoch-like to Box Hill-like. These results on their own are of interest, for they suggest that the differences between the finer details of the colour patterns between these two distant populations are due to polygenic modifiers. However, they throw up one insurmountable problem, for if the differences between f. *nigricata* and f. *nigra* are not retained in the progeny between these forms, it becomes impossible to determine whether the same or two different alleles of the same locus controlled the two melanic forms. That said, the question becomes somewhat trivial if the differences between these forms are controlled polygenically. Furthermore, the more crucial question of whether the *nigricata* and *nigra* alleles are from the same or different genes, could still be addressed.

Crosses of moths from within the Rannoch population showed that f. *nigricata* is controlled by a single allele that is dominant to the typical allele. The same was true of f. *nigra* when crosses were made between moths from Box Hill. The question was, were the melanic alleles from the two populations the same. To distinguish between the two hypotheses, I needed to cross melanics from the two populations and then obtain large numbers of families by mating F_1 melanic progeny together from families producing a 3:1 ratio of melanic: non-melanics. Figure 7.15 shows the expected outcome from such a procedure on the basis of each of the two hypotheses. The crucial ratios being sought are 15:1, or 7:1 of melanics: non-melanics, for while these could result were the melanics controlled by alleles of different genes, they could not be produced if the alleles were the same, or were different alleles of the same gene. However, to be sure that the lack of 15:1, or 7:1 ratios could reliably be interpreted as showing that only one gene was involved, two safeguards had to be considered. First, analysis of the possible genotypic combinations from the two gene model, indicates that only one in nine crosses between melanic F_1s from appropriate broods would give a 15:1 ratio. Of the other eight, half would give a 7:1 ratio, the other half giving a 3:1 ratio. Second, Ford (1937) has shown that heterozygotes of f. *nigricata* are fitter than non-melanics, particularly under harsh conditions. It was thus imperative to rear sufficient broods to minimise the possibility that the lack of any 15:1 or 7:1 ratios was due to chance, and to rear offspring in optimum conditions, keeping a record of mortality, thus reducing any physiological advantage afforded heterozygous melanics, which might bias progenic ratios. From each family, therefore, I selected randomly precisely 100 one day old neonate larvae to rear up. This allowed me to keep numbers within check, and to monitor mortality more easily. None of the F_2 families produced either 15:1 or 7:1 ratios. I am therefore of the opinion that a single gene is responsible for the control of full melanism in both the Rannoch and Box Hill populations. Similar tests, using stocks from Dyfed, Cambridge, and Sheffield, gave similar results, leading to the conclusion that the inheritance of f. *nigricata* and f. *nigra* is based on a single gene through-

male	melanic homozygous	melanic heterozygous
female		
melanic homozygous	All melanic	All melanic
melanic heterozygous	All melanic	3 melanic:1 typical

(b) If both loci are heterozygous in both parents, the cross is:

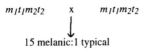

15 melanic:1 typical

If one parent is heterozygous for both alleles, the other being heterozygous at one locus and homozygous for the typical allele at the other, the cross is, e.g.:

7 melanic:1 typical

Fig. 7.15 The expected outcome of test crosses of the mottled beauty to determine whether the *nigra* and *nigricata* forms are controlled by the same or different gene loci.

out Britain (Majerus in preparation *e*). Furthermore, this finding leaves the speculative hypothesis for the evolution of the habitat preferences seen in this species, in both industrial and rural districts, in place.

Anti-search image melanism

The final subdivision of non-industrial melanism is additional to those designated by Kettlewell (1973), but it is clear that he recognised that common species, by producing additional forms, might avoid predation by birds that form search images. The likelihood of a bird forming a search image for a particular prey type depends on the abundance of the prey, its conspicuousness, and its profitability as a food item. Consider a common monomorphic species of cryptic moth. Even if it is well camouflaged, the abundance of the species may itself be detrimental if predators begin to search actively for them. In such circumstances, a new form, that is different enough from the old form for predators not to recognise it, will be advantageous and begin to spread. The frequency to which the new form spreads will depend mainly on its relative crypsis compared with the old form. However, the new form is very unlikely to become fixed, for its initial selective advantage will be negatively frequency dependent (p. 91). Once an equilibrium is reached, additional novel forms may arise and spread, again because they initially gain an advantage, by being different and rare. Consequently, in this category the benefit gained by melanic forms is not specifically in being dark, but in being different. This scenario assumes, of course, that the colour pattern of the

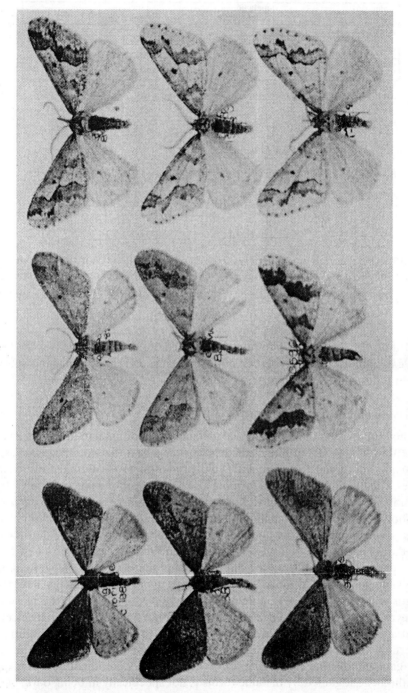

Fig. 7.16 Variation in the mottled umber may be a result of search image avoidance.

species is not important in species recognition or courtship. However, in most moths olfactory rather than visual stimuli are used in finding conspecific mates.

Several of the commonest species of British moth have a wide variety of different forms, including melanics. They include: the common marbled carpet; grey pine carpet; July highflyer; mottled umber (Plate 5b and Fig. 7.16); large yellow underwing; clouded drab, *Orthosia incerta* (Hufnagel); beaded chestnut, *Agrochola lychnidis* (Denis and Schiffermüller); dunbar, *Cosmia trapezina* (Linnaeus); common rustic *Mespamea secalis* (Linnaeus), and lesser common rustic, *Mespamea secalella* (Remm). In all of these species, melanic forms occur in non-industrial regions, although in some, such as the common marbled carpet and clouded drab, melanics occur at higher frequency in industrial regions, probably as a result of increased crypsis.

Little evidence exists for the role of search image avoidance in the maintenance of melanic polymorphism in adult Lepidoptera. Only in the recent work on the decline of melanism in the peppered moth on the Wirral is there an indication that, as the *carbonaria* form becomes rare, the selection against it is beginning to diminish (p. 155). On the other hand, it has been shown that the existence of different larval forms of the angleshades moth, which include a dark brown form, is partly a result of search image avoidance (p. 92).

Industrial melanism

Full industrial melanic polymorphism

The classic example of full industrial melanic polymorphism is, of course, the peppered moth. The number of other examples of British moths that fall into this category is rather small. These include a number of species in which melanic forms had been described from single individuals caught in non-industrial regions, before the establishment of true genetic polymorphisms (*sensu* Ford 1940*a*). A case in point is that of the great prominent, *Peridea anceps* Goeze, in which melanic form, f. *fusca*, has been recorded from Westmoreland (now Cumbria). This form appears to have been inherited as a unifactorial recessive. More recently, a phenotypically similar form, inherited as a unifactorial dominant, has established as an industrial melanic in southern England, particularly around London, where the frequency of the melanic form reached over 40 per cent, in some localities, in the 1970s.

Of course, it is not always easy to be absolutely certain that species with melanic polymorphisms, in areas affected by industrialisation, did not have such forms maintained at low frequencies prior to industrialisation. However, the history of butterfly and moth collecting in Britain over the last 200 years does give a strong degree of likelihood for many species. The confidence with which we may state that a species shows full industrial melanic polymorphism, depends on the likelihood that melanic polymorphism would have been detected before it became associated with industrialisation. Two factors are thus crucial to the assessment of whether a species shows full or partial industrial melanic polymorphism. First, the rate at which a species is, and has previously been, encountered by entomologists, which in turn is a function of its abundance, its distribution, and its appearance, and second, how recently melanic polymorphism was initially recorded in industrial areas.

It is perhaps pertinent to point out that one could classify industrial melanism in another more precise way. There is no doubt that in some species, melanic forms that existed as polymorphisms before the widespread industrialisation, have increased in industrial areas subsequently. However, in other cases, new melanic forms have arisen by mutation and then spread in industrial regions of Britain, despite the existence of melanic forms in other parts of Britain. It could therefore be argued that the best classification of melanism would be one that depended on the genes controlling melanism. In practice, there would be several problems with such a classification. First, in many species, the inheritance of melanism has not been investigated, even for the commoner melanic forms occurring in industrial regions. The crosses necessary to show whether melanics from different parts of the country are genetically the same, would have to continue for at least two, and in many cases three or four generations, in optimal conditions, to ensure that differences in the physiological fitness of genotypes did not produce biased progenic ratios. This has been achieved for the melanic forms of the mottled beauty (p. 193), but such examples are few. Second, it must be realised that some misclassifications might still be made in instances when homologous mutations occurred in different locations.

In a few species, industrial melanism has developed very recently, within the last half of this century. This is true of both the sprawler, *Brachionycha sphinx* (Hufnagel), and the early grey, *Xylocampa areola* (Esper), in which melanic forms have increased considerably over the last three decades. The reasons why industrial melanism did not develop in these species earlier is probably serendipitous: the relevant mutation simply did not arise in the right place at the right time.

These recent cases of industrial melanism emphasise an important feature of evolution by natural selection. Natural selection cannot cause change unless it has phenotypic variation to work upon. If a melanic form does not arise by mutation (or recombination), in an area where it would be at an advantage, selection has nothing to act upon.

The case of the peppered moth's closest British relative, the oak beauty, *Biston strataria* (Hufnagel) (Fig. 7.17), is instructive in this regard. This species has a melanic form, f. *melanaria*, which is a recent and common industrial melanic in the Netherlands, but has never occurred at appreciable frequencies in Britain. To date, f. *melanaria* has only been found here as a rare mutation, from Kent, Surrey, Hampshire, and Cambridge. The ecology and behaviour of the oak beauty is

Fig. 7.17 The oak beauty.

similar to that of the peppered moth in many ways. However, the melanic mutation in the oak beauty seems never to have arisen in this country in favourable
circumstances, with the consequence that it has not successfully established itself
here.

In the Netherlands, f. *melanaria* seems to be analogous to f. *carbonaria*. Other
forms of the oak beauty also occur in the Netherlands. Two of these, f. *nigricans*
and f. *robiniaria*, both have wings suffused with dark brown, and perhaps parallel the darker forms of the *insularia* complex in the peppered moth. Both these
forms occur in Britain, but never at high frequency.

The relatively recent spread of f. *melanaria* in the Netherlands, in conditions
that appear to mirror those experienced in Britain, and the lack of industrial polymorphism in this species in Britain, emphasises the importance of mutational
events occurring in the right place and time. Given the reversal of the fortunes
of melanic peppered moths since the 1960s, it is unlikely that the *melanaria* form
of the oak beauty will establish itself in Britain in the foreseeable future.

In at least one moth, industrial melanism in Britain is the result of migration
from the continent. The melanic form, f. *fusca* Cockayne, of the figure of eighty
(Plate 4a and b) appeared in southern and eastern England in the mid-1940s, and
has increased rapidly since, so that over much of its range in England, f. *fusca* is
currently more common than the non-melanic form. Although f. *fusca* is
found in Wales and the Channel Islands, it has yet to be recorded in Scotland or
Ireland.

Fig. 7.18 Typical, half melanic, and full melanic, f. *rebeli*, forms of the willow beauty.

The group of fully industrial melanics also includes strongly melanic forms of some species, which previously had less phenotypically extreme melanic forms, of non-industrial origin. For example, the willow beauty had a non-industrial melanic form, f. *perfumaria*, which has increased in frequency around urban conurbations. However, in some localities this morph has been partly or completely replaced by another, more extreme, form, f. *rebeli* Aigner, which is only known from polluted regions (Fig. 7.18).

Fig. 7.19   Non-melanic (above) and melanic (below) forms of the lobster moth. At several sites in England, the melanic form is currently declining.

The decline in the frequencies of melanic forms in Britain over the last 30 years is not confined to the peppered moth. Although published data are sparse, the melanic frequencies of several other species are currently in decline. For example, in the south-east of England, the melanic form of the lobster moth, *Stauropus fagi* (Linnaeus) f. *obscura* Rebel (Fig. 7.19), has declined significantly at four sites since the 1970s (Majerus in preparation *f*). The same is true of f. *nigra* of the poplar grey, *Acronicta megacephala* (Denis and Schiffermüller), in north Staffordshire. One extraordinary case is that of the May highflyer, *Hydriomena impluviata* (Denis and Schiffermüller), in Dartford, Kent. West (1992), reports that between 1969 and 1989, all captures of this moth at light in his garden were of the melanic form, f. *obsoletaria* Schille. In 1989, one of the five specimens taken was of the normal grey form. A second individual of this form was taken the following year, in a slightly larger sample. No May highflyers were caught at the site in 1991, but in 1992, all 15 moths taken were of the typical form!

Although there appears to be a general decline in the frequency of truly industrial melanics in Britain at present, this is not true in all other parts of the world. In Finland, for example, Mikkola (1989) discovered a new melanic form, f.

inferna, of the noctuid *Xestia gelida* (Sparre Schneider), which he interprets as the first case of industrial melanism in the subarctic, and which he associated with the intense pollution emanating from the Kola Peninsula. It is likely that other novel cases will be discovered in countries which are only now developing heavy industry.

Partial industrial melanic polymorphism

Perhaps more species of British moth fall within this category of melanism than any other. Most species that had melanic forms prior to the industrial revolution fall into this class, if their distribution includes major industrial regions, for in the majority of such cases, the frequency of melanics increased in the nineteenth and first half of the twentieth centuries.

The case of the willow beauty, mentioned above, demonstrates this principle. Originally a non-industrial melanic, f. *perfumaria*, rapidly increased in frequency in and around London, in the second half of the nineteenth century, appearing in other industrial heartlands such as Sheffield in 1902, Newcastle upon Tyne in 1914 and Burnley in 1916 (C. Jones 1993). Other similar cases include the green brindled crescent (Plate 4e and f), the clouded-bordered brindle, *Apamea crenata* (Hufnagel) (Plate 4g and h), and the satin beauty, *Deileptenia ribeata* (Clerck), as well as most of those species showing morph specific habitat preferences discussed previously (p. 184). In addition, the majority of species showing search image avoidance melanism (p. 195), also have higher melanic frequencies in industrial regions than elsewhere. This is to be expected, as the likelihood of a predator forming a search image for a particular type of prey depends on the apparent abundance of the prey to the bird, which in turn is a function of both the true abundance of the prey, and the ease with which it is initially found. Melanic forms will thus appear less abundant in polluted habitats than in unpolluted ones, as a result of their improved crypsis, even if the melanics are at the same density in the two habitats.

Often, when non-industrial melanics have spread into urban areas, the forms have darkened in appearance, with time. Kettlewell (1973) suggests that geographic/ancient melanic forms have only occasionally acted as reservoirs for the increase in industrial melanism. The rarity of such occurrences Kettlewell accounts for by noting that many non-industrial melanics are mainly found far from industrial centres. However, Kettlewell's analysis is usually based on the phenotypic similarity of urban and rural melanics. Unless the melanic forms taken in industrial regions are phenotypically identical to those from rural areas, Kettlewell assumes the forms have arisen independently. This ignores the possibility that non-industrial melanics may have been acted upon by natural selection in a novel, urban, and polluted environment. For example, Kettlewell argues that the genes controlling the similar, but distinguishable, non-industrial and industrial melanic forms of the mottled beauty, f. *nigricata* and f. *nigra*, respectively, are unlikely to be genetically identical. However, as I have shown (p. 194) genetic analysis shows that they are at least allelic, and may be controlled by the same allele.

Rather few of the species that fall into this class of melanism have been investigated in any depth. Exceptions are the pale brindled beauty (Lees 1971, 1974, 1975) (Plate 4a and b), the mottled beauty (Kettlewell 1973; C. Jones *et al.* 1993) and the green brindled crescent (Steward 1977*d*). All three of these are trunk resters, and the increase in melanics in industrial regions is attributed to increased crypsis.

Many species have received virtually no attention, and their status is certainly open to conjecture. I will give just a handful of examples, which would repay attention.

Among the Noctuids, the brindled green, *Dryobotodes eremita* (Fabricius), has a recurring melanic form around London, and Skinner (1984) mentions that the darker form occurs frequently in the Midlands. Old collections exhibit considerable variation in this species, with melanics also occurring in Argyll, Rannoch,

Fig. 7.20 Typical (above) and melanic (below) forms of the common heath.

and Cumberland around the turn of the century. Similarly, the lead coloured drab, *Orthosia populeti* (Fabricius), has an industrial melanic, f. *nigra*, in London, with melanics also recorded from Rannoch, Herefordshire, and near Worcester. In these cases, the old moths from rural areas were not as dark as more recently caught melanics from southern industrial regions. However, this may be because the old specimens have 'faded' in storage. Here then, it is not possible to be sure whether non-industrial melanics acted as a reservoir for the current industrial melanics. Such cases would certainly repay comparative genetic analysis of the melanic forms.

Other species, which may be worthy of close consideration, are: the common marbled carpet; the common heath, *Ematurga atomaria* (Linnaeus) (Fig. 7.20); the engrailed; the small engrailed, and the grey pine carpet, for in all of these the relationship between the industrial and non-industrial melanic forms is unclear.

It would also be interesting to see how species that have long had melanic forms, but in which these forms have increased since the industrial revolution, respond to the recent decline in pollution levels. Appendix B in Kettlewell (1973), provides melanic frequencies of many species that fall within this category, from around the time that pollution levels began to decline. These are of extreme value, for they may be used as baseline data, to assess the effect of reduced soot fall-out on melanic frequencies of species in this class. All that is now needed are current data sets, from the same sites, for comparison.

Polygenic industrial melanism

This final category of melanism is the least considered, and perhaps the most difficult to address. Examination of material collected over the last 150 years, reveals that in many species there appears to have been a general darkening of the colours of populations coming from industrial regions, irrespective of morph. Some of this is due to the fading of museum specimens captured well over a century ago, yet it is difficult to ascribe all of the differences to this one factor alone. Recently, I compared the ground colour of series of specimens, of six species, from rural and industrial regions, collected over 80 years ago, with those collected over the last 4 years. The comparisons showed that the ground colour had darkened more in industrial regions than in the rural areas. As the fading should be the same for moths from both types of region, these results seem to support the idea that there has been a gradual darkening in the ground colour of many moths in industrial localities.

Many species show changes of this kind. Such changes are probably the result of selection acting on polygenic variation in respect of ground colour, and novel major mutations are not implicated. Again, it is likely that selective predation of less cryptic forms, in regions affected by particulate air pollution, is primarily responsible. If this speculation is correct, the sharp drop in airborne pollution since the Clean Air legislation of the 1950s should lead to a reversal of this trend, with ground colours gradually becoming paler and patterns more obvious again.

Fig. 7.21   Typical (above) and melanic, f. *nigra* (below), forms of the brindled beauty.

This should be monitored over the next few decades. The species that perhaps could most usefully be studied in this regard are, I think: the common marbled carpet; dark marbled carpet, *Chloroclysta citrata* (Linnaeus); figure of eighty; grey pine carpet; July highflyer; November moth; winter moth; scalloped hazel; pale brindled beauty; willow beauty; mottled beauty; engrailed; small engrailed; iron prominent, *Notodonta dromedarius* (Linnaeus); turnip moth, *Agrotis segetum* (Denis and Schiffermüller); dark arches; clouded drab; Hebrew character; sycamore, *Acronicta aceris* (Linnaeus); poplar grey; lunar underwing; and the large yellow underwing. The brindled beauty should also be monitored in this regard, because although it has a well known melanic form, f. *nigra* (Fig 7.21), which unusually for an industrial melanic is controlled by a single recessive allele, assessments of Cambridgeshire non-melanics over the last 16 years, suggest that this species is already getting paler. Novel methods of measuring the spectral reflectance of surfaces, and storing the information digitally, should allow such monitoring to be achieved without reliance on museum specimens or photographs, both of which may fade with time.

Concluding remarks

In this chapter, I have, I hope, shown that even in a strongly limited group of organisms, the night flying moths that rely on crypsis for protection during the day, the factors responsible for melanism, and the form of the melanism that

results, are very variable. Because a great variety of factors may promote
melanism, it may be misleading to extrapolate from one population to another,
let alone from one species to another. Even within one class of melanism, the relative influence of different aspects of a species biology will vary from one species
to another. Each species which has evolved melanic forms, will have done so in
the presence of a variety of different intrinsic and extrinsic circumstances. The
variety of different factors which may play a part in the evolution of melanism,
the different parts played by particular factors, such as selective predation, migration, non-visual selection and the underlying genetics of melanic variation in different species, and the complexity of systems that have been uncovered, in even
the few well studied species, suggest that there is still enormous scope for original research into this phenomenon.

8 Melanism in conspicuous Lepidoptera

Introduction

In the previous three chapters, the majority of Lepidoptera discussed have relied on some form of camouflage to avoid predation, and the melanic forms that they exhibit have often evolved to maximise predator avoidance, either when at rest, or, in some instances, when in flight. However, many organisms have evolved conspicuous coloration for one or other of a variety of reasons. Principal among these are: as warning coloration to warn potential predators of some defensive property; to mimic thus protected species; or to attract mates or deter rivals. Melanism in conspicuous species occurs in a wide range of taxa, some of which have already been discussed in Chapter 2. However, in this chapter I will again confine myself to the Lepidoptera, for the main principles involved in the evolution of melanism in brightly coloured species can all be related using examples from this order. I will continue this theme in Chapter 9, in which I will discuss melanism in a brightly coloured family of beetles, the ladybirds, in which warning coloration, mimicry, sexual selection, and other factors, such as thermal melanism, may all have an effect on the evolution and maintenance of melanism, sometimes within a single species.

Melanism in Lepidoptera showing warning coloration

At the outset, it is important to realise that butterflies and moths showing warning coloration can be broadly split into two categories: those in which the warning colours are honest signals to potential predators, and those in which the warning coloration is a sham. Species in the former group are said to show aposematic coloration, meaning that they are memorably coloured and that the colours serve to advertise additional defences of one sort or another. In the Lepidoptera, these defences usually take the form of foul smelling and distasteful chemicals, although other defences such as urticating hairs, or stinging spines also occur. The shammers are those species that have evolved warning coloration to resemble, or mimic, aposematic species. Such mimetic species are said to show Batesian mimicry (p. 212). As black is possibly the most commonly used colour in warning patterns, it is not surprising that melanins form a significant part of the colour patterns of both aposematic and mimetic species, and in both groups, melanic forms are known. I will deal with the former first.

Melanism in aposematic Lepidoptera

Aposematic colour patterns work, because they usually consist of two, or occasionally more, strongly contrasting, memorable colours. Naive predators, that try to eat an aposematic prey, discover the less obvious defences, such as distastefulness, or sting, and begin to associate their unpleasant experience with the striking colour pattern of the prey they have just attacked. Many aposematic species are extremely resistant to injury and survive initial attacks. If so, their coloration will help protect them from subsequent attack, at least by that particular predator. However, their memorable coloration will also deter the predator from attacking other individuals with the same colour pattern. These will often be relatives, as most members of a family tend to live relatively close together, so the selective advantage of having aposematic coloration, will fall, at least partly, upon kin. This then, is an example of kin selection. The insect that was initially attacked will, of course, also benefit when other birds learn from attacking other insects that show the same aposematic coloration that it does. From this, it becomes obvious that in such species, there is an advantage in monomorphism, so that the number of individuals that predators can learn from is as large as possible. Here is a case in which it is better to be one of many, rather than one of a few. It is perhaps not surprising then, that melanic polymorphism is not common in aposematic Lepidoptera.

Despite this, melanic forms of many aposematic species are known. In many instances, these melanics occur as rare mutations that are not maintained for long in natural populations. From the British Lepidoptera, several examples can be cited in the burnet moth (Zygaenidae) and tiger moth (Arctiidae) families. The majority of members of both of these families are distasteful to birds, and many are toxic. In both the 5 spot burnet, *Zygaena trifolii* Verity, and the 6 spot burnet, *Z. filipendulae* Dupont, which normally have deep blue-black forewings spotted with red, and predominantly red hindwings, forms with the red replaced by black occur (Plate 7f). These forms are inherited as unifactorial recessives, and occur at very low frequency. Similarly, in several of our tiger moths, such as the garden tiger, *Arctia caja* Linnaeus, the cream-spot tiger, *Arctia villica* Oberthür, and the scarlet tiger, completely black specimens have occasionally been reared from wild stock. Again, these forms are generally inherited as simple recessives. Kettlewell (1973) has speculated that the recessive nature of the alleles controlling such forms, is a consequence of the recurrent nature of mutation and the low fitness of these forms. Following Fisher's theory of the evolution of dominance (p. 90), Kettlewell argues that selection will tend to promote modifiers that reduce the expression of these 'black' alleles in heterozygotes. It is certainly true that some of these melanic forms have lower viability than their respective typical forms. For example, the *nigra* form of the scarlet tiger moth only rarely manages to eclose successfully. Furthermore, given the general theory of the selective advantage of aposematic coloration, it is likely that completely black forms of these species will be maladaptive in the wild, because of their rarity. Birds that have learnt to avoid the common typical forms of these species are unlikely to avoid forms that are so different, without at least trying them.

Despite the fact that few aposematic species have common melanic forms, a few cases do occur. Most occur in the Arctiidae, and five examples should suffice to show the type of factors that may be involved in allowing such forms to flourish. The first case is that of the melanic polymorphism in the scarlet tiger moth, at Cothill, in Berkshire. I have already discussed both the genetics of this species (p. 43), the balance of mating advantage to the rarer forms and viability advantage to the common form (p. 86), which are involved in maintaining this polymorphism. Less is known about the other cases. The ruby tiger moth, *Phragmatobia fuliginosa* Linnaeus, typically has reddish-brown forewings and bright red abdomen and hindwings, the latter being more or less lightly marked with dark blue. However, throughout much of Scotland, Sweden, and Norway, a darker form, f. *borealis* Staudinger (Plate 7b), in which the forewings are darker brown, and the hindwings are almost totally suffused with dark grey, replaces the nominate form. Skinner (1984) notes that this is not a case of polymorphism, for the dark northern form slowly grades into the southern form, specimens in the north of England being intermediate between the two.

Kettlewell (1973) argues that the northern form of this day flying species is specialised for flying over heathland, where the dusky form is very difficult to see on the wing. He describes how the moths, when netted, go into a state of catalepsy, curving forward their red abdomen to expose its aposematic colours. This is surely the case, and it is interesting that this species appears to employ both crypsis and warning coloration in its defence. However, I have also been intrigued to note the increase in the proportion of moths with the majority of the hindwing dark blue or grey, in samples taken in south-east England and East Anglia, over the last three decades. I am at a loss to explain this recent change, although I have shown that it is genetic and controlled polygenically. The species should certainly be monitored, in the southern part of Britain, in the future.

A difficulty in being seen when in flight may not be the only advantage gained by northern ruby tiger moths by being darker than the norm. It is also possible that thermal melanism is involved in the evolution of this form. Such speculation is not based on any empirical evidence, but on the fact that a second species of the family, the white ermine, *Spilosoma lubricipeda* Linnaeus, also shows melanism in northern latitudes (Plate 7d). In Scotland and Ireland, the normally white colour of the forewings of this species are more or less suffused by brown. It is difficult to understand how such a change might have a defensive function, but in these cool and cloudy regions, the possibility that a thermal advantage results from having dark forewings, which are exposed when the moth is at rest, should at least be considered.

While melanism in the ruby tiger occurs over a wide area, that in the buff ermine, *Spilosoma luteum* Hufnagel, is a much more local phenomenon. The buff ermine is normally yellow, lightly spotted on both wings and abdomen, with black. It is relatively unpalatable, but is not as toxic as some other members of the genus (Rothschild 1963). In a few coastal sand-dune locations, on the east of England, the continental Channel coast and some of the islands off Germany, a melanic form, f. *zatima*, occurs as a polymorphism (Fig. 8.1). This form has most of the intervenic regions of all the wings dark brown, with just the veins and small areas of the forewing retaining the normal yellow coloration. The *zatima* form is controlled by a single allele, showing incomplete dominance, the het-

Fig. 8.1 Typical (above) and melanic, f. *zatima*, forms of the buff ermine.

erozygote being less extreme than the homozygote. Although little work has been conducted on the adaptive significance of this form, it is probable that in the locations where it occurs, its colours are cryptic. On the Yorkshire and Lincolnshire dunes, it appears to rest on the dead trunks and branches of elder, against which it is very well camouflaged. Such cases, in which aposematic coloration has been forfeited for the advantages of not being detected by predators, are rare.

The final example is perhaps the most remarkable. Over most of its range, the muslin moth, *Diaphora mendica* (Clerck), has white females marked with small black dots, while the male is dark, smoky grey, again with black dots. However, in Ireland, males are white like the females, or are pale buff, being gently flushed with brown (Plate 7e). Genetic analysis shows that the Irish race is due to an allele of a single gene, that is incompletely recessive to that producing the dark male race, and that the heterozygote has a rather variable expression (Onslow 1921; Adkin 1927*a*,*b*). This then is a case of a sex-limited trait. Where males are black and females white, the sexes have different flight times. Males fly late during the night, seeking females that 'call' pheromonally, and mating takes place just before or at dawn. Females are active in the late afternoon, flying in the sunshine, to seek oviposition sites. Both sexes are distasteful, and have the characteristic behaviour of the ermines, in that they do not fly when disturbed, but curve the abdomen forward to reveal the warning colours on the underside of the body and legs.

Here then, it appears that the muslin moth shows sexual dimorphism, resulting from differences in the selective factors to which males and females are exposed, because of differences in their habits. This is commonly the case, in moths in which the sexes find and recognise each other primarily by scent. In butterflies (and a few moths, such as the ghost moth, p. 176), in which at least part of the mate identification is the result of visual recognition, sexual dimorphism is more common, and is often attributed to a combination of sexual selection acting on one sex and natural selection acting on the other.

Unfortunately, I can find no information in the literature on when male Irish muslin moths fly. It is tempting to expect that they will fly during the day, but I have no confirmation of this. If any lepidopterists working in Ireland could clarify this point, they would certainly be making a most useful contribution.

As an aside, it is notable that the time of flight of nocturnal moths has received very little systematic attention. The recent development, by a team at Anglia Polytechnic University, of a mechanised moth light trap, that rotates over a series of six holding chambers at fixed periods through the night, should allow this gap in our knowledge to be filled.

From the above instances of melanism in aposematic species, it is again apparent that many factors can affect the evolution of melanism, and that the balance between these will vary not only from species to species, but within a species, from location to location.

Melanism in mimetic species

In North America, larvae of the pipe-vine swallowtail, *Battus philenor* (Linnaeus), a large black butterfly with yellow spots towards and at the edges of the wings, feed on plants of the genus *Aristolochia*. From these plants, the larvae sequester defensive chemicals, which are passed on to the adult stage. The adults are thus aposematic. Indeed, Jane van Z. Brower (1958*a,b*) has shown that a single encounter with one of these butterflies is sufficient to deter Florida scrub jays, *Cyanositta coerulescens*, from attacking butterflies with this pattern. Several other species—such as the Diana fritillary, *Speyeria diana* (Cramer); the parsnip swallowtail, *Papilio polyxenes* (Fabricius); the red-spotted purple, *Limenitis arthemis* (Fabricius); and the tiger swallow-tail, *Papilio glaucus* Linnaeus—all of which appear to be perfectly edible, have melanic females that closely resemble the pipe-vine swallowtail. The males of all these species are quite different from the females, their colour patterns being constrained, because females respond to males that they recognise by sight. The females have evolved melanic forms, mimicking the pipe-vine swallowtail, to take advantage of the fact that birds that have encountered a pipe-vine swallowtail will not attack other butterflies with the same pattern. Interestingly, the females of one of these mimetic species is polymorphic, having both the melanic mimetic form, and a yellow form similar to males of this species. The black form only occurs at appreciable frequency where *B. philenor* flies.

This, and many other cases of mimicry, have intrigued biologists for over a hundred years. Before discussing the evolution of such melanic mimetic forms, and the maintenance of polymorphisms in some, a few introductory words on mimicry are worthwhile, to give context. This is because the study of mimicry, particularly in the Lepidoptera, has a special place in unravelling some of the more intractable problems of evolution.

Mimicry is defined as the resemblance of one organism, or part of an organism, to another, for protective or, more rarely, aggressive purposes. In some instances mimics are themselves harmless, or relatively unprotected, and gain protection by resembling an organism (the model) that has strong defences, such as being toxic, tasting or smelling obnoxious, or possessing a sting, poisonous spines, or a powerful bite. This type of mimicry is termed Batesian mimicry after H.W. Bates, who first described it in 1862. Batesian mimicry may be contrasted with the second main class of defensive mimicry, Müllerian mimicry, in which a number of well protected species come to resemble each other, to reduce the harm caused to each species by inexperienced predators while learning of their unpalatability (Müller 1878).

Conventionally, then, in Batesian mimicry, the mimic gains at the expense of the model, without having to pay the energetic cost of bearing the protection. In Müllerian mimicry all bear the costs of manufacturing protection, but all also gain from their resemblances.

A number of different points may be expected in examples of each of these two types of mimicry. In Batesian mimicry:

(a) The model must be relatively inedible or otherwise protected.

(b) The model must have some conspicuous feature, most usually a warning colour pattern.

(c) The model must be common compared with the mimic. This condition is obviously necessary, because, should the model be rare, the predator will have little chance of learning of its harmful properties, and will therefore not avoid the mimic.

(d) Both model and mimic must usually be found together at the same time.

(e) The mimic, unlike the model, will not be particularly resistant to injury.

(f) The mimic should bear a very close resemblance to the model. Selection will always tend to increase the resemblance, because the closer it is, the less chance a predator has of distinguishing between the model and the mimic.

(g) The resemblance is only likely to extend to visible structures, colour patterns, behaviours, scents, sounds, or other criteria, by which a predator may identify prey. It will not extend to basic anatomical make-up, developmental processes, metabolism, or biochemistry.

In Müllerian mimicry:

(a) All the species must be protected.

(b) All the species must be relatively conspicuous.

(c) All species may be equally common, but the abundance of any one species will not be limited by its mimicry.

(d) The species are likely to be relatively resistant to injury during initial attack.

(e) The resemblance between species need not be very exact, but must simply involve a feature which reminds a potential predator of an unpleasant previous experience.

Many spectacular examples of both Batesian and Müllerian mimicry are provided by the Lepidoptera. Just from the British Lepidoptera one might think of species; such as the hornet clearwing, *Sesia apiformis* (Clerck), lunar hornet moth, *Sesia bembeciformis* (Hübner), narrow-bordered bee hawk moth *Hemaris fuciformis* (Linnaeus), and broad-bordered bee hawk moth *Hemaris tityus* (Linnaeus), as classical examples of Batesian mimics, and the burnet moths (Zygaenidae) as Müllerian mimics. Further afield, females of many palatable swallowtail butterflies (e.g. *Papilio dardanus*, *P. memnon* Linnaeus, *P. polytes* Linnaeus) mimic a variety of noxious or toxic species, in consequence having melanic forms, while larvae of some harmless South American hawk moths (Sphingidae) resemble small snakes.

Mimicry has long fascinated biologists from an evolutionary standpoint. Indeed, Darwin (1859, 1871) used examples of the close resemblance of edible butterflies to poisonous ones, as one of his finest examples of the role of natural selection, the resemblance conferring a survival advantage on the mimic. Batesian mimics have one unique feature, in terms of the understanding of evolutionary mechanisms. The end-point of an evolutionary process may be predicted. The process of evolution by natural selection is unpredictable, to the extent that natural selection has no foresight. It is the fittest now that survive, not those that may be most fit in the future. We cannot tell, in most instances, which variations will arise by mutation or genetic reassortment, so we cannot predict the future course of evolution (although many evolutionary biologists try to do so) and we usually cannot have an image of the end-point of evolution. However, in examples of Batesian mimicry, we do have such an image. That image is the model species, for selection will favour closer and closer resemblance of a mimic to its model, until, at least to potential predators, the two cannot be distinguished.

Batesian and Müllerian mimicry are usually described in a way that suggests that the distinction between them is precise and clear-cut. This is not necessarily the case, and it is often difficult to be sure whether two warningly coloured species that resemble one another, but are not related, are a pair of Müllerian mimics, or comprise a Batesian mimic and its model. Furthermore, edibility is a subjective trait, depending on a predator's immune system, taste and scent sensors, age, experience (we know we 'get a taste' for something), and state of hunger, all of which are variable. This means that an organism that is distasteful to one predator may be perfectly edible to another. This variation in palatability means that we should not regard possible prey species simply as edible or inedible, but think of them in terms of what Turner (1984) has termed a 'palatability spectrum', from species that are very well protected and eaten by only a few specialist predators, to highly palatable prey, that are appetising to most predators of appropriate size.

I will discuss the evolution of Müllerian mimicry in Chapter 9, when I discuss melanic polymorphism in ladybirds. Here I will confine myself, in the main, to a discussion of melanism in Batesian mimetic Lepidoptera.

One of the features of Batesian mimicry is that it may produce negative frequency dependent selection, thus leading to polymorphism. This is easy to understand, if one considers the relative abundance of a mimic and its model. If the mimic is rare compared with the model, inexperienced predators are more likely to come across the model before the mimic, and so will learn to associate its pattern with unpalatability. However, if the mimic is more common than the model, the chances are that inexperienced predators will first come across, and attempt to eat, mimics. This they will do readily, as the mimics are palatable, having no second line of defence, such as unpleasant smell or taste, to reveal. Consequently, the rarer a Batesian mimic is, relative to its model, the greater the selective advantage accruing from its mimicry. One way in which a species can circumvent the problem that its mimicry may become disadvantageous should the species become too common, is to spread its mimicry among a number of different models. This is the case in a number of species of swallowtail butterfly, perhaps most famously in the African swallowtail, *Papilio dardanus*.

Considerable attention has been paid to this, and a number of related species, showing mimetic polymorphism. This was only partly because these examples provided workers with a view of the optimum phenotype that should be the outcome of the evolutionary process. Perhaps, more importantly, early studies of these polymorphic Batesian mimics confronted evolutionists with what appeared to be two contradictory facts. On the one hand, the resemblances of the mimics to the models were very exact. On the other hand, the differences between forms of a species appeared to be controlled by a multiple allelomorphic series of a single gene. The resolution of this apparent biological contradiction has been of great importance in the development of population and ecological genetic thinking.

I will illustrate the evolution of mimetic polymorphism with two of the butterflies which have received most attention in this regard, *P. dardanus* and *P. memnon*, for in these the genetics underlying the different mimetic and non-mimetic forms has been worked out. However, before doing so, one more introductory point should be stressed. Most texts which discuss the evolution of Batesian mimicry give the impression that the expected outcome of Batesian mimicry is polymorphism. This is not true. In most instances natural selection will lead Batesian mimics to be monomorphic. This is because of the following.

(a) The models of most Batesian mimics are themselves components of Müllerian mimic complexes. This means that the effective population size of models may be the sum total of all the species in the Müllerian complex. In most cases, before the population density of a Batesian mimic can increase enough for its advantage to be lost, because it is too common relative to the model, other factors, such as food availability, will limit its population size;

(b) Some predators may have innate avoidance behaviours of some warning patterns. In this situation the frequency dependent selection pressure which may otherwise lead to the evolution of mimetic polymorphism will be irrelevant.

(c) The number of potential models is often limited.

That most Batesian mimics are monomorphic may be easily appreciated if the assemblage of yellow and black insects in Britain are considered. These include bees, wasps, hoverflies, robber flies, beetles, adult moths, and moth larvae. Many of these, such as the clearwing moths (Sesiidae) and the bee hawk moths, are Batesian mimics and are monomorphic in respect of colour pattern. Indeed, in the case of the British Lepidoptera, I can think of only one example of polymorphism which may involve Batesian mimicry. This is the melanic larva of the death's-head hawk moth, *Acherontia atropos* (Linnaeus), which may be a snake mimic, although to my mind, not a very good one (Plate 2g and h).

The number of cases of polymorphic Batesian mimicry is rather small. Yet, given the number of conditions that have to be met, it is remarkable that this type of situation has evolved as often as it has. These are that:

(a) the morphological changes involved must not be detrimental;

(b) the frequency of the model (compared with the mimic) must act as a major limiting factor on the population size of mimetic forms;

(c) there must be more than one model pattern available to the mimic. Of course, some potential models may not be available as they already have Batesian mimics at equilibrium densities;

(d) the morphological change which initiates the evolution of the mimicry must be sufficient to confer on the new mimic some selective advantage as a consequence of its mimetic resemblance;

Table 8.1 The genetics of polymorphism in *Papilio dardanus* (After Nijhout 1991.)

Female form	Principal allele	Known genotype(s)*	Models
Mimics			
hippocoonides	h	hh	*Amauris niavius*
cenea	H^c	H^cH^c, H^ch, H^cH^{na}	*Amauris albimaculata*
			A. echeria
trophonius	H^T	H^TH^T, H^Th, H^TH^c	*Danaus chrysippus*
planemoides	H^{Pl}	$H^{Pl}H^{Pl}$, $H^{Pl}h$	*Bematistes poggei*
Non-mimics			
natalica	H^{na}	$H^{na}H^{na}$, $H^{na}h$	
leighi	H^L	H^LH^L, H^LH^{na}, H^LH^c, H^Lh	
niobe	H^{Ni}	$H^{Ni}H^{Ni}$, $H^{Ni}h$, $H^{Ni}H^c$, H^TH^{Pl}	
Yellow	H^y	H^yH^y, H^yh, H^yH^c	
Pale-poultoni	H^{pp}	$H^{pp}H^{pp}$, $H^{pp}h$	
Bright-poultoni	H^{bp}	$H^{bp}H^{bp}$, $H^{bp}h$, $H^{pp}H^T$	
salaami	—	H^LH^T	
Imperfect heterozygotic mimics			
cenea-like	—	H^cH^{bp}, H^cH^{pp}, H^cH^{Ni}	
trophonius-like	—	H^TH^{na}	

Apparent dominance relations†

$H^c > H^{na}$	$H^T \approx H^{na}$
$H^T > H^c$	$H^{bp} \approx H^c$
$H^{pp} > H^T$	$H^{pp} \approx H^c$
$H^L > H^c, H^{na}$	$H^T \approx H^{Pl}$
$H^y > H^c$	$H^{Ni} \approx H^c$

Note: The known genotypes of the named female forms are shown, together with the species mimicked by each. Most of the phenotypes can be produced by several different genotypes. Not all forms of *P. dardanus* are known mimics.
Source: After Clarke and Sheppard (1960a,b).
* The phenotypic effects of these genes are given in Table 8.2.
† The symbol > indicates dominance of allele at left over allele at right; the symbol ≈ indicates additive or codominant alleles. allele h appears to be recessive to all the others.

(e) the enormous limitations which derive from basic genetic mechanisms (e.g. coadapted gene complexes, dominance effects, gene interactions, and disruptions of pre-existing linkage groups) must be circumvented.

This last problem is perhaps the least considered, but the most interesting. Consider, as an example, the African swallowtail *Papilio dardanus*. This butterfly is sexually dimorphic over most of its range, males being yellow and black, while the females are either similar to the males, or very different because they mimic several different species of unpalatable butterflies. The changes from the ancestral non-mimetic form necessary to produce these mimetic females may involve several different elements, including: pigments; distribution of pigments on wings and body; presence or absence and length of tails; shape of wings; flight pattern; resting behaviour; and oviposition behaviour. Yet, apparently, all these changes appear to be produced by a single gene, the different mimetic forms being controlled by a multiple allelomorphic series. The same is true of most of the mimetic forms of other *Papilio* species.

While it seemed inconceivable that a single gene mutation could be responsible for the many changes which are involved in the production of a perfect Batesian mimetic form, early genetic analysis of the various female forms revealed this to be the case for the majority of the traits involved. In *Papilio dardanus*, the gene involved is the *H* locus which has 10 different alleles (Table 8.1) (Ford 1936; Clarke and Sheppard 1960a,b).

A theory, which in modified form is now generally accepted, was put forward by Fisher. He suggested that when a mutation arose which gave a non-mimetic species a slight resemblance to a potential model, this would begin to spread. Selection would then favour any increase in the accuracy of the mimicry, as long as the genetic factors which were being selected were only expressed with the initial mutation. Two mechanisms which would lead to this limitation of expression were postulated. Either the initial mutation could act as a switch gene, switching on the expression of a series of modifiers; or, all the factors involved could become tightly linked into a supergene, so that recombination between the loci involved occurred only very rarely.

In different parts of the sub-Saharan range of this butterfly, the forms present, and their frequencies, vary. Analysis of crosses between forms from different regions has provided evidence for the switch gene modifier system. Considering just three of the most melanic mimetic forms, *hippocoon* from West Africa, and *hippocoonides* and *cenea* from East Africa, all of which have considerable extensions to the black patterning, compared with the ancestral non-mimic, the main alleles which control these forms are:

hippocoon – h

hippocoonides – h

cenea – H^c

where H^c is dominant to h.

Although both *hippocoon* and *hippocoonides* are controlled by the same alleles, they look different, and mimic different models. Results of crosses show that this is because different sets of modifier genes have been favoured in West and East Africa (see below). So, the differences between these forms are polygenically controlled.

(a) *hippocoonides* (East Africa) × *hippocoon* (West Africa) gave intermediate F_1, and a continuous range of forms in the F_2, with different modifier genes segregating out independently.

(b) *hippocoonides* (East Africa) × *cenea* (East Africa) gave all *cenea* F_1 (or 1 *cenea*:1 *hippocoonides*), and a 3:1 ratio of *cenea* to *hippocoonides* in the F_2, as expected if *cenea* is dominant to *hippocoonides*.

(c) *hippocoon* (West Africa) × *cenea* (East Africa) gave intermediate F_1, and a continuous range of forms in the F_2, showing that the dominance modifiers for H^c over h are not present in West African genomes.

Nijhout (1991) has considered the way the alleles of the H locus affect the development of the different elements of the patterns of the various mimetic and non-mimetic forms. The pattern diversity of the mimetic female forms results from two basic types of pattern formation. The melanic elements of the colour pattern can increase or decrease in size, or the background colour of four different areas of the wings may vary. In addition, intervenic melanic stripes on the hindwings vary in length and breadth. Changes in the width of the pattern elements affect the amount of ground colour that shows through, and can have dramatic effects on the overall pattern and the precision of the mimicry. The forewing seems to consist of three areas in which the ground colour is under independent genetic control. Interestingly, the boundaries of these regions do not correspond to any known 'landmarks' of Lepidoptera wings. The extent of each of the melanic pattern elements is also controlled independently.

The specific effects of the alleles of the H locus on the various elements are summarised in Table 8.2. Analysis of these phenotypic effects suggests that the alleles act quantitatively. The background colour varies from white to orange-brown, with the intermediate shades simply resulting from lower concentrations of the darker pigment. One exception is the yellow of males and non-mimetic females, which is due to a different pigment. This can easily be shown, because this yellow fluoresces in ultraviolet (UV) light, while the other ground colours absorb UV (Clarke and Sheppard 1963). Nijhout (1991) notes that the pattern of the ancestral form is a very abbreviated version of the pattern theme that is controlled by the other nine alleles of the H locus. It seems, therefore, that the evolution of mimicry in this species has involved additions to the ancestral pattern, not merely modifications of it. Indeed, several of the pattern elements on the forewings of mimics have no counterparts in the ancestral form.

Table 8.2 Phenotypic effects of various alleles of the H locus in *Papilio dardanus*.
(After Nijhout 1991.)

Pattern	Expression by major allele									
	h	H^c	H^T	H^L	H^{na}	$H^T H^L$	H^{pp}	H^{bp}	H^{ni}	H^{Pi}
Background colour*										
Forewing discal bar	1	1	1	3	2	1	1	2	1	1
Forewing band	1	1	1	3	2	3	2	5	4	1
Forewing posterior blotch	1	2	4	3	3	4	2	5	4	1
Forewing submarginal spots	1	3	4	2	2	4	2	5	4	1
Hind wing	1	3	4	2	2	4	2	5	4	1
'Pattern' on forewing†										
Discal bar	M	M	M	M	L	S	S	S	M	L
Band	L	S	L	L	L	M	L	L	L	L
Posterior blotch	L	S	L	S	L	L	L	L	L	M
Submarginal spots	L	M	M	M	S	S	S	S	M	L
'Pattern' on hind wing†										
Rays	L	L	M	L	L	S	S	S	M	L
Black band	M	L	S	L	L	S	S	S	S	L
Submarginal spots	M	M	S	M	M	M	M	M	S	L

Note: Each allele controls a combination of background colours and pattern sizes simultaneously. This tabulation allows for a convenient comparison of the differential effects of each allele on various features of the colour pattern. Dominance relations of the alleles, insofar as they are known, are given in Table 8.1.
Sources: Clarke and Sheppard (1959, 1960*a,b*, 1963).
* Arbitrary colour scale: 1 (white) to 5 (dark orange-brown).
† Arbitrary size scale, relative to mean: S = small, M = medium, L = large. All these patterns, except the hindwing rays and black band, actually represent background colour. When such a coloured band is large, for example, it means that the (black) pattern elements that flank it are small or narrow.

Although many authorities have suggested that the H locus is a supergene, largely on theoretical grounds, there is virtually no evidence to support this view. Only one putative recombinant has been reported, this being an f. *salaami* phenotype, produced from a homozygote rather than a heterozygote (Clarke *et al.* 1968). Of course, if the supergene was very tight, recombinants might be so rare that evidence based on their occurrence might be difficult, or impossible, to obtain. Yet given the number of broods of *Papilio dardanus* that have been reared for genetic analysis, it seems probable that more cross-over products would have been obtained were the H locus a supergene. Much more probable, is that the alleles of the H locus merely switch on the expression of genes elsewhere in the genome that are responsible for the various components of the different mimetic and non-mimetic forms.

However, evidence for Fisher's supergene mechanism of Batesian mimicry has been obtained for another swallowtail, *Papilio memnon*. The occurrence of rare recombinants, in crosses of different forms of this species, has shown that at least eight loci are involved in the polymorphism. Of these six are tightly linked into a supergene and usually segregate together (Clarke *et al.* 1968; Clarke and Sheppard 1971; Turner 1984).

So, in many respects Fisher's ideas are vindicated. However, Fisher appears to have been wrong in two respects. First, he envisaged that the initial mutation may have been very slight. We now know that for selection to favour the initial mutation, the degree of resemblance to the model must be quite considerable.

Secondly, the selective coefficients involved in these mimetic systems are probably not strong enough 'to pull' genes from different chromosomes, or otherwise unlinked genes, into a supergene. For a supergene to evolve the genes responsible for the different components of the mimicry probably need to be at least loosely linked already (Turner 1984).

It is unlikely that the loci responsible for the components of the mimicry just happened to be appropriately linked, and in most cases they are not, which is why examples of spectacularly polymorphic Batesian mimicry are so rare. Only a few species have the basic requirement of clusters of appropriate loci that control wing pigments and the extent of the melanic pattern elements. These are the species that have attracted attention, and because a biological paradox exhibited by these has been explained, the fact that these are rare special cases has been obscured.

In conclusion, melanism is seen in many Lepidoptera that may be considered conspicuous. In many of these cases, there is both a defensive and a sexual component to the evolution of the conspicuous coloration and the melanic forms that so many species exhibit. The polymorphic Batesian mimetic *Papilio* species perhaps show this interaction best, by their extreme sexual dimorphism. Similar interactions between warning coloration and sexual selection are seen in some of the species of ladybird that show melanic polymorphism. It is these that I shall explore in Chapter 9.

9 Melanism in ladybirds

Introduction

The Lepidoptera have been pre-eminent in the study of the evolution and maintenance of melanism. However, as mentioned in Chapter 1, there have been many studies of melanism in other groups. Some, such as the work on the spittlebug, *Philaneus spunarius*, do little more than confirm that evolutionary phenomena, first described in moth or butterfly species, are also relevant to other taxa. However, one other group of insects in which melanism is common, is of particular note, because this group demonstrates a number of features related to melanism not seen in Lepidoptera. These are the ladybirds.

Melanic varieties are a feature of many species of ladybird. In some species, darker or blacker forms occur as rarities, in others, the amount of black patterning varies with geographical location, while in still others, melanic and non-melanic forms occur together as genetic polymorphisms.

The existence of melanic polymorphism in many ladybirds is problematic. Most species of coccinellid are aposematic, i.e. they are both memorably coloured and are armed with an array of defensive chemicals making them unpalatable to many potential predators. Most species that have true warning coloration are monomorphic, at least within a particular population. This is expected on the basis of evolutionary theory, because warning colours operate best when they are shared by many individuals. Uniformity of appearance is, of course, expected in aposematic species in any case. The more protected individuals that share the same memorable colour pattern, the smaller the proportion that suffer injury, or even death, while inexperienced predators learn to associate a particular colour pattern with unpalatability. The idea can be thought of as a form of intraspecific Müllerian mimicry. The theory of Müllerian mimicry (Müller 1878) is that two or more different species that are protected from predation by some form of armour, weaponry, or chemical defence, come to resemble one another, thereby reducing the inroads made upon them by naive predators learning to avoid prey items bearing particular colour patterns. Expectation is that different species should evolve to look similar, and that within a species all individuals should look the same.

That uniformity is the expected outcome of Müllerian mimicry is easy to see, if the evolutionary pathway leading to the production of a Müllerian mimicry complex is considered. The pathway is simple. Suppose two protected species with different colour patterns coexist in the same place, but that one species, 'A', is much commoner than the other, 'B'. If a naive predator needs the same number of trials to learn to avoid the colour pattern of each species, the rarer species, B, will obviously suffer disproportionately more than A. Therefore, if a mutation occurs which makes a member of B resemble A sufficiently, predators that have learned to avoid those with species A's pattern avoid it too, hence, it will gain a selective advantage, for it will then be one of many, rather than one of a few. This mutation will increase in frequency, the appearance of species B converging on to that of species A. Other protected species may later be drawn into the complex. Of course, the counterside of this proposition is that if an individual differs in its appearance sufficiently from the norm, so that it is not recognised by predators as being unpalatable, it will be at a selective disadvantage. The result is a tendency to general uniformity both between the species in the complex, and more pertinently, within the species in the complex.

Yet ladybirds do not all look the same. Not only is there great variation in the colours and patterns of different species, but there is considerable variation within many species of ladybird. In just the 24 British species of ladybird, some are red or orange with black spots, others are black with red spots, or yellow and black, or brown and black, or orange or maroon with white spots, or chestnut red with pale yellow stripes. One species, the eyed ladybird, *Anatis ocellata* (Linnaeus), commonly has three colours, being dark red with black spots surrounded by cream rings. Conversely, in Britain, the larch ladybird, *Aphidecta obliterata* (Linnaeus), is commonly just brown with no spots at all. Elsewhere, the range of colour combinations is greater still. The ground colours cover almost the whole colour spectrum, including blues, lilacs, and turquoise, but, I think, with the exception of green, and the colours of markings are almost as diverse. Among all this colour and pattern diversity within the family, one phenomenon is common to all species which have been studied to any significant degree. Melanic forms are known in virtually every case. I have discussed some of the reasons why ladybirds are not all alike elsewhere (Majerus 1994a). Here I will focus on the possible reasons why melanism is such a common phenomenon in this aposematic group.

Types of melanic forms in the ladybirds

In general, there are two different ways in which a ladybird may be considered melanic. Most coccinellids have a two-colour pattern, essentially a ground colour with a pattern of spots or other markings of a quite different colour. The spots may be darker than the ground colour as in the red and black 7 spot ladybird,

Coccinella 7-punctata Linnaeus, or lighter as in the striped ladybird, *Myzia oblon-goguttata* (Linnaeus), which usually has rich chestnut brown elytra sporting cream or pale yellow stripes and spots. Thus, it is possible to increase the overall darkness of an individual by darkening the ground colour, or by altering the ratio of the two colour elements in favour of the darker colour.

Examples of both types occur. Several species of ladybird have melanic forms in which the ground colour is darkened, the pale patterning being unaffected. For example, the striped ladybird has a form, f. *lignicolor*, controlled by a single reces-sive allele, in which the ground colour is dark chocolate brown (Majerus 1993). Similarly, the cream-spot ladybird, *Calvia 14-guttata* (Linnaeus), has a form f. *nigripennis* in which the normal maroon-brown ground colour is replaced by black (Mader 1926–37). In a number of usually red species, recessive mutants are known which cause the normal bright red colour to gradually darken through purple to almost black. Both of the commonest British species of ladybird, the 7 spot and the 2 spot, have rare aberrant forms of this type, which occur as extreme rarities in the wild (Plate 8a) (Majerus 1994a; Majerus *et al.* in preparation). This type of gradual deepening of the ground colour to almost black is not rare in all species. In one American ladybird, *Anatis labiculata* (Say), all individuals appear to go through this transition. Within 48 hours of eclosion, this ladybird is a handsome morning suit grey colour, patterned with 15 black spots. However, the ground colour continues to develop over the ensuing weeks, the grey first becoming brown, and eventually, after about 3 months, a dark rich purple, inclin-ing to black, so that by the time the ladybird is 4 or 5 months old, the black spots are virtually indiscernible (Fig. 9.1).

Examples of melanic forms apparently produced as a result of an increase in the extent of dark patterning are even more common. These include several species in which fully melanic forms occur only as extreme rarities: such as the 24 spot ladybird, *Subcoccinella 24-punctata* (Linnaeus) f. *nigra*; the 16 spot *Tyt-thaspis 16-punctata* (Linnaeus) f. *poweri* (Majerus 1991; Revels and Majerus 1997), and f. *hebraea* of the eyed ladybird.

The reason for the existence of melanic forms that occur only at very low fre-quency and are controlled by major genes, such as f. *nigripennis* of the cream-spot ladybird, or f. *lignicolor* of the striped ladybird, need not concern us unduly. Such forms occur as a result of recurrent mutation and are probably deleterious, eventually being eliminated by selection. Although the inheritance of the rare melanic forms of *Subcoccinella 14-punctata* f. *nigra*, and *Tytthaspis 16-punctata* f. *poweri*, have not been elucidated, it is probable that these are also due to major genes, and may be considered as deleterious mutations eliminated by selection, soon after they arise.

In other species, the black spots enlarge to such an extent that they fuse together, producing blotches or grid patterns. For example, Mader (1926–37) depicts 26 pattern varieties of *Aiolocaria mirabilis* Motschulsky, from completely

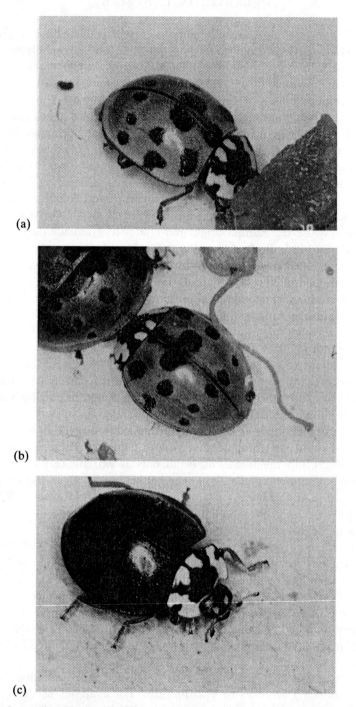

Fig. 9.1 Increasing pigment laydown with age in *Anatis labiculata*: (a) 2 days old; (b) 6 weeks old; (c) 9 months old.

unmarked on the pronotum, apart from a black scutellary spot, to completely black, each variety shown having the black areas slightly enlarged compared with the previous one.

The most melanic forms of some species occur rarely in some regions, but predominate in others, forming clines in the number and extent of black spots. These include Adonis' ladybird, *Adonia variegata* (Goeze), the 13 spot, *Hippodamia 13-punctata* (Linnaeus), the 7 spot, and the 5 spot, *Coccinella 5-punctata* Linnaeus (Dobrzhanskii 1922*a*,*b*). These species are widely spread through Europe, Asia, and northern Africa. In general the palest, i.e. the reddest, populations occur in central Asia, with the extent of melanic patterning on the elytra increasing in populations to the north, east, and west. In the 7 spot, this increase is achieved primarily through an increase in spot size, while in Adonis' ladybird, the 5 spot and the 13 spot, it is the number of spots that varies. In three of these species there is considerable variation between populations but little variation within populations, except in the size of spots. The exception is Adonis' ladybird in which most populations contain a considerable range of spot numbers (Dobzhansky 1933; Majerus 1994*a*) (Figs 9.2 and 9.3).

The eyed ladybird, has 18 or 20 black spots, more or less ringed with cream or yellow. One, several or all of the black spots may be lacking, leaving pale spots, or the spots may be fused into longitudinal black stripes (Mader 1926–37; Majerus 1994*a*). In Europe, striped individuals are exceedingly rare. However, in Siberia, spot fusions become more frequent to the east, and east and north-east of Lake Baikal, striped forms predominate (Dobzhansky 1933). Farther east still, towards the Pacific coast, less melanic forms increase in frequency again.

In these species, and a range of others, such as: *Synharmonia conglobata* (Linnaeus); *Coccinella transversoguttata* Faldermann; *Coccinella trifasciata* Linnaeus; the scarce 7 spot, *Coccinella magnifica* Redtenbacher; and *Coccinella 9-notata* Herbst, more melanic forms are found primarily to the north and east,

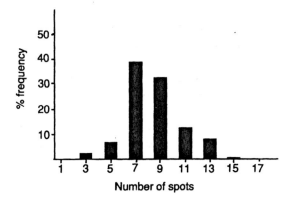

Fig. 9.2 The percentage of individuals of Adonis' ladybird with different spot numbers from a Staffordshire population. (From Majerus 1994*a*.)

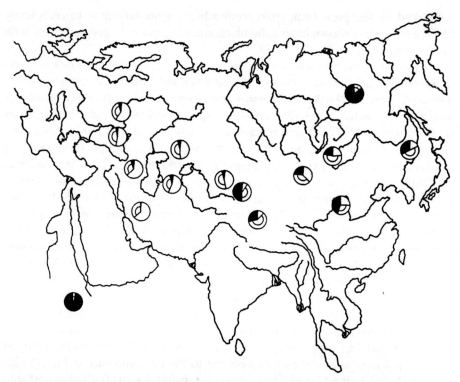

Fig. 9.3 Geographic variation in the proportion of Asian populations of Adonis'
ladybird with none to nine spots (white segments), 11–15 spots (white with central
stripe), and 15 fused spots. (After Dobzhansky 1933.)

with less marked forms having highest frequency to the south and west. This
broad generalisation applies to both the Eurasia and North American continents.
The areas of least melanism for these species are the arid, hot, desert regions of
Turkestan and Iran in Eurasia, and California, Arizona, and New Mexico in North
America. Highest levels of melanism occur in northern Russia and Siberia, and
in Canada, where climates tend to be cool and humid (Dobzhansky 1933). The
differences in the degree of melanisation within a species are probably genetic
in the majority of these species. Although not all the species have been studied
using controlled crosses, in all those that have been analysed, spot numbers, spot
size, and spot fusions have been found to have high heritability. Furthermore,
rearing coccinellids at different temperatures has little effect on the pattern produced (Johnson 1910; Dobzhansky 1933). The deduction is that selection promotes melanism in cool, humid climes, but acts to reduce melanism in hot, arid
regions. Appropriate experiments have not been performed on most of these
species to determine why blacker individuals are selected in cool, humid condi-

tions and pale ones in hot regions, but it is likely that thermal melanism is involved.

The trends discussed above occur over vast geographic distances, and only show a course general pattern. In many instances, deviations from this general pattern occur in particular localities, presumably because other local selective factors outweigh the broad advantages, or disadvantages, induced by the thermal properties of being black. For example, it is not clear why intra-population variation in spot number is so considerable in Adonis' ladybird, yet so little in the 13 spot or 7 spot ladybirds.

In the case of the 14 spot ladybird, the proportion of the elytra that is black varies tremendously within many European populations. For example, in British populations, the proportion of the elytra that is black varies from less than 10 to over 80 per cent. In the palest individuals, the spots are fully discrete. However, in individuals in which more than 30 per cent of the elytral surface is black, the spots begin to fuse, and in those with more than 70 per cent black, the impression is of a black ladybird with yellow spots. Despite the intra-population variation in the amount of black in this species, there is also a latitudinal cline, the mean proportion of black increasing northwards from populations in the Mediterranean region towards Scandinavia (Majerus unpublished data). The variation in the amount of black in this species has a genetic component (heritability = 78 per cent for British populations, p. 64). It is probable that the differences in the average blackness of populations from different latitudes are the result of selection favouring different optima, the tightness of the distributions reflecting different intensities of selection around the mean. The factors that may favour different optima in the proportion of the elytra that is black have not been studied in the 14 spot, but it is probable that thermal properties are involved, the lighter forms being prevalent in warmer climes, while the blacker forms are favoured in cooler Scandinavia. Why the variance is greater in British populations than elsewhere in Europe is not obvious, but the changeability of the British climate from year to year may be influential, with the result that the optimum changes between quite wide limits from year to year.

The same arguments may apply to the ladybird *Epilachna 28-punctata*. This species is highly variable in terms of spot number and size (Dieke 1947; Richards 1983; Abbas *et al.* 1988; Katakura *et al.* 1988). In Sumatera Barat, Indonesia, Abbas *et al.* (1988) found that the increase in spot number, and the amount of fusion between spots, was correlated with increasing altitude.

These cases, and my own studies of the Lapland ladybirds, which tend to be darker than those from more southerly populations, either because of a darker ground colour (e.g. cream-spot ladybird), or because of an increase in black patterning (e.g. 2 spot ladybird), support the theory that melanism in ladybirds is affected by the thermal properties of dark versus light surfaces. However, there is a danger that the influence of thermal factors may be overemphasised, simply because it is easy to understand.

Lesley Stewart and Tony Dixon (1989) discussed the thermal properties of 2 spot and 7 spot ladybirds in a paper entitled 'Why big species of ladybird beetles are not black'. While the paper is interesting, giving valuable information on the heat excesses experienced by the two species in artificial conditions, the title is somewhat unfortunate. The trouble is that a number of big ladybirds are black. I have already described melanic forms of the eyed ladybird, and the striped ladybird also has melanic forms in Asia (Mader 1926–37). Both of these species are larger than the 7 spot. Many other examples from other parts of the world could also be quoted (e.g. *Axion tripustulatum* (Degeer), *Hippodamia glacialis* Mulsant, *Hippodamia moesta* Leconte, *A. labiculata*, *H. axyridis*). Stewart and Dixon argue that the reason that big ladybirds, such as the 7 spot, are not black, is that given their large body size, they would suffer heat stress when foraging on hot sunny days. However, these other species of temperate coccinellid, all of which have melanic forms, or are normally black, cope with this problem. The way they do so is by avoiding heat when it is likely to be excessive. Some species do this simply by moving into the shade during the hottest part of the day (Hodek and Honek 1996). Others aestivate during the hottest months of the year. In Mediterranean countries, the 7 spot behaves in this way. Indeed, several species of ladybird, including the 7 spot and 2 spot, have been reported to become dormant during heat waves in Britain (Majerus 1986*b*, 1994*a*). Finally, Stewart and Dixon's proposition is completely undermined by the fact that the 7 spot ladybird does have melanic forms, such as f. *confusa*, in India (Rhamhalinghan 1988). While it is highly likely that thermal considerations do play a significant part in determining the extent of melanisation of many coccinellids, these are not the only considerations.

The variation in amount of black on the elytra of many ladybird species is continuous, and follows a more or less normal distribution. This is not true of either the larch ladybird or the hieroglyphic ladybird, *Coccinella hieroglyphica* Linnaeus.

The typical form of the hieroglyphic ladybird is brown, with a black streak running backwards from the front of each elytron, and one or 2 spots behind these. These markings may be reduced or extended, so that a range of forms from very lightly marked to almost completely black occurs (Fig. 9.4) (Edwards 1914). However, although the variation is continuous, some patterns are much more common than others, and if the percentage black is plotted for a sample, the distribution is strongly bimodal, with one mode at about 15 per cent and the other at about 98 per cent black. Indeed, if the variation is divided into three classes, typical, intermediate, and melanic, having 0–25 per cent, 25–75 per cent, and 75–100 per cent of the elytra coloured black, respectively, 67.1 per cent fall into the typical class, 23.3 per cent into the melanic class and only 9.5 per cent into the intermediate class (Fig. 9.5) (Majerus 1994*a*).

A somewhat similar situation is seen in the larch ladybird. Although British populations of this species show relatively little variation, almost all individuals

Fig. 9.4 Melanic and non-melanic forms of the hieroglyphic ladybird.

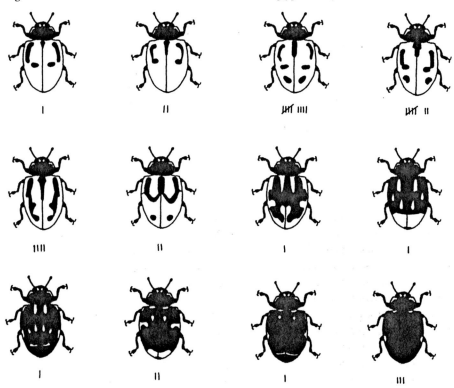

Fig. 9.5 Variation in patterning of a sample of hieroglyphic ladybirds from Chobham Common, Surrey. (From Majerus 1994a.)

being brown and either unspotted, or with just a small oblique black mark near the back of the elytra, many Continental populations have wide range of forms, from unspotted to completely black (Fig. 9.6). Here again, the unspotted/small oblique dash forms, and almost fully black forms, predominate over intermediates. Although intermediates do occur, they are always in a considerable minority. Breeding experiments with Luxembourg material has shown that the full melanic form is controlled by a single allele that is recessive to the allele controlling the unmarked form (Majerus unpublished data).

It is possible that the bimodal nature of the variation in the hieroglyphic ladybird is the result of selection acting on a polygenic system. The expected outcome of disruptive selection, acting on continuous variation controlled by a polygenic system, is the evolution of some mechanism by which the production of less fit intermediates, between the two (or more) optima, is reduced. This may be by assortative mating, which would be the case if typicals preferred to mate with typicals and melanics with melanics. Such assortative mating might ultimately lead to speciation, as discussed previously (p. 146). Alternatively, certain combinations of genes in the polygenic system, which together produce one of the optimum phenotypes, may become tightly linked into a supergene, or may come under the control of a switch gene. In either instance, the result is a reduction in the number of intermediates produced, compared with those expected if all the genes in the polygenic system segregated independently and were expressed. In this way, a polygenic system evolves to become polymorphic, with the majority

Fig. 9.6 Melanic and non-melanic forms of the larch ladybird from Luxembourg.

of individuals in the population being one, or other, of a small number of forms, apparently showing discontinuous, or strongly bi- or multimodal distributions.

Unfortunately, due to problems in breeding the hieroglyphic ladybird in captivity, in particular inducing matings of this species in captivity, possible explanations of this bimodal distribution of colour pattern have yet to be tested.

Melanic polymorphism in ladybirds

Most of the cases of melanism in ladybirds discussed above cannot be considered as true genetic polymorphisms (*sensu* Ford 1940*a*), either because the melanic forms occur only very rarely, or because the variation is continuous. Only the hieroglyphic and larch ladybirds can be considered to show true genetic melanic polymorphism. However, there are many other ladybird species in which melanic polymorphism has been reported (e.g. *Coleophora inaequalis* (Fabricius), Houston and Hales 1980; Ng 1986 (Fig. 9.7); *Menochilus 6-maculatus* (Fabricius), Chapin 1965; Subramaniam 1923; Ng 1986; *Harmonia arcuata* (Fabricius), Ng 1986). Three species, *Harmonia axyridis*, the 2 spot ladybird, and the 10 spot

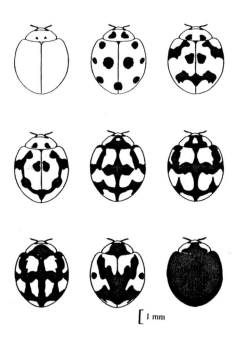

Fig. 9.7 Colour pattern variation in *Coelophora inaequalis*. (From Ng 1986.)

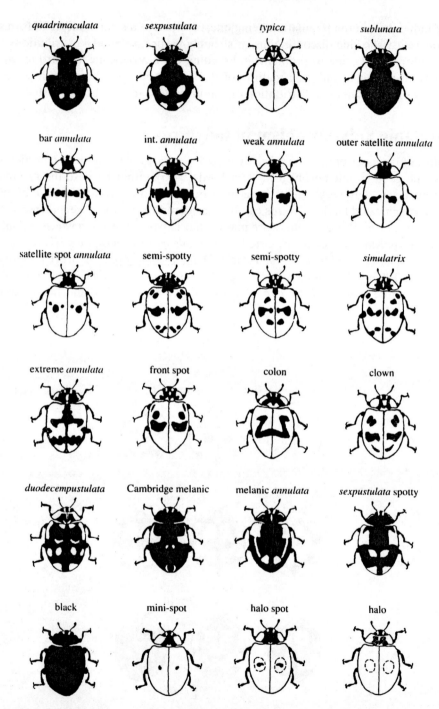

Fig. 9.8 Colour pattern variants of the 2 spot ladybird.

ladybird have received considerable attention, both in respect of the genetics of melanism, and the underlying reasons for the evolution and maintenance of melanic polymorphism. All three are extremely variable with respect to the colours and patterns of their pronota and elytra.

Taking the 2 spot first, Mader (1926–37) described and named over a hundred 2 spot varieties from Europe and Asia, and many additional pattern variations have been illustrated by other authors (Palmer 1911, 1917; Hawkes 1920; Lus 1928, 1932; O'Donald and Majerus 1989; Majerus and Kearns 1989; Majerus 1989, 1994a, 1995; Majerus, T. *et al.* in preparation *a*). In Cambridge, for example, we have recorded over 600 varieties from British populations alone. Some of the forms mentioned by name are shown in Fig. 9.8.

Despite this enormous wealth of variety, in most populations only a small number of forms reach frequencies above 1 per cent. It is outside the scope of this discussion to give a detailed description of the varietal composition of populations of 2 spots from around the world. In brief, however, European populations tend to be either predominantly *typica*, or polymorphic for *typica* and two melanic forms, f. *quadrimaculata* and f. *sexpustulatus* (Plate 8b). The only other forms to reach significant frequencies are weak *annulata*, a variety which is genetically heterogeneous, comprising heterozygotes between the *typica* allele and at least three comparatively rare alleles (*simulatrix*, *duodecempustulata*, and extreme *annulata*), and f. *frigidda* in the north of Scandinavia. This latter form is similar to f. *typica*, but with the black spot extended into a bar, and sometimes additional spots on the elytra. Some authors have raised f. *frigidda* to the status of subspecies. However, these diagnoses have been based on morphological features, which may have more to do with adaptation to the specialised circumstances facing organisms in sub-polar regions, than with reproductive isolation or genetic distance. Molecular genetic work and information on the viability and fertility of hybrids between f. *frigidda* and forms from farther south, are needed to resolve this question.

In central and eastern Asia, the *typica* form predominates in most populations, sometimes together with a different melanic f. *sublunata*, while *quadrimaculata* and *sexpustulata* are absent. In some Asian populations, other forms, such as f. *simulatrix*, and f. *duodecempustulata*, reach significant frequencies in some populations (Zakarov personal communication).

In American populations, five forms attain frequencies in excess of 1 per cent. These range from f. *melanopleura* which has unpatterned red elytra, through *typica*, *coloradensis*, and *annectans*, which are all red with varying numbers of black spots, to f. *humeralis*, a melanic form, which, according to published description, is indistinguishable from f. *quadrimaculata* (Fig. 9.9). As we shall see, despite their apparent phenotypic similarity, the genetic control of f. *humeralis* from the United States, f. *quadrimaculata* from Britain and western Europe, and f. *quadrimaculata* from Asia, are different.

Less survey work has been conducted on colour pattern variation in the 10 spot

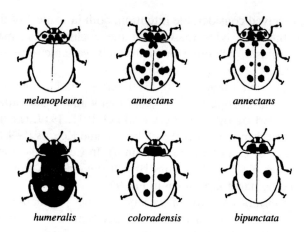

melanopleura   *annectans*   *annectans*

humeralis   *coloradensis*   *bipunctata*

Fig. 9.9 Variation in the 2 spot ladybird from North America. (Adapted from Palmer 1911, 1917.)

ladybird, but in Eurasia, this species shows even more variation than the 2 spot. This is because the 10 spot shows greater variation in both the ground colour and the colour of its markings. Essentially the 10 spot has three basic morphs (Fig. 9.10): *decempunctata* is orange with anything from zero to 15 dark dots; *decempustulata* (Plate 8d) has a dark grid-like pattern on a paler ground colour; and *bimaculata* (Plate 8d) is dark all over the elytra, with the exception of a pale flash at the anterior outer margin of each elytron. Within the constraints of each of these basic forms there are many variations on each theme. The pale colour involved may vary from cream or pale beige through a range of dull yellow, orange, various light tans and browns to bright red, while the dark colour varies from mid tans, maroon, and purple, through darker shades, to black. The precise strength of the markings also varies. In addition, the rate at which pigments are laid down can be very fast, or extremely slow, even when temperatures are controlled (Majerus 1994). If just the three basic forms are considered, the pattern of temporal and spatial variation is very different to that in the 2 spot, for all populations appear to support all three forms at significant frequencies, *decempunctata* usually being the commonest form and *bimaculata* the least frequent. Furthermore, the frequencies of these forms do not vary very greatly, either temporally or spatially, and changes that do occur take place slowly (Brakefield and Lees 1987; Majerus 1994a).

The Asian ladybird, *Harmonia axyridis*, is as variable as either of the others. The patterns on the elytra of this species tend to be very bold combinations of black and orange or black and red. The form that might be considered typical is orange with 18 black spots. However, some or all of these spots may be absent, or they may be extended so that some are fused together. In western populations, such as those from the Altai Mountains. in Mongolia, a chequered patterned form

Fig. 9.10 The basic forms of the 10 spot ladybird. From left to right, lightly spotted f *decempunctata*, heavily spotted f. *decempunctata*, f. *decempustulata*, and f. *bimaculata*.

Table 9.1 Relative per cent frequencies of the different colour pattern forms in *Harmonia axyridis*, in Asia. (From Dobzhansky 1933.)

	siccoma	frigida	19-signata	axyridis	spectabilis	conspicua	aulica	n
Altai Mountains			0.05	99.95				4013
Yeniseisk Province		0.9		99.1				116
Irkutsk Province			15.1	84.9				73
Transbaikalia (western part)		4.9	45.9	49.2				61
Amur Province	7.3	29.3	41.5					41
Maritime Province (Khabarovsk)	18.6	18.1	38.7	0.2	13.4	10.7	0.3	597
Maritime Province (Vladivostok)	16.9	31.1	37.6	0.8	6.0	6.8	0.8	765
Manchuria	12.9	32.8	34.0		11.2	8.6	0.5	232
Korea	28.1	26.6	26.6		6.2	12.5		64
Japan		16.4	3.0	4.5	16.4	59.7		67
China (Chi-Li, Shan-Si, Shan-Tung)	36.8	12.5	27.0		12.5	10.5	0.7	152
China (Kan-Su, Sze-Chuan)	3.7	26.0	40.7		11.1	14.8	3.7	54

is the commonest morph, although this form is rare in more easterly populations. A wide variety of forms which seem to be variations on a melanic theme occur. These are characterised by having most, or all, of the lateral margins of the elytra black, with the interior of the elytra sporting various bold orange or red markings. Some of these forms are shown in Plate 8e and f.

Survey work on *Harmonia axyridis* has been hampered because much of the distribution of this interesting genetic species is in the former Soviet Union, and work on genetic polymorphism was frowned upon for political reasons for almost half of the twentieth century. However, Dobzhansky did survey many populations in the 1920s, demonstrating that this species shows great geographic variation in the frequencies of the various forms (Table 9.1 and Fig. 9.11). Furthermore, in the species' range outside the former Soviet Union, surveys have also indicated considerable spatial variation in form frequencies; for example, Komai (1956) noting that the frequency of melanics is greater in southern locations, suggests *H. axyridis* exhibits climatic adaptation. Variation in this species has the added component that melanic frequencies alter seasonally (Tan 1949).

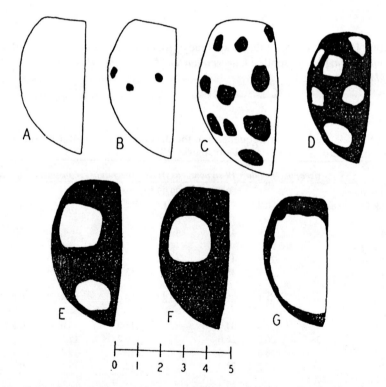

Fig. 9.11   Forms of *Harmonia axyridis* found in Dobzhansky's survey of the species in Asia (see Table 9.1). A: f. *siccoma*, B: f. *frigida*, C: f. *19-signata*, D: f. *axyridis*, E: f. *spectabilis*, F: f. *conspicua*, G: f. *aulica*. (From Dobzhansky 1933.)

Without going into detail, the inheritance of many of the colour pattern forms of these three species has been analysed.

(a) Two spot: Palmer (1911, 1917), Hawkes (1920), Lus (1928, 1932), Creed (1966), Majerus and Kearns (1989), Majerus (1994a), and Majerus T. *et al.* (in preparation *a*).

(b) Ten spot: Lus (1928) and Majerus (1994a).

(c) *Harmonia axyridis*: Hosino (1933, 1936), Tan and Li (1934), Komai (1956), and Sasaji (1971).

In all of them, a single locus with multiple alleles appears to control the majority of forms. For example, using European and Asian stocks of 2 spot, Lus identified 13 alleles of a single locus which followed a relatively straight-forward dominance hierarchy: s^i (the top dominant), s^{lu}, s^m, s^p, s^i, s^a, s^{il}, s^t, s^o, s^{st}, s^z, s^a, s^d the bottom recessive) (for allelic notation see Fig. 9.12a and b). In general the melanic forms are dominant to the non-melanics, but the correspondence is not exact. The line of decreasing melanism in respect of homozygous forms is: *sublunata, lunigera, quadrimaculata, sexpustulata, duodecempunctata, annulata, ocellata, simulatrix 1, simulatrix 2, typica, impunctata 2*, and *impunctata 1*. The dominance hierarchy of the multiple allelic series in *Harmonia axyridis* is similar, with melanic forms being dominant over non-melanics. Conversely, in the 10 spot ladybird, the top dominant allele is the non-melanic, followed by f. *decempustulata*, with the melanic f. *bimaculata* being the bottom recessive (Lus 1928; Majerus 1994a).

There is good evidence that in the 2 spot ladybird this main colour pattern locus is a supergene, containing at least four, and possibly more, very tightly linked genes (Majerus 1994a). The mechanism by which this supergene has arisen is not known. It is possible that it is the result of selection acting to draw relevant genes together from different parts of the genome. However, the fact that three of the four genes within the supergene affect the same trait, elytral colour pattern, opens the possibility that these loci have arisen by gene duplication. Unequal crossing-over during meiosis, in which homologous chromosomes do not align absolutely precisely, may give rise to two or more copies of the same gene, lying next to each other, on a chromosome. These duplicate genes may then evolve independently to one another with novel alleles arising by mutation at each locus within the supergene. Although three of the genes within the 2 spot colour pattern supergene may have evolved through gene duplication, it is unlikely that the fourth did, for this gene affects pronotal, rather than elytral, pattern (Lus 1932).

Recent work on the 10 spot ladybird and *Harmonia axyridis* has provided some evidence to suggest that a supergene is involved in these species as well, in both species rare phenotypes appearing in laboratory bred crosses, which may be the result of cross-overs within the supergenes (Majerus and T. Majerus unpublished data).

Although one main gene/supergene is responsible for most of the colour

(a)

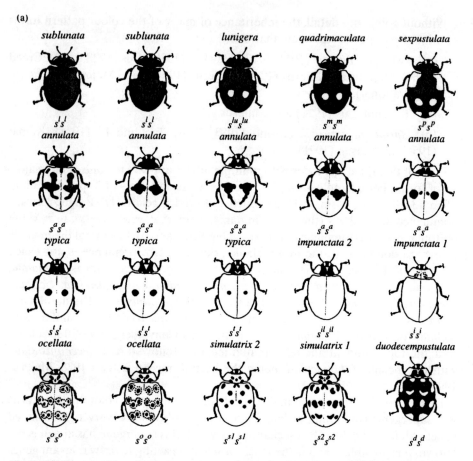

Fig. 9.12 The genetics of colour pattern variation in the 2 spot ladybird from Lus'
1928 and 1932 analyses: (a) homozygous forms; (b) heterozygous forms. (After Hodek
1973.)

pattern varieties within each of these species, in all of them, some other unlinked
loci have affects on some of the varieties. In the 2 spot, the recessive allele of an
unlinked gene increases the extent of melanic markings of some forms, converting extreme *annulata* into melanic *annulata*, and spotty and strong spotty into
sexpustulata spotty. A recessive allele at another locus causes black patterning to
be extended still further, producing the blackest of all the forms of the 2 spot, f.
nigra (Majerus and Kearns 1989).

There is also evidence that modifiers are involved in the dominance of several
of the alleles in the supergene. So, for example, while *quadrimaculata* is dominant over *typica* in Europe (Hawkes 1920), the phenotypically indistinguishable
humeralis in America is recessive to *typica* there (Palmer 1911, 1917). Further-

(b)

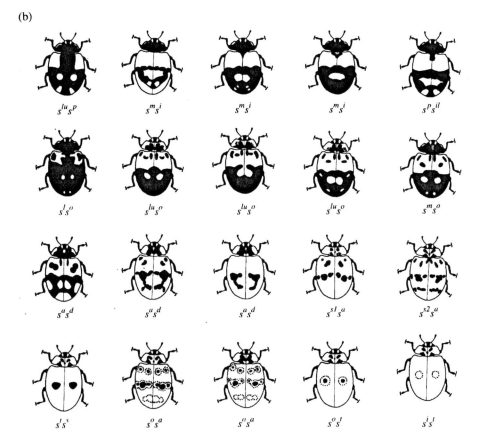

Fig. 9.12 *Continued.*

more, crosses between Scottish and southern English, or Parisienne 2 spots, produce breakdown of dominance, suggesting that different sets of dominance modifiers have been selected for in these different populations (Majerus and Kearns 1989). One intriguing finding in this respect concerns melanic 2 spots from New Zealand. The 2 spot ladybird was introduced into New Zealand in the early days of colonialism, probably from Europe. Melanic 2 spots supplied by Dr Anthony Harris from Dunedin, were of two basic types. Some were indistinguishable from British f. *quadrimaculata* and f. *sexpustulata*, while others were similar to melanic *annulata*. Genetic analysis revealed both types of melanic to be controlled by a single allele of the main colour pattern locus, with the melanic allele recessive to *typica*. Assuming that the original introductions were from Britain or Europe, these findings present two alternative possibilities. First, the melanics in New Zealand may be the product of selection acting on recessive *annulata*-type alleles of the supergene and their modifiers, in the absence of the *quadrimaculata* allele, to create a stable melanic form. Second, selection may have acted against the expression of melanic alleles when in the heterozygous state, gradually reversing the dominance of the alleles involved. This latter possibility seems unlikely, because, were the melanic form selected against, being originally dominant, the probability is that it would have simply been selected out of the population (T. Majerus *et al.* in preparation *a*).

In the 10 spot ladybird, the number of dark dots on the *typica* form is controlled, in part, by the major colour pattern locus, and partly by a polygenic system. The two loci controlling rate of laydown of carotenoid and melanic pigments, respectively, are closely linked to one another, but unlinked to the colour pattern locus. For both loci, the fast laydown alleles are dominant to the slow laydown alleles (Majerus 1994*a*).

The maintenance of melanic polymorphism in ladybirds

Considerable research effort has been devoted to unravelling the factors that have played a part in the evolution and maintenance of melanism in these ladybirds, with most effort being expended on the 2 spot. Our continued uncertainty with regard to the reasons why the frequencies of common melanic forms vary both temporally and spatially are not a result of lack of scrutiny. Rather, they stem from the equivocal and contradictory nature of a mass of data, from both field and experimental studies, and in some cases, lack of crucial details, such as the exact patterning of melanics found. The literature on melanic polymorphisms in the 2 spot abounds with speculative explanations of the evolution and maintenance of the phenomenon.

The amount of evidence in support of these suggestions varies considerably. Broadly, explanations can be split into four categories: involving mimicry; the thermal properties of black surfaces; the reproductive behaviour of the morphs; and the effect of changes in selective forces on a co-adapted gene complex. These explanations are not mutually exclusive, but it is easiest to recount the evidence for each by discussing them in turn, before considering ways in which they may interact.

Before doing so, it is important to be specific about the proportion of the great wealth of colour pattern variation in the 2 spot that one is trying to explain. If we were to follow Kettlewell's implied definition of a melanic form, i.e. one in which the amount of black or dark coloration was in excess of that occurring in the typical form, the vast majority of forms of the 2 spot would have to be considered melanic, the only exceptions being forms such as *typica* itself, the American *melanlopleura*, the Asian *impunctata*, and the rare British varieties 'mini-spot' and 'halo' (Majerus 1994*a*). For this discussion such a categorisation is unhelpful. Consequently, I make a totally arbitrary split that those forms in which more than 50 per cent of the elytra is black should be designated melanic, while those in which red covers more than half are considered non-melanic. With this designation in mind, in Britain we have two categories of melanics defined on the basis of the frequencies that they achieve. These are the common melanics which comprise more than 5 per cent of some populations and include just the forms *quadrimaculata* and *sexpustulata*. The other class, the rare melanics include forms such as *duodecempustulata*, strong spotty, extreme *annulata*, and melanic *annulata*, among others. None of these have been recorded as comprising even 1 per cent of any British population, and it is doubtful whether any specific genotype in the weak *annulata* complex reaches even this frequency. Consequently, *typica* is considered the only common non-melanic. The plethora of other non-melanics, including weak *annulata*, bar *annulata*, intermediate *annulata*, front-spot, halo, semi-spotty, *simulatrix*, new *duodecempustulata*, and many others, are considered rare non-melanics.

Thus, for British 2 spots, the following questions need to be addressed.

(a) Was the ancestral form of the 2 spot, in Britain, non-melanic or melanic?

(b) What causes the geographic variation in the frequencies of the common melanics relative to the common non-melanic?

(c) What causes temporal variation in the frequencies of the common forms?

(d) Why do so many other forms exist at low frequencies?

The same questions could be asked for populations of the 2 spot in other parts of the world, but with different other forms comprising the common melanic and common non-melanic classes.

To consider these questions, we need to investigate the various factors that, to date, have been implicated in the evolution and maintenance of melanic polymorphism in ladybirds.

Is the 2 spot ladybird a Batesian mimic, a Müllerian mimic, or neither?

At the outset, the problem of an allegedly aposematic species being polymorphic, must be addressed. If the 2 spot is not protected from predation by a combination of defensive chemicals and memorable colours, it may be a Batesian mimic rather than a Müllerian mimic, in which case polymorphism would not be unexpected. This possibility was first proposed by Brakefield (1985). Brakefield (1985) argued that the 2 spot may not be toxic to birds. This proposition has been confirmed by comparative tests, feeding 2 spots to blue tit chicks, using 7 spot ladybirds as a control (Marples *et al.* 1989; Marples 1990). While the 7 spot ladybirds were lethal to the blue tit chicks, the 2 spot ladybirds were not. Brakefield (1985) speculates that if 2 spots are palatable to birds, the non-melanic forms are

Batesian mimics of black on red species such as the 7 spot, while the melanic forms are Batesian mimics of red on black species such as the pine ladybird, *Exochomus 4-pustulatus* (Linnaeus). However, although 2 spots may not be toxic to birds, this does not mean that they are palatable. Indeed, Marples (1990) showed that 2 spots did have a deterrent effect. Recognising this possibility, Brakefield (1985) also speculates that the 2 spot may be a Müllerian mimic, being involved in two Müllerian mimicry complexes. The question of whether the 2 spot is aposematic thus becomes crucial.

Is the 2 spot ladybird palatable to birds?

Evidence on this point is equivocal. In the case of most ladybirds, there can be little doubt that the bright colours are not used in a deceptive way. They are truthful advertisements. Ladybirds do smell foul, taste horrible, and some are toxic, at least to some predators. Much research has been carried out on the chemical defence systems that the bright colours of ladybirds advertise. At the centre of coccinellid chemical defence lies a group of related nitrogen containing molecules, known as alkaloids. These include: N-oxide coccinelline, and its corresponding free base precoccinelline, first extracted from the 7 spot ladybird (Tursch *et al.* 1971) and also occurring in other species of the genus *Coccinella*, such as the 5 spot, 11 spot, scarce 7 spot and hieroglyphic ladybirds; N-oxide convergine and its free base hippodamine from the convergent ladybird, *Hippodamia convergens* Guerin (Tursch *et al.* 1974); myrrhine from the 18 spot ladybird *Myrrha 18-guttata* (Linnaeus) (Tursch *et al.* 1975), and propyleine and adaline yielded by the 14 spot and 2 spot ladybirds, respectively (Tursch *et al.* 1972, 1973).

These chemicals appear to be synthesised by the ladybirds that bear them, and are found in all life-history stages (Tursch *et al.* 1971, 1975). Many coccinellids also contain relatively high concentrations of histamine (Frazer and Rothschild 1960). The 11 spot ladybird, when feeding on the oleander aphid (*Aphis nerii* Boyer de Fonscolombe), sequesters cardiac glucosides (Rothschild *et al.* 1973; Rothschild and Reichstein 1976). Other volatile chemicals, such as quinolenes and pyrazines, have also been extracted from some coccinellids (Rothschild 1961).

An elegant series of experiments, carried out by Paul Brakefield's team, at Leiden, Holland, on the reflex blood exuded by 7 spots and 2 spots (de Jong *et al.* 1991; Holloway *et al.* 1991), has demonstrated that the defensive chemicals are distributed throughout all parts of adult ladybirds, but occur at much higher concentrations in reflex blood, the yellow fluid exuded from joints between the femur and tibia of the legs, than elsewhere. All six legs may reflex bleed simultaneously or independently (Fig. 9.13). The amount of reflex blood, and the concentration of defensive chemicals therein, varies, both between individuals, and with season, less fluid being produced during the winter, when fluid retention is crucial to survival through the months of dormancy.

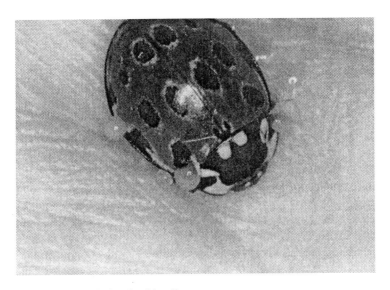

Fig. 9.13   An eyed ladybird reflex bleeding.

Thus, there is evidence that 2 spots contain defensive chemicals. There is also evidence from feeding experiments to suggest that 2 spot ladybirds are distasteful to many, although not all, vertebrates. For example, Frazer and Rothschild (1960) offered 7 spots, 11 spots and 2 spots to mice, voles, hedgehogs, bats, lizards, terrapins, a toad, and six species of bird. The ladybirds offered were given the highest rating of unacceptability. Conversely, several authors have reported observations of birds eating ladybirds in the wild (Heikertinger 1932; Kristin 1984, 1986, 1991; Majerus and Majerus 1997), or found the remains of ladybirds in the stomachs of wild birds, sometimes in significant numbers (Mizer 1970; Netshayev and Kuznetsov 1973). Interestingly, 2 spots were not found in significant numbers in any of these studies, although the allegedly more toxic 7 spot was (Mizer 1970; Netshayev and Kuznetsov 1973; Kristin 1984, 1986, 1991; Majerus and Majerus 1997).

Muggleton notes the contradictory nature of data on whether ladybirds are aposematic, and suggests that although the colour patterns and chemistry of ladybirds undoubtedly afford ladybirds some degree of protection from predation by vertebrates, previous authors may have overemphasised the protection. To support his case, Muggleton presents the results of experiments in which frozen 2 spot ladybirds were glued to trees or bushes, in 'life-like' positions, and left for 24 hours. Between 4 and 28 per cent of the ladybirds disappeared, in different runs. These results seem to suggest that levels of bird predation on ladybirds may be highly significant, if, that is, the disappearances were due to bird predation, which was not demonstrated. However, Muggleton's experiments are open to the criticism that dead ladybirds do not have the defensive abilities of live ladybirds, in that they cannot reflex bleed. This criticism has some support, for Majerus and Majerus (1997) have observed crows eating dead ladybirds in the tide-line on the Norfolk coast, while ignoring huge numbers of living ladybirds on the beach. In addition, Brakefield (1984), conducting similar experiments in which dead lady-birds were glued to rose leaves, obtained considerably lower rates of disappear-ance (0 and 13 per cent, in two runs).

Marples and Brakefield's work with blue tit chicks shows that 2 spot ladybirds are not actually toxic to the chicks. Recent observations of wild birds feeding on ladybirds, suggest that a variety of bird species will feed on ladybirds in specific circumstances, such as when the birds are starving and other food is short (Majerus and Majerus 1997). Perhaps more relevantly, observations of robins feeding ladybirds to their chicks were recorded. These observations are of par-ticular interest for two reasons. First, the parent robins were not seen to eat any ladybirds themselves; they only fed them to their chicks. Second, all the ladybirds brought to the nest were yellow and black or beige and black species (14 spot, 16 spot and 22 spot *Thea 22-punctata* (L.)). No red and black species were caught despite both 2 spots and 7 spots being much more abundant than any of the three species that were taken.

Majerus and Majerus (1997) conclude their series of observations on the pre-dation of ladybirds by wild birds with the general deductions that:

i) most, if not all, ladybirds are distasteful to most birds;

ii) some ladybirds may be toxic to some birds, but most are not toxic to most birds, and few, if any, are toxic to all birds;

iii) some species of bird which habitually feed on the wing, are able to eat ladybirds without obvious detriment;

iv) birds that are starving may eat ladybirds as a last resort;

v) adult birds feeding nestlings may include some species of ladybird in the food they bring to their young, and these are acceptable to the nestlings at least for a time;

vi) some birds take ladybirds selectively, preying on certain species to the exclusion of others;

vii) the reactions of birds, particularly wild birds, to ladybirds (whether they eat them or not) may depend crucially on the ladybird's ability to reflex bleed;

viii) great caution should be taken if extrapolating from the results of exper-iments using either captive birds, or, perhaps more critically, dead ladybirds, to the role of bird predation in the evolution of ladybird chemical defence and warning coloration;

ix) in our opinion, the view that the bright contrasting colours of ladybirds are aposematic is justified.

Some birds can, and do, eat 2 spot ladybirds, others do not, or only do so under conditions of hardship. The published literature on bird predation of ladybirds does not present a clear picture. Sadly, few reports contain all the information necessary to clarify the matter, that is to say, species of bird, species of ladybird, likely condition of predator, abundance of ladybirds, likely abundance of other suitable prey, and number of ladybirds eaten in a given time. Careful field observations are urgently needed to obtain this type of data.

So what of Brakefield's ideas on the different morphs of the 2 spot ladybird, and indeed the different morphs of the 10 spot, being parts of either Batesian or Müllerian mimetic complexes? Although, it is not possible to say with any certainty that advantageous resemblance to well protected aposematic species was not an influence in the evolution of the colour patterns in the 2 spot, in the past, irrespective of whether the 2 spot is itself unpalatable, it is highly improbable that such resemblance has any significant influence on the frequencies of the various forms of the 2 spot now.

Were melanic 2 spots to be Müllerian or Batesian mimics of other red on black ladybirds, one would expect the frequencies of melanic 2 spots to be highest in those regions and habitats where species such as the pine and/or kidney-spot ladybirds (*Chilocorus renipustulatus* (Rossi)) were most common. But, in Britain and all parts of western Europe that I have surveyed, this is not the case. In Britain, the reverse is true. High melanic frequencies are most often found in 2 spot populations from the industrial Midlands, into the north-west of England, and the industrial heartlands of Scotland. In these regions the pine ladybird occurs at relatively low density, if at all, while the kidney-spot, although locally common in parts of northern England, is very rare in Scotland. Furthermore, both species are rather host-plant specific, the kidney-spot favouring deciduous trees such as ash, willows, sallows, poplars, and birches, and the pine ladybird, conifers and ash. On none of these trees, with the exceptions of willows and sallows, do 2 spots occur commonly. In the south of England and East Anglia, where the kidney-spot is common, and the pine ladybird is often extremely abundant on conifers, melanic 2 spots are comparatively rare, or absent.

Interestingly, Brakefield's speculations on the possibility of mimicry being involved in the evolution of the diverse array of bright colour patterns in the Coccinellidae highlights one of the differences between Lepidoptera and many other groups of insects, and emphasises one of the drawbacks of working with museum specimens, without a close knowledge of the living insects.

Brakefield proposes that British ladybirds can be divided into five colour pattern categories (see Table 9.2). Consideration of these strongly suggests that they have been concocted on the basis of dead museum specimens. Whereas the colours of most Lepidoptera remain reasonably unchanged after death when kept away from the light, the colours of many other insects, including beetles, alter substantially. This is particularly true of some paler colours, such as bright reds and oranges, which tend to lose lustre, becoming a rather dull orangey brown, often termed testaceous.

If some of Brakefield's Müllerian mimic complexes are considered, they only make sense, in terms of showing resemblance, if the ladybirds were seen when long dead. The species involved in some of the groupings show little similarity when alive. For example, one proposed group includes the cream-spot ladybird, the orange ladybird, *Halyzia 16-guttata* (Linnaeus), the 18 spot ladybird, and the cream-streaked ladybird, *Harmonia 4-punctata* (Pontoppidan), under the description 'brown with yellow marks'. This description makes little sense if these species are examined when alive. Certainly, there is a resemblance between the cream-spot and 18 spot ladybirds. Both are a maroon-brown colour, but their markings are white or off-white, not yellow. The orange ladybird is bright orange with brightly white spots, while the cream-streaked ladybird is yellow, light brown, pink, or red, variably streaked with cream, and it has black spots (most usually either four or 16).

This case illustrates the difficulties inherent in working with dead material, without checking that death itself does not cause changes in the appearance of the material. The same may be true of work on chemically defended prey species. Offering dead material to predators may not be the same as offering live material (p. 244).

Taking all these, admittedly somewhat insubstantial threads together, my view is that the case for mimicry being strongly influential in the evolution of melanic polymorphism in ladybirds has not been made.

Table 9.2 Brakefield's proposed Müllerian mimetic rings in ladybirds. (Based on Brakefield 1985.)

Red or orange-red with black marks	Black with red marks	Yellow with black spots	Brown with yellow marks	Red-orange, pink or yellow, with black lattice pattern
7 spot	Pine	16 spot	Cream-spot	14 spot
5 spot	Kidney-spot	22 spot	Orange	Hierogylphic
11 spot	Heather	Water	Cream-streaked	10 spot, f.
13 spot	2 spot (melanics)		18 spot	*decempustulata*
Eyed	10 spot, f.			
Adonis'	*bimaculata*			
2 spot (non-melanic)				
10 spot, f.				
decempunctata				

Evidence for thermal melanism

The most significant early studies of factors affecting melanic polymorphism in the 2 spot were those of Timofeeff-Ressovsky (1940), who recorded the numbers of melanics and non-melanics, in Berlin, from 1929 to 1940. Counts were made in the autumn and spring. The proportions of melanics and non-melanics are given in Fig. 9.14, together with the frequency of the melanic alleles (assuming all melanic alleles are allelic and are dominant to the *typica* allele). The figure shows that the frequency of the melanics declines during the winter, and increases

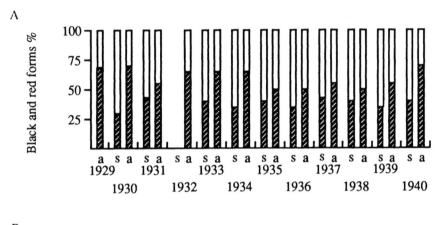

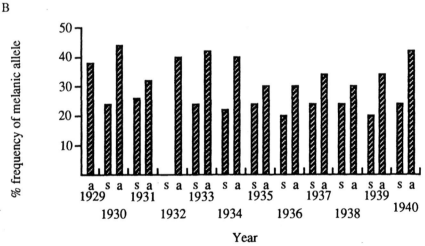

Fig. 9.14 (A) The proportion of melanics and non-melanics in spring and autumn samples of 2 spot ladybirds taken in Berlin from 1929 to 1940. (B) The estimated frequency of the melanic allele (After Majerus 1994, from data in Timofeeff-Ressovsky 1940.)

Table 9.3 The mortality of different morphs of the two-spot ladybird during the winter

(a)

Location	Composition of group: October		January		March		Number known to have died		Number which dispersed or disappeared before March	
	melanics	non-melanics	melanics	non-melanics	melanics	non-melanics	melanics	non-melanics	melanics	non-melanics
Keele, Staffs.	6	38	6	33	5	31	1	3	0	7
Keele, Staffs.	4	32	4	30	4	26	0	3	0	3
Keele, Staffs.	18	164	16	131	16	117	1	24	1	23
Keele, Staffs.	2	37	2	31	1	26	0	5	1	6
Keele Staffs.	5	14	5	13	4	11	0	3	1	0
Keele, Staffs.	1	26	1	24	0	7	0	2	1	17
Keele, Staffs.	2	8	2	7	2	6	0	2	0	0
Keele, Staffs.	15	49	13	42	12	35	1	7	2	7
Stoke, Staffs.	3	33	3	22	3	21	0	10	0	2
Stoke, Staffs.	7	41	6	39	5	37	1	2	1	2
Stoke, Staffs.	9	79	9	79	9	74	0	0	0	5
Stoke, Staffs.	10	133	6	84	3	30	6	84	1	19
Stoke, Staffs.	4	20	4	18	4	18	0	2	0	0
Stoke, Staffs.	1	6	1	6	1	5	0	1	0	0
Stoke, Staffs.	2	41	2	41	2	28	0	13	0	0
Stoke, Staffs.	4	12	4	10	1	9	3	1	0	2
Stoke, Staffs.	17	39	15	36	12	30	3	6	2	3
Stoke, Staffs.	7	49	7	47	5	31	2	16	0	2
Manchester	12	41	11	38	8	21	4	16	0	4
Manchester	3	6	3	6	3	6	0	0	0	0
Manchester	18	22	18	22	18	21	0	1	0	0
Manchester	24	34	15	23	5	9	10	14	9	11
Manchester	6	18	6	18	2	7	4	11	0	0
Manchester	29	24	28	24	26	22	2	2	1	0

Manchester	4	7	3	7	2	7	1	0	1	0
Manchester	13	30	12	23	4	10	8	15	1	5
Total	226	1003	202	854	157	645	47	243	22	118

(a) The composition (melanics and non-melanics) of 26 overwintering groups in October, January, and early March, with the number of melanics known to have died at the sites between October and March, and the number of each form which dispersed or disappeared from these groups
% of melanics known to have died at overwinter sites = 20.80.
% non-melanics known to have died at overwinter sites = 24.2.

(b)

Sample number	Number of ladybirds surviving	
	quadrimaculata	*typica*
1	24	31
2	43	47
3	41	32
4	18	26
5	47	39
6	34	33
7	37	42
8	28	19
9	49	44
10	49	50
11	42	48
12	31	28
13	8	10
14	42	30
15	44	32
16	50	47

(b) The mortality of melanics (*quadrimaculata*) and non-melanics (*typica*) in 16 overwintering groups of 50 melanics and 50 non-melanics placed into an insectory in October and removed in late March
Total melanics that died = 213 (26.6%) of those put out.
Total non-melanics that died = 242 (30.3%) of those put out.

during the summer. The explanation of this cyclical change in frequencies is based on the thermal properties of dark, compared with light, surfaces. Brakefield and Willmer (1985) have shown, by placing minute thermocouples on, and under, the elytra of 2 spots, that black areas heat up, and cool down, more rapidly than do red areas. The melanic forms may therefore warm up more quickly on spring and summer mornings, giving them a competitive advantage, in replenishing exhausted food and fluid reserves, finding mates, and laying eggs (Benham *et al.* 1974). As the 2 spot produces three generations a year in Berlin (Timofeeff-Ressovsky 1940), the advantage to melanics through the summer may be substantial, leading to the observed increase in the frequencies of melanics during this period. However, the same thermal properties of black surfaces will mean that, in the winter, melanics will be subject to greater fluctuations in temperature than non-melanics. It is known that overwintering ladybirds are particularly susceptible to rapid changes of temperature (Hodek 1973), so it is likely that melanics will suffer higher winter mortality than non-melanics, again as observed. This, then, is a case of thermal melanism.

It is symptomatic of the problems associated with the understanding of melanic polymorphism in the 2 spot, that other workers who have studied this polymorphism (for example, in Britain, Creed 1966, 1971*a,b*; Muggleton 1979; Brakefield and Lees, 1987; Majerus and Kearns 1989; Majerus 1989*a*, 1994*a*, 1995; Finland, Mikkola and Albrecht 1988; Norway, Bengtson and Hagen 1975, 1977; Italy, Scali and Creed 1975; Holland, Brakefield 1984; Germany, Schummer 1983; Czechoslovakia, Honek 1975; Russia, Zakharov and Sergievski 1980) have not found such seasonal changes in the frequencies of the forms. Brakefield's work in Holland illustrates the difficulties well, for while he found some evidence to support the thermal melanism hypothesis (melanics mated earlier in the spring), he also showed that during the winter non-melanics suffered significantly higher mortality than melanics. This is precisely the opposite of Timofeeff-Resovsky's findings. Furthermore, in a protracted series of studies of the mortality rate of *typica* and *quadrimaculata*, at overwintering sites in Britain, no morph specific differences were detected (Table 9.3) (Majerus 1994*a*).

The lack of concordance between the different studies does not greatly undermine the thermal melanism hypothesis, for most of them have provided evidence that high melanic frequencies occur in regions with lower sunshine levels, or low temperatures. However, this is not always the case. For example, in northern-central Italy, Scali and Creed (1975) found that melanic frequency decreases with increasing altitude, concluding that melanic frequency was positively correlated with temperature. Furthermore, Lusis (1973) records the melanic f. *sublunata*, to comprise up to 96 per cent of some populations in extremely hot regions of central Asia.

Creed (1966, 1971*a*) demonstrated that high melanic frequency was correlated with the level of smoke pollution (Fig. 9.15). Further, he showed that melanics had declined in frequency by about 15 per cent, at six sites near Birmingham,

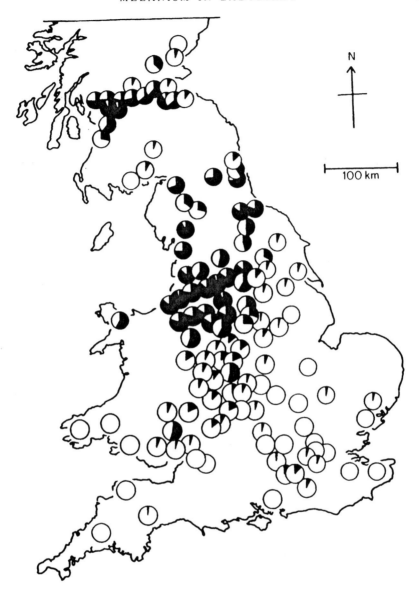

Fig. 9.15 Distribution of melanic (black segments) and non-melanic (open segments) of the 2 spot ladybird in Britain in the late 1960s. (After Lees 1981, from data in Creed 1971*a*.)

during the 1960s, noting that this decline followed on shortly after the clean air legislation of 1956, and the consequent introduction of smokeless zones (Creed 1971*b*). Brakefield and Lees (1987) added to this data set for Creed's Birmingham sites, showing that the decline in melanics had continued into the 1980s

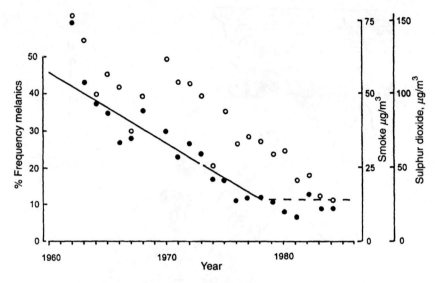

Fig. 9.16 The decline in 2 spot ladybird melanic frequencies, in Birmingham, since 1960. (After Brakefield and Lees 1987.)

Table 9.4 Numbers of non-melanic (n-m) and melanic (mel) 2 spot ladybird in annual samples from sites in the Birmingham conurbation from 1960 to 1986. (Data from Brakefield and Lees 1987.)

Year	Aston Church n-m	mel	Castle Bromwich n-m	mel	Egbaston n-m	mel	Hall Green n-m	mel	Maw Green n-m	mel
1960	—	—	47	21	55	64	—	—	—	—
1961	—	—	—	—	15	39	5	6	27	18
1962	—	—	—	—	51	65	44	42	51	28
1963	—	—	—	—	2	5	—	—	—	—
1964	36	26	51	17	—	—	30	17	109	71
1965	15	12	—	—	—	—	33	18	41	23
1967	244	77	108	23	—	—	132	58	119	54
1968	4	3	48	5	—	—	54	21	106	38
1969	106	41	40	12	87	91	61	29	97	32
1970	128	33	31	5	—	—	70	25	—	—
1971	29	8	—	—	—	—	51	13	—	—
1972	70	22	186	35	—	—	64	13	—	—
1974	131	19	—	—	—	—	150	36	—	—
1975	—	—	—	—	4	6	—	—	—	—
1976	207	25	—	—	19	6	131	15	71	14
1977	154	24	19	0	114	48	271	33	—	—
1978	31	1	—	—	27	13	—	—	—	—
1986	46	5	2	0	251	84	167	12	85	9

(Table 9.4 and Fig. 9.16). Bishop *et al.* (1978*b*) demonstrated similar declines for eight sites in north-west England. Creed (1971*a,b*) implied that smoke has a direct and detrimental selective effect on non-melanics, compared with melanics. Muggleton *et al.* (1975) and Brakefield and Willmer (1985) have suggested that the correlation between smoke pollution and melanic frequency could be the result of soot particles in the air reducing sunshine strength at ground level. The melanics then gain an advantage through thermal melanism. In this they are echoing Lusis (1961), who found melanic frequencies to be highest in industrial areas, and, to a lesser extent, areas with a humid maritime climate.

Whether melanic frequencies are positively correlated with smoke levels, or negatively correlated with sunshine levels, it must be stressed that such correlations provide no more than circumstantial support for the thermal melanism hypothesis. They should not be taken as proof of either cause or effect. Further, as indicated above, some evidence of a contradictory nature exists (e.g. Scali and Creed 1975; Bengston and Hagen 1975, 1977; Honek 1975; Muggleton 1978).

An extensive survey in Britain between 1984 and 1994 has shown that the frequencies of melanics, while still in decline in the Midlands and the north of England, are increasing in many parts of southern England, and East Anglia (Majerus 1995). In some districts the increase in melanic frequency has been dramatic, e.g. from 0 to 30 per cent, between 1985 and 1988, at one site in south Devon (Majerus 1994*a*). It is difficult to envisage why the frequencies of melanic forms should be moving in opposite directions, if only thermal factors were affecting the morph frequencies.

The stability of melanic polymorphism in the 10 spot ladybird

Melanic polymorphism in the 10 spot ladybird is very different to that in the 2 spot. First, the melanic form *bimaculata* is recessive to the other common forms. Second, the three main forms show relatively little in the way of spatial or temporal variation in their frequencies. In Britain, all three forms occur in all populations. Generally, *decempunctata* is the commonest morph, followed by *decempustulata*, with *bimaculata* the least frequent, but still having a frequency of 5–20 per cent (Fig. 9.17). Brakefield (1985) has shown a similar lack of geographic variation in morph frequencies in the Netherlands. Third, there is little evidence to suggest temporal changes in the frequencies of the forms. Rather the reverse, for Brakefield and Lees (1987), comparing frequency data at five Birmingham sites, from the 1960s, 1970s, and 1980s, did not find reduced melanic frequencies to parallel those seen in the 2 spot (Table 9.5).

The reasons why the 10 spot does not have the same pattern of variation in melanic frequencies as the 2 spot, are unknown, but neither thermal melanism,

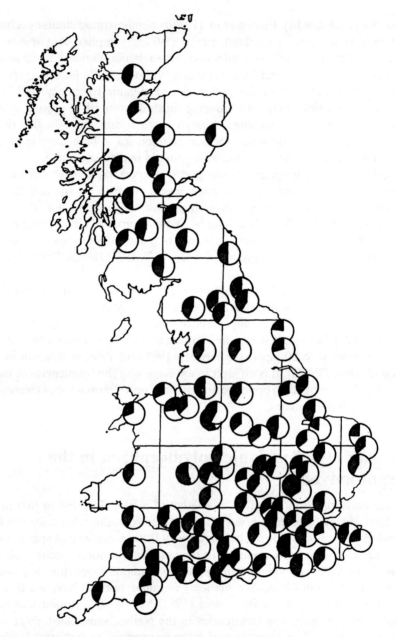

Fig. 9.17 The frequencies of the *decempunctata* (open segments), *decempustulata* (stippled segments), and *bimaculata* (black segments) forms of the 10 spot ladybird in Britain.

Table 9.5 Numbers and percentage frequencies of the *decempunctata* (*typ.*), *decempustulata* (*dec.*), and *bimaculata* (*bim.*) forms of the 10 spot ladybird in combined samples for each of three decades from sites in the Birmingham conurbation. (From Brakefield and Lees 1987.)

Site		1960–69			1970–78			1986		
		typ.	*dec.*	*bim.*	*typ.*	*dec.*	*bim.*	*typ.*	*dec.*	*bim.*
Aston Church	n	15	5	4	59	19	6	5	2	1
	%	51.7	17.2	13.8	70.2	22.6	7.1	62.5	25.0	12.5
Castle	n	16	7	3	10	5	1	—	—	—
Bromwich	%	61.5	26.9	11.5	62.5	31.2	6.2	—	—	—
Edgbaston	n	18	9	4	14	9	4	19	9	8
	%	58.1	29.0	12.9	51.9	33.3	14.8	52.8	25.0	22.2
Hall Green	n	42	25	11	89	40	7	4	3	4
	%	53.8	32.1	14.1	65.4	29.4	5.1	36.4	27.3	36.4
Maw Green	n	5	3	3	5	1	0	1	1	0
	%	45.5	27.3	27.3	83.3	16.7	0.0	—	—	—

nor mating preferences are thought to be important. Brakefield (1985), noting the 10 spot ladybird's more arboreal habits, suggests that in more shaded situations, thermal melanism is unlikely to have any significant influence on adult reproduction. Banbura and Majerus (unpublished data) conducted mating tests using the three forms of the 10 spot, and found that mating was random with respect to form.

In the light of this lack of support of any selective factor which might maintain the polymorphism in the 10 spot, the spatial and temporal constancy of the frequencies of the three forms is remarkable. A concerted programme of research on this species would certainly be worthwhile. Indeed, this is just the type of species that Ford would have pointed to as a prime candidate for research, for he would have regarded the constancy of the frequencies of the forms as indicative of very strong balancing selective pressures maintaining the polymorphism.

Non-random mating in ladybirds

Other factors have been implicated in the evolution and maintenance of melanic polymorphism in the 2 spot. Lus (1961) provided evidence that mating in 2 spot ladybirds is not random with respect to morph. Muggleton (1979), using Lus' data and further data of his own, showed that the mating biases were frequency dependent, the rare morph, regardless of colour, mating more frequently than the common morph. It has been shown by O'Donald and Muggleton (1979) that this type of non-random mating may be sufficient to maintain the polymorphism. Lees (1981), noting this fact, suggests that it is possible that the principal factor which ensures the maintenance of melanic polymorphism in the 2 spot, is unrelated to any environmental factor, and that these simply affect the frequency at which a stable equilibrium, maintained by the frequency-dependent selective mating, is reached.

Investigations into the mating preferences shown by 2 spot ladybirds, since Muggleton's 1979 paper, have been considerable. Work initiated by Peter O'Donald in Cambridge, has demonstrated that 2 spots from different populations show different types of mating preference. For example, *annulata* phenotypes from Cambridge show assortative mating (O'Donald *et al.* 1984), while in populations from Staffordshire, Glasgow, and the Loire Valley, France, some females prefer to mate with black rather than red males, irrespective of their own genotype (Majerus *et al.* 1982*a,b*; O'Donald and Majerus 1992). This latter mating preference has received particular attention, because a mating preference for a specific phenotype will necessarily be negatively frequency dependent (O'Donald 1980), and so may maintain polymorphism.

Selection experiments for increased or reduced levels of preference showed that the mating preference was a case of genetically controlled female choice, rather than male competition, and suggested a single dominant gene controlled the preference (O'Donald and Majerus 1985; Majerus *et al.* 1986). The female mating preference for melanic males is itself a genetic polymorphism, some females having the preference, others lacking it. Indeed, the polymorphism in respect of mate preference seems to be as variable temporally and spatially as that in respect of colour pattern. For example, the frequency of the preference gene in Cambridge populations is zero, whereas in north Staffordshire, it had a frequency of about 0.13 in the early 1980s (O'Donald *et al.* 1984, O'Donald and Majerus 1985). Furthermore, the frequency of the preference allele in one north Staffordshire population appears to have declined rapidly in the mid-1980s, this decline being correlated with a sharp decline in the melanic frequency of the population in question (Kearns *et al.* 1990).

The variability in mating preferences is not confined to the 2 spot ladybird. Naoya Osawa and Takayoshi Nishida (1992), in an excellent series of studies involving both field observations and laboratory experiments, showed that both sexes of *Harmonia axyridis* have mating preferences, but to different degrees. Female mating preferences were stronger than those of males. Moreover, the preferences shown changed with the season. In the spring, both melanic and non-melanic females chose to mate with non-melanic males, while in the summer, melanics were over-represented in matings. By manipulation experiments, Osawa and Nishida showed that the patterns of mating were directed mainly by female choice, but that in the summer generation there was some element of male choice as well. These experiments also demonstrated that female *H. axyridis* use male colour pattern in making their choice, although other features, such as chemically or physiologically based attractiveness were also implicated.

Why do rarer melanic forms recur?

Melanic polymorphism in the 2 spot ladybird appears to be a very fluid system, which is acted upon by a variety of selective factors. These undoubtedly differ in their directions, strengths, and relative importance, from one place to another, and from one time to another. Furthermore, the genetic control system that lies at the core of this polymorphism, the main colour pattern supergene, is itself dynamic. Lus (1932) notes that melanics occur at significant frequencies, both in Peterhoff and Semirjetche. In Peterhoff, the melanics were entirely of the forms *lunigera, quadrimaculata*, and *sexpustulata*, while in Semirjetche all melanics were of the *sublunata* form. In Britain, Creed (1971*a*) notes that in Birmingham and Scotland *sexpustulata* is much commoner than *quadrimaculata*, the situation being reversed in Durham and Manchester.

In most work on melanic polymorphism in the 2 spot, all predominantly black forms are lumped together as melanics, and all predominantly red forms are lumped together as non-melanics. If we are to resolve the difficulties in explaining melanic polymorphism in this ladybird, this is surely a fundamental error. In many populations, the uncommon forms together reach frequencies well in excess of 10 per cent. Yet, these genotypes are neglected entirely. Consequently, we are in danger of ignoring the underlying genetics of the system, the fluidity of the colour pattern supergene.

Very little experimental attention has been paid to the rarer forms. The explanation of the presence of these rare forms in so many populations may be a consequence of the colour pattern of the 2 spot being primarily under the control of a supergene. It is likely that many of the alternative alleles of the main colour pattern supergene are the products of recombination, or rearrangements, within the supergene. The frequencies at which recombinants are produced, will be a function of the distances between the various genes within the supergene, and the selective factors that act upon them each generation.

Sergievsky and Zakharov (1989) have proposed that the maintenance of a wide range of genetic variants may be crucial to the survival of 2 spot populations during periods of environmental stress. They have postulated that the initial response to a variety of types of environmental stress is an increase in melanic frequencies. This increase results from the destabilisation of a previously stable co-adapted gene complex, by novel selective factors. Subsequently, melanic frequencies decline again, as the genome as a whole re-establishes stability, finely adapted to the new environmental conditions. While I do not see why the initial response should be an increase specifically in melanic phenotypes, this idea has some very interesting features. Not least of these is that it creates a role for the maintenance of a fluid genome. There is one problem with the idea. That is, it seems to suggest that deleterious alleles may be maintained in the population in case they become beneficial if the environment changes. This is contrary to accepted ideas on the way that selection operates, because selection is not predictive. Selection acts now, as a result of current circumstances. Yet, as we know that the 2 spot carries an exceptionally high proportion of lethal recessive alleles (Lusis 1947; Majerus 1994a), which selection should purge from populations, there is perhaps an inkling that we need to display some degree of unconventional or lateral thinking if we are to explain melanic polymorphism in these ladybirds.

Melanic polymorphism in ladybirds: we do not have the answers yet?

The melanic polymorphisms in the three ladybirds I have discussed in detail are complex. Data from population surveys and laboratory experiments on these species lead to contradictory interpretations of the causes of the evolution and maintenance of the polymorphisms. The consensus view appears to be that thermal (and possibly humidity) conditions are important in the 2 spot ladybird, although the evidence is not unequivocal. Yet, thermal considerations are not thought to be important in the closely related 10 spot ladybird (Brakefield 1987), and in *Harmonia axyridis*, the clines in melanic frequency are precisely the reverse of those one would expect on the basis of thermal melanism (Osawa and Nishida 1992).

The data on mating preferences also show contradictions within one species (cf. O'Donald and Majerus 1985; Kearns *et al.* 1992; O'Donald and Majerus 1992) and different patterns in others.

The presence of female-biased sex ratios in many populations of two of these species, due to the presence of male-killing bacteria, may be partly responsible for the fluidity of the mating preferences found. The principal cause of the prevalence of male-killing bacteria is the resource reallocation from dead males to sibling females, when the neonate female larvae eat the eggs containing their dead male siblings (Hurst *et al.* 1992; Hurst and Majerus 1993). Furthermore, the densities and size structures of aphid populations available to young larvae are likely to influence the advantage gained by the females in male-killed clutches. Consequently, it is possible that the population sex ratio may change considerably over fairly short periods. Certainly, in some populations of both the 2 spot and *Harmonia axyridis*, natural adult sex ratios approaching two females per male have recently been found (Zakharov *et al.* 1996; T. Majerus *et al.* in preparation *b*).

The role of mimicry in these polymorphic species is also controversial. The analyses to date, both in respect of the palatability of ladybirds to birds, and the resemblances between species, are somewhat flawed, because of the artificiality of the experiments, and the use of dead ladybirds, both for comparisons, and in predation experiments. Furthermore, two of the species discussed herein, the 10 spot and *Harmonia axyridis* have hardly been assessed. Certainly, if mimicry does play a significant part in the maintenance of melanic polymorphism in these, or other coccinellids, more attention must be paid to the species of ladybird, and their densities, found within the specific forage areas of birds that may attempt to prey upon them. The resemblance between an exclusively arboreal coccinellid, and one that lives just on low vegetation, can be of little mimetic significance if a different set of predators forage in each habitat. Furthermore, while predation experiments using captive birds are useful in assessing possibilities, such as the variation in learning abilities and risk taking when seeking food, assessments of the level of palatability of a prey species must be verified in the field using live prey, preferably at different times of the year, with predators in a range of conditions, and different ranges of other prey species available. While this may be a tall order, I think that we will continue to obtain unconvincing and contradictory data if we persist in carrying out experiments in conditions that have little, or no, relevance to the natural situations in which melanic polymorphisms in these ladybirds originally evolved and are now maintained.

Melanic polymorphism in ladybirds: a thesis

The underlying genetics of melanic polymorphism in ladybirds, and the contradictory nature of the data collected from observations and experiments on a range of possible evolutionary explanations for the existence of these polymorphism, suggest that these systems are complex and fluid. We may envisage each case as being, at any one time and place, either in a delicate state of equilibrium, or out of equilibrium and so moving slowly, or rapidly, towards some new equilibrium state. If transient, the rate of change towards a new equilibrium will depend upon the type and strengths of the selective factors acting upon the particular population at that time. The situation is perhaps like having a ball-bearing on a very gently undulating smooth piece of thin card, held by hand. While it is possible to hold the card still enough to keep the ball-bearing in one of the shallow depressions for some time, even small movements of the card may shift the ball-bearing out of that depression, so that it rolls off into some other depression. Given the range of potential variation held in the colour pattern supergene, a range of different phenotypes may be available for selection to act upon every generation. Following our analogy, this means that there will be many alternative depressions on our piece of card.

Ford argued, with great insight, that the most temporally stable polymorphisms would be the ones maintained by the strongest and most consistent balancing selection. In two of the species discussed, considerable temporal variations in melanic frequencies have been recorded. Consequently, in these species, our difficulties in identifying the selective agents involved may emanate from the fact that the selective pressures are not strong: they are very weak or very variable. The relative stability of melanic polymorphism in the 10 spot ladybird, compared with the 2 spot and *Harmonia axyridis* may mean that this species is the most suitable for studies into the factors that maintain balanced polymorphisms in ladybirds, for it will be in this species that the selective pressures are strongest and most constant. In the 2 spot and *H. axyridis*, the polymorphisms seem to be continually varying and readjusting in response to a complex set of extrinsic influences. I have little doubt that climatic factors, and the colour patterns and population densities of other aposematic insects that live in close proximity with these species, are two of these extrinsic factors, in virtually all populations. In addition, mating preferences may play a significant part in many populations, and may be crucial in maintaining some forms for substantial periods, because of the negative frequency dependence of the selection they produce. Other factors that play a part almost certainly await identification.

Recent work on the mating behaviour of 2 spot ladybirds has provided some circumstantial support for the suggestion that the life history of this species may have evolved to cope with a highly fluid and changing environment. This ladybird is highly promiscuous, females mating on average about 10 times more often than is necessary to maintain high egg fertility (Majerus 1994*b*). This is unexpected from an evolutionary point of view, because of the costs of mating. These costs include mating time, which may last several hours, virtually every day during the breeding period. While mating, females cannot lay eggs, and their feeding efficiency is considerably reduced. There are energy costs in terms of initial rejection behaviour (Majerus 1994*a*), carrying the male around, and the mechanics of

Fig. 9.18   Individuals of the sexually transmitted mite, *Coccipolipus hippodamiae*, on the underside of an elytron of a 2 spot ladybird.

copulation itself. In addition, many populations of 2 spots have a sexually trans-mitted disease in the form of a mite, *Coccopilipus hippodamiae*, which is severely detrimental to the fecundity and fertility of female 2 spots (Hurst *et al.* 1995) (Fig. 9.18). Given these costs, what counteracting advantage is there to the promiscu-ous behaviour of females? Two possibilities have been proposed (Majerus 1997). First, by mating with many male partners, the females may be providing the potential for sperm competition (Parker 1970). Second, by mating with many dif-ferent partners females may be bet-hedging. By this I mean that the promiscu-ous behaviour of females may have evolved to ensure that their offspring are as genetically diverse as possible, so that at least some of the progeny have a genetic make-up that is appropriate to the conditions they have to face. The ladybirds in question feed on aphids species, which have volatile and unpredictable popula-tion increases and decreases. Consequently, the conditions to which a new gen-eration of ladybirds will be exposed will depend upon which aphids happen to be doing well at a particular time. When a female is mating, she cannot predict which aphids will be available to her offspring at the crucial time when they are breeding the following year. In one year, the common aphids may be on birch, the next, on nettles, the next on lime trees, and in a fourth year there may be a surfeit of aphids on a whole range of hosts. Given the potential temporal vari-ability of favourable habitats for a female's future generations, it is perhaps not surprising that the strategy of a female ladybird is to produce very large numbers of eggs (up to 2000, Majerus 1994*a*), and to make these as genetically variable as possible, by mating with many different males.

The melanic polymorphism in the 2 spot and *Harmonia axyridis* may also be the outcome of strong, but inconsistent, selective pressures. Given the rapidity of some of the temporal and spatial changes in melanic frequencies that have been recorded, the suggestion of Sergievsky and Zakarov (1989) (see also Zakharov 1992), that such changes are transitions between fragile stable states is the only explanation that I find persuasive. Here then we may have a case that comes close to Professor Sewall Wright's (1982) 'shifting balance' theory, but with the adap-tive topology being only slowly undulating, without sharp and high adaptive peaks.

I believe that it will only be by protracted and continuous monitoring of natural populations of these ladybirds, with field and laboratory experiments being carried out as appropriate, that the conundrums of melanic polymorphisms in this group of aposematic beetles will be resolved. Having accumulated 17 years of data I am just beginning to think that I almost know the right questions to ask. I hope, over the next 17 years, the Ladybird Research Group in Cambridge will be able to answer some of them.

10 The future of research into melanism

The contribution of the study of melanism to biological understanding

One way to begin a summary of this book is to list the ways in which study of melanism has contributed to our understanding of biological, and in particular evolutionary or genetic, phenomena. Having taught undergraduates for over 20 years, I am aware that such summaries are popular and indeed useful in two ways. First, they give those students or pupils, who are set essays on melanism, a quick resumé of the more important applications of the subject to wider biological thought, without having to plough through 80 000 other words. Second, and I hope more importantly, it will help to focus on the contribution that studies of melanism have made to my own field, that of evolutionary genetics, and to other spheres of biology, and may allow a sensible course of future work to be plotted.

The simplest way to produce a synopsis of the major contributions of the study of melanism is perhaps a list of statements, and this is the style I will follow, describing the nature of the contribution and giving the organism(s) involved in brackets. I am aware that the list is not exhaustive, and is probably rather subjective. Other biologists would undoubtedly feel that studies of melanism have been influential in other ways. Furthermore, the illustrative species listed are also not chosen randomly, for I concentrate on those species I have discussed most fully in the previous text, which, not surprisingly, have tended to be those I am most familiar with. However, this may not be disadvantageous, for it is my belief that the more one knows about a species and its way of life, the greater will be the understanding of the way particular features of that organism fit into the larger picture of its interactions with others of its own kind and those of other kinds. In this I disagree with E.B. Ford, at least to some extent. Ford was an advocate of having a question to be answered and then looking through the diversity of life to find the ideal organism to study, in order to answer the question. There is obvious merit in this approach. Indeed I am currently following it, in my work with Greg Hurst and my wife Tamsin on male-killing endosymbiotic bacteria, for we have recently turned our attention from male-killers in ladybirds to seek them in *Drosophila*. The reason for this is that, if we are successful in finding such, genetical questions which are likely to be intractable in ladybirds, given the current paucity in knowledge of ladybird genetics, should become open to investigation, for the *Drosophila* have been subjected to more genetical research than any other group. However, having worked with ladybirds for 16 years and with Lepidoptera for much longer, I am coming more and more to realise that many of the most interesting phenomena and questions only come to light as one begins to know an organism, or group of related organisms, intimately. Furthermore, I also believe that I am a much more efficient and effective scientific questioner now than formerly, as long as I stick to those species whose natural history I have gained some 'feeling for', as a result of long hours observing them in captivity, or more usefully in nature, and thinking about them and their ways.

So, to a listing of the contributions that the study of melanism has made to biological understanding.

(a) First among these must be the demonstration that natural selection, the primary mechanism of evolution put forward by Charles Darwin, actually happens. As Ford (1955) put it: 'The industrial melanism of the Lepidoptera is the most striking example of evolution which has ever actually been witnessed in any organism, animal or plant' (e.g. the peppered moth, Chapters 5 and 6).

(b) Resolution of the controversy over whether selection or drift are responsible for the existence of conspicuous polymorphisms. In the main, such polymorphisms have been shown to be maintained by a balance of selective advantage and disadvantage (e.g. the scarlet tiger moth, p. 86).

(c) Balanced polymorphisms may be maintained not only by the balance between the selective advantages and disadvantages of different genotypes, but also by a balance between selective advantages to certain genotypes and migration into the population of other genotypes that are favoured elsewhere, thus producing a selection/migration balance (e.g. the peppered moth, p. 132).

(d) As demonstrated mathematically by J.B.S. Haldane (1924), the fitness differences between two commonly occurring morphs of a species may be much greater than Darwin or his early advocates thought likely (e.g. the peppered moth, p. 102).

(e) Natural selection can only act on the variation available within a population (e.g. the oak beauty moth, p. 199).

(f) Females of a species may have mating preferences between genetically different males thus confirming part of Darwin's second mechanism of evolution, sexual selection through female choice (e.g. *Harmonia axyridis*, p. 256; the 2 spot ladybird, p. 256; the Arctic skua, p. 94).

(g) Female mating preferences can be genetically controlled, thus confirming one key element in Sir Ronald Fisher's theory of the evolution of sexual selection by female choice (e.g. the 2 spot ladybird, p. 256).

(h) The genetic dominance of one allelomorph over another, in respect of a particular phenotypic effect, can evolve through natural selection acting upon modifier genes (e.g. the peppered moth, p. 119; *Papilio dardanus*, p. 217; the 2 spot ladybird, p. 238).

(i) Evolution is not a one-way process. If environmental factors change, forms previously at an advantage may become disadvantageous and decline (e.g. the peppered moth, p. 151; the 2 spot ladybird, p. 250).

(j) Balanced polymorphisms may be maintained when fitness differences between morphs are dependent upon the frequencies of forms in populations (e.g. the scarlet tiger moth, p. 86; angleshades moth larvae, p. 92).

(k) Negative frequency-dependent selection is a demonstrable consequence of search image formation by predators (angleshades moth larvae, p. 92).

(l) The thermal attributes of dark versus light surfaces may give advantage to different forms of a species in different environments (e.g. various moths, p. 162), or altitudes (the viper, p. 20), or latitudes, (e.g. the 14 spot ladybird, p. 227), or seasons (e.g. *Harmonia axyridis*, p. 236).

(m) Cuticular pigments may evolve for a variety of protective purposes, such as screening ultraviolet (UV) (*Daphnia*, p. 28), or protection against abrasion (e.g. bird feathers, p. 30; desert beetles, p. 29).

(n) In many species, melanism is shown in a variety of different ecological circumstances, the presence of melanic forms in each circumstance often depending upon different selective factors (e.g. industrial and non-industrial melanism in the mottled beauty, p. 193; the 2 spot ladybird, p. 258).

The factors which affect the evolution and maintenance of melanism in a species are usually complex and vary both in time and space, with the result that in no case can we feel certain that we yet know the whole story.

The future of research into melanism

Having given my opinion of the major contributions that the study of melanism has made to biological understanding, it is now perhaps useful to discuss some of the directions that future work into this phenomenon might take. At the outset, it is worth pointing out that future work may be split into two types, which may be broadly defined as 'more of the same', and new approaches based upon emergent technologies.

Dealing with the 'more of the same' class first, by this I do not mean to be derogatory; quite the reverse, for I mean that the techniques used to study melanism in species such as the peppered moth, the 2 spot ladybird, the viper, and the Arctic skua, should be used to scrutinise melanism in other species. As described in Chapter 2, melanism is a common phenomenon throughout the animal kingdom. Yet this phenomenon has been closely studied in only a few handfuls of species. The result is that explanations of the existence of melanism in many species are based, not on research into that species, but on extrapolation from research on other species. This may be valid, but it is not necessarily so. For example, the frequency of melanic forms in both the peppered moth and the 2 spot ladybird has been declining over the last 30 years, as a result of reductions in pollution levels. Were it not obviously the case that crypsis is not the function of the colour pattern of the 2 spot ladybird, it might be supposed, from research on the peppered moth, that the cause of the decline in melanism in the 2 spot was a consequence of changes in the nature of the substrates that the 2 spot rests upon. Fortunately, the thermal dimension in respect of melanism in the 2 spot has been recognised, but there may be many species showing melanism, in which the critical factors affecting melanism have not been determined, because it is assumed that these are the same as those in the well worked species. Consequently, my first plea is for work to be carried out on a much wider array of species, preferably from a range of different taxonomic groups.

The first approach, and one in which the amateur naturalist may make a tremendous contribution, is to monitor the frequencies in the forms of a species over time and/or space. Some years ago I wrote an article aimed at amateur lepidopterists, entitled: 'The importance of form frequency data to ecological genetics' (Majerus 1990). The aim of this article was to encourage lepidopterists to collect form frequency data on British moths and to publish it. The paper was partly written out of frustration. On a number of occasions I have spent considerable time breeding a species of moth to work out the genetics of a form, or collected data on the frequency of a particular form in a specific region, only to receive a letter after publication of the results stating that another lepidopterist had done similar work, giving the same results some time earlier. The frustration is that had the previous work been published I could have saved myself considerable time which could have been devoted to other questions. In respect of form frequency data, it is probable that much data exists in the unpublished record books of lepidopterists throughout Britain. This is likely to be particularly true since the 1950s, when mercury vapour moth traps were developed. Since then, many lepidopterists have regularly run moth traps, diligently recording the numbers of different species caught and the number of each form taken. Such data would be of great value to evolutionary biologists interested in polymorphism, and particularly those interested in melanic polymorphism, for the period from the 1950s to the present covers the time when pollution levels have been declining. Yet, despite the value of such data, little has been published by amateur lepidopterists, and data that are not published, and made available to other scientists, are of no value at all. I have said that the next 20 or 30 years are likely to be a critical period in the history of industrial melanism, for during this period melanic forms of some species may become extinct. Therefore, I urge all readers of this book with an interest in moths to start collecting form frequency data on species which have melanic forms, and to publish such data regularly. And to any lepidopterists who already have such data, please collate it and make it available to the scientific community by publication as soon as possible. There is, perhaps, no other field in which the amateur lepidopterist can contribute more usefully to scientific research with so little effort beyond that which they normally devote to their hobby.

Other lists could have been compiled on different groups of organism. For example, the same type of data would be useful for those species of ladybird that have melanic forms (2 spot, 10 spot, hieroglyphic, striped), although here, through the activities of members of the Cambridge Ladybird Survey, we already have considerable data sets on melanic frequencies (Majerus 1994, 1995). Other groups which would repay attention in Britain include hoverflies (Syrphidae) in which many species show melanism, land snails, particularly of the genus *Cepaea*, and some species of true bug (Hemiptera).

In addition to form frequency data, the genetics of melanism is unknown in a great many species, and here again amateur entomologists can make significant contributions if they breed the insects that they are interested in, as so many now do. All that needs to be done is to keep families separately, to think about which adults should be crossed to produce subsequent generations, to apply the basic genetical rules outlined in Chapter 3 (or in Ford, 1945*a*) to the results, and then to publish said results.

Other types of research where amateurs can make significant contributions, include observations of the basic natural history and ecology of species. Of particular value would be details of the natural resting sites of species, any changes in distribution, particularly where changes in food plant or habitat are involved, any significant causes of mortality, and any records of predation, either through witnessed observation, or examination of bird droppings. It should be stressed that such information may be of value, not only in respect of species showing melanism, but also in those that do not, for the corollary to the question of why melanism has evolved in such and such a species, is why it has not evolved in others, and this corollary has rarely been addressed.

For many of the other studies that are needed, rather more resources and equipment may be required. In addition, some understanding of experimental design and the use of appropriate controls are probably required. Consequently, predation experiments in a range of circumstances, tests to investigate resting site and habitat preferences, studies of the relative fitnesses and the dispersal rates of forms are probably the territory of professional scientists, or the very serious amateur. Again, attention needs to be turned to species of moth and other organisms that have, as yet, received little attention, to assess whether it is reasonable to extrapolate from species such as the peppered moth to other cases.

Having said that one of the main thrusts of future research should be to investigate melanism in species that have previously received little attention, I would also advocate that efforts devoted to the well worked species are increased. The reasons for this are twofold. First, as stated earlier in this chapter, the factors which affect the evolution and maintenance of melanism in any species are extremely complex, with the result that there is no case in which we are confident that we know the whole story. If we are to understand any case fully, it would obviously be most efficient to extend work on those species that we already know most about. The type of work that is still necessary can best be illustrated by pointing to areas of research on the two most well worked examples of melanic polymorphism, the peppered moth and the 2 spot ladybird, that may be addressed using traditional methods.

For the peppered moth in Britain, the crucial studies are the monitoring of the decline in form frequencies over the next 20 or 30 years, more detailed assessment of its natural resting sites; predation experiments on moths placed in natural positions; reassessment of dispersal rates taking wind blown larval dispersal into account; and a whole range of studies on the *insularia* complex, to see where these forms fit into the picture. Any studies of melanism in the peppered moth from other countries would also be of inestimable value, as comparitors with the situation in Britain.

In the 2 spot ladybird, we need to know more about the natural rates of mortality due to predators that find prey by sight; the relationship between temperature and activity in the wild, over a range of climatic conditions, including very hot conditions, and the consequences this has on survival and reproduction; how mating preferences vary both in time and space, and, perhaps most crucially, full details of the genetic control of melanism throughout the species range. Of course, in both these species, further investigations may throw up new insights which may turn the research in directions that have not been considered before.

I now turn my attention to examples of future work which could not have been conducted before, because the requisite technology was not available. Two examples, one essentially genetical, the other ecological, should suffice to give a flavour of the advances that might be made in our understanding.

Taking the genetical case first, in no field of biology has the period since Kettlewell's book on melanism seen more advance. New molecular genetic developments, from the mapping of genes and the sequencing of the first gene, via DNA fingerprinting, the amplification of single strands of DNA using the polymerase chain reaction and mini- and microsatellite DNA analysis, to the development of automated gene sequence technologies, all permit huge amounts of often previously unobtainable data to be collected quickly. These technological advances provide the evolutionary biologist with an array of enormously powerful tools. However, these novel techniques have yet to be applied to the study of melanism in a meaningful way. One crucial investigation would be to find and sequence the gene(s) responsible for melanism in species such as the peppered moth and the 2 spot ladybird. If this was done, as it has been in some bacteria, fungi, fruit-flies, and mammals, a whole range of new avenues would be open for investigation. By taking the sequence of a melanic gene in one species, molecular probes could be constructed to determine whether the genes causing melanism in different populations or different species were the same or different. In the 2 spot ladybird for example, it would be possible to discover whether the supergene that controls most of the colour pattern polymorphism arose as a result of gene duplication, or as a result of selection favouring translocations that brought appropriate genes together into tight linkage.

DNA fingerprinting techniques, which allow paternities to be assigned with more confidence in wild populations than do behavioural observations, could be used to study the effects of mating preferences, and multiple mating, on the relative success of melanics and non-melanics in mating. The same technique could be used in studies of sperm competition. Analysis of the nucleotide sequences of genes in the mitochondria, which in most species is inherited just from females, may allow assessments of gene-flow between populations. The list of potential studies using these recent genetic advances goes on and on, and there is no reason to suppose the next decade or two will not witness further technological improvements, increasing our ability to unravel the mysteries of past evolution with ever greater precision.

For an area in which a recent advance offers the possibility of greater insight into the ecology of melanic organisms, I have chosen the discovery that birds see UV light to which humans are almost completely blind. I have already described the problems this finding poses to the interpretation of experimental work, designed with a human eye, into melanism in moths. One of the problems in including the UV dimension into experiments is that we cannot see it ourselves. However, with the development of reflectance spectrometers covering UV wavelengths, and advances in video cameras and camera filters, the UV component can now be taken into account. Indeed, research projects are currently underway in Bristol and Cambridge to assess the role of UV light in the defensive patterns of insects, including the Lepidoptera.

To it: 'the loving care of its numerous admirers'

I hope that in this book I have shown that while the study of melanism has advanced considerably over the last half century, and that the discoveries made have had far reaching effects on biological understanding, there is still a great deal to be learnt about those species which exhibit this phenomenon. Again, I want to stress two points, which I believe are crucial to the understanding of this phenomenon, and which have often been given rather scant regard in writings on melanism. First, the factors that affect the success, or otherwise, of melanic forms will be many, and they will vary in strength and importance from habitat to habitat and from year to year. Second, the precise causes of, and influences on, melanism in each species in which the phenomenon occurs, will depend on the genetics, behaviour, and ecology of that species, which may in turn be influenced and evolve simply as the result of the evolution and spread of melanism in the species. There has, I believe, been some tendency for workers in this field, myself not excluded, at times to fall into the trap eluded to by Tutt (1891) in the first book specifically on the subject of melanism. There he wrote: 'There are so many probable causes to be worked out in a subject of this kind that one is apt to emphasise some one special condition to the exclusion of others'.

Thus, we should be wary of extrapolating from one species to another, and even from one population of a species to another. This is not a novel warning, for a similar warning was made over a hundred years ago by Lord Walsingham. In his address to the Fellows of the Entomological Society of London in 1890 (cited in Tutt 1891) he discussed explanations of the features of the morphology of Lepidoptera, including melanism, that may be adaptive and be the result of natural selection. In his final paragraph he states:

> It cannot be too freely admitted that in all cases of supposed 'natural selection', accompanied by advantage to the species, such advantage is probably by no means the sole and exclusive cause of, or inducement to selection—all the special conditions under which the species exists must be taken into consideration, and any inclination to overrate the active value of one special condition should carefully be discounted. The study of such supposed causes and effects is yet in its infancy, and although the promising child has 'grown apace' under the loving care of its numerous admirers, it has by no means arrived at maturity; on the other hand, no jealous or disparaging critic can at present be justified in putting it down as an 'ill weed'.

Appendix

Table A1 Details of habitat types at field sites involved in investigations of habitat-related differences in morph frequencies over short distances

Location	Close canopy habitat	Open canopy habitat	Nature of boundary
Linford, Hants.	Douglas fir plantation	Mature oak woodland	Rough scrub, bramble
Juniper Bottom, Box Hill, Surrey	Mature yew woodland	Chalk grassland	Thick mixed hedge
Kidnalls Enclosure, Pillowell, Forest of Dean	Mature Scots pine plantation	Bracken and domestic garden	Bracken
Ynys-Hir, Dyfed	Douglas fir plantation	Mature open oak woodland	Rough scrub, bramble
King's Forest, Suffolk	Mature Scots and Corsican pine plantation	Open grassland with young Scots and Corsican pine and gorse scrub	Mixed open deciduous saplings, mainly sycamore
Springfield, Coton, Cambs.	Mature apple orchard	Rough grassland	Rough grassland
Madingley Wood, Cambs.	Mature oak–ash woodland	Arable farmland	Mature mixed deciduous hedge

Table A2 The percentage frequencies of morphs of the 14 species of Lepidoptera, caught in moth traps set approximately 20 m apart in sharply contrasting habitats. Data on five species showed no morph specific differences in their use of habitats. (Data on these species are given in Majerus in preparation *a*.)

	Close canopy			Open canopy		
	melanic	banded	typical	melanic	banded	typical
Mottled beauty						
Dyfed, 1984	3.4	19.0	77.6	0	0	100
Surrey, 1989	37.1	7.1	55.8	17.0	3.0	80.0
Surrey, 1990	53.7	8.5	37.8	17.8	6.7	75.5
Surrey, 1991	39.8	9.2	51.0	21.0	5.7	73.3
Surrey, 1992	35.3	7.6	57.2	21.4	6.1	72.5
Surrey, 1993	39.2	8.5	52.3	19.5	6.3	74.3
Hants., 1991	21.1	7.9	71.0	6.4	2.4	91.2
Hants., 1992	19.7	11.1	69.2	9.2	3.0	87.8
Hants., 1993	18.2	10.1	71.7	8.2	3.2	88.7
Suffolk, 1991	28.4	7.2	64.4	9.7	5.2	85.2

Table A2 *Continued*

	Close canopy			Open canopy		
	melanic	banded	typical	melanic	banded	typical
Suffolk, 1992	32.5	6.4	61.1	16.4	4.9	78.7
Suffolk, 1993	26.7	6.1	67.2	13.5	4.2	82.3
Dyfed, 1991	1.6	16.3	82.2	0.9	9.1	90.0
Dyfed, 1992	4.7	21.4	73.9	1.2	7.2	91.6
Dyfed, 1993	1.9	18.9	79.2	0	7.4	92.6
Grey pine carpet						
Glos., 1991	42.9	32.2	24.9	34.8	19.9	45.3
Suffolk, 1991	51.2	28.7	20.1	43.9	21.7	34.4
Suffolk, 1992	53.4	31.7	14.8	37.5	23.2	39.4
Suffolk, 1993	61.7	29.4	8.9	39.6	22.7	37.7

	melanic	non-melanic	melanic	non-melanic
Willow beauty				
Surrey, 1989	33.9	66.1	15.8	84.2
Surrey, 1990	35.5	64.5	6.0	94.0
Surrey, 1991	34.2	65.8	12.4	87.6
Surrey, 1992	43.7	56.3	14.8	85.2
Surrey, 1993	32.4	67.6	16.3	83.7
Hants., 1991	21.1	78.9	7.2	92.8
Hants., 1992	24.7	75.3	9.4	90.6
Hants., 1993	20.1	79.9	11.1	88.9
Suffolk, 1991	30.5	69.5	14.1	85.9
Suffolk, 1992	32.5	67.5	11.7	88.3
Suffolk, 1993	34.1	65.9	18.4	81.6
Dyfed, 1991	8.2	91.8	1.3	98.7
Dyfed, 1992	6.4	93.6	5.2	94.8
Dyfed, 1993	9.1	90.9	3.9	96.1
Tawny-barred angle				
Dyfed, 1984	76.9	23.1	12.5	87.5
Hants., 1991	8.4	91.6	0	100
Hants., 1992	7.2	92.8	0	100
Dyfed, 1991	64.8	35.2	18.2	81.8
Dyfed, 1992	70.2	29.8	14.1	85.9
Dyfed, 1993	61.9	38.1	9.3	90.7
Pale brindled beauty				
Cambs., 1991	28.4	71.6	16.3	83.8
Cambs., 1992	21.5	78.5	17.1	82.9
Cambs., 1993	30.2	69.8	19.2	80.7
Hants., 1991	18.8	81.2	13.5	86.5
Hants., 1992	14.1	85.9	13.9	86.0
Dotted border				
Hants., 1989i	12.7	87.3	1.5	98.5
Hants., 1989ii	14.3	85.7	2.9	97.1
Hants., 1991	13.1	86.9	3.1	96.9
Hants., 1992	11.0	89.0	1.2	98.8

Table A2 *Continued*

	Close canopy		Open canopy	
	melanic	non-melanic	melanic	non-melanic
Cambs., 1991	9.2	90.9	2.0	98.0
Cambs., 1992	8.6	91.4	2.8	87.2
Cambs., 1993	10.4	89.6	1.4	98.6
Common pug				
Cambs., Aug., 1991	86.9	13.1	79.3	20.7
Cambs., Spring, 1992	92.2	7.8	75.6	24.4
Cambs., Aug., 1992	88.7	11.3	80.6	19.4
Engrailed				
Cambs., 1991	15.2	84.8	4.9	95.1
Cambs., 1992	19.8	80.2	7.3	92.7
Dyfed, 1992	26.8	73.2	14.2	85.8
Dyfed, 1993	31.2	68.8	12.7	87.3
Surrey, 1992	39.3	60.7	16.7	83.3
Small engrailed				
Surrey, 1991	23.1	76.9	15.8	84.2
Surrey, 1992	29.4	70.6	12.7	87.3
Surrey, 1993	21.7	78.3	13.2	86.8
Cambs., 1991	18.6	81.4	7.3	92.7
Cambs., 1992	21.4	78.6	6.4	93.6
Dyfed, 1992	7.2	92.8	8.1	91.9
Dyfed, 1993	6.4	93.6	0	100
Riband wave				
Surrey, 1990	34.6	65.4	11.5	88.5
Surrey, 1991	31.4	68.6	17.1	82.9
Surrey, 1992	27.7	72.3	15	85
Surrey, 1993	25.9	74.1	18.2	81.8
Cambs., 1992	26.7	73.3	21.4	78.6
Cambs., 1993	29.4	70.6	17.9	82.1
Common marbled carpet				
Glos., 1991	44.6	55.4	22.9	78.1
Cambs., 1992	79.2	20.8	69.6	30.4
Cambs., 1993	75.8	24.2	69.2	30.8
Pine carpet				
Glos., 1991	43.8	56.2	30.2	69.8
Large yellow underwing				
Glos., 1991	59.1	40.9	0	100
Cambs., 1992	46.2	53.8	41.7	58.3
Lunar underwing				
Cambs., 1991	15.5	84.5	8.4	91.6
Cambs., 1992	18.1	81.9	8.7	91.3

Table A3 Resting positions of 14 species of moth recorded in the wild since 1966. (Data on additional species given in Majerus in preparation b.)

Species	Exposed on trunks	Hidden on trunks	Under branches	Above branches	Twigs	Tree foliage	Ground herbage	Other	Total
Mottled beauty	217	31	62	5	8	24	17	14	428
Willow beauty	189	41	53	2	14	31	30	7	367
Tawny-barred angle	27	6	9	0	9	20	27	3	101
Pale brindled beauty	51	27	26	1	13	1	1	2	122
Dotted border	10	7	19	4	10	1	12	6	69
Common pug	4	1	2	1	3	11	5	1	28
Engrailed	130	49	54	2	26	22	4	4	291
Small engrailed	104	23	41	5	31	18	7	2	231
Riband wave	20	4	9	0	3	22	31	3	65
Common marbled carpet	31	6	12	1	11	10	16	2	89
Grey pine carpet	44	16	16	1	9	9	2	1	98
Pine carpet	18	4	16	0	7	13	2	0	60
Large yellow underwing	1	5	0	0	0	3	104	87	200
Lunar underwing	0	0	0	0	0	2	19	2	23

Table A4 Resting positions taken up by moths released at, or soon after, dawn. Moths were released from pill boxes in one of four habitats: close canopy coniferous woodland with little ground vegetation (CCCW), open deciduous woodland with mixed forbe ground vegetation (PDW), apple orchard with mixed herb ground vegetation (AO) and chalk grassland (CG). The moths released were either laboratory bred (LB) (approximately equal numbers of males and females used), or moth trap caught and retained for 24 hours (MTC) (only males used).

	Habitat	Source	Trunks	Branches	Twigs	Foliage	Ground	Lost	Total
Mottled beauty	CCCP	LB	32	15	12	5	1	38	103
	ODW	LB	43	30	17	12	3	34	139
	CCCP	MTC	56	26	20	8	2	41	153
	ODW	MTC	89	51	27	17	7	58	249
	CG	MTC	74	35	29	17	15	38	208
Willow beauty	CCCP	LB	58	21	7	5	0	48	139
	ODW	LB	72	27	13	9	10	71	202
	CCCP	MTC	64	29	9	6	1	60	169
	ODW	MTC	61	9	19	10	13	38	160
	CG	MTC	73	27	13	11	39	52	215
Tawny-barred angle	CCCP	MTC	39	10	7	15	0	21	92
	ODW	MTC	32	7	13	11	8	30	101
Pale brindled beauty	CCP	MTC	53	11	12	5	0	18	99
	ODW	MTC	138	48	29	3	1	47	266
	AO	MTC	51	39	27	2	4	13	136
Dotted border	CCCP	MTC	92	38	27	4	23	21	205
	ODW	MTC	128	34	39	3	59	28	300
Common pug	CCCP	LB	20	14	7	2	1	7	51
	ODW	LB	16	17	13	10	9	9	74
	CCCP	MTC	22	12	9	3	1	4	51
	ODW	MTC	25	19	15	17	5	7	88

Engrailed	CCCP	MTC	61	22	12	13	0	26	134
	ODW	MTC	79	37	26	24	7	35	208
Small engrailed	CCCP	MTC	38	22	7	5	0	31	103
	ODW	MTC	41	36	19	18	3	27	144
Riband wave	CCCP	MTC	21	18	7	13	1	20	80
	ODW	MTC	31	17	10	26	11	15	110
	AO	MTC	12	23	17	12	18	17	99
	CG	MTC	8	6	4	8	29	21	76
Common marbled carpet	CCCP	MTC	26	9	10	12	2	7	66
	CG	MTC	11	17	24	29	24	21	126
Grey pine carpet	CCCP	MTC	42	11	8	10	3	21	95
	CG	MTC	19	25	14	28	23	29	138
Pine carpet	CCCP	MTC	21	8	7	3	1	17	57
	CG	MTC	7	16	13	19	21	39	115
Large yellow underwing	ODW	MTC	8	4	2	5	227	58	304
	AO	MTC	3	1	1	7	194	52	258
Lunar underwing	AO	MTC	2	1	1	3	89	12	108

Table A5 Resting positions taken up by different morphs of four species of moth released at, or soon after, dawn, at the boundary between two contrasting habitat types. (Moths that failed to fly are omitted.)

Surrey		Yew woodland				Boundary vegetation	Chalk grassland			Lost	Tatal
		Tk	Br	F	G		Tr	Sh	Gf		
Mottled beauty	melanic	58	21	17	1	13	4	2	2	29	147
	banded	32	18	14	0	10	5	3	0	20	102
	typical	39	24	18	2	31	15	10	3	60	202
Willow beauty	melanic	28	12	3	1	6	2	1	0	17	70
	intermediate	63	16	3	3	19	15	10	3	38	170
	typical	64	20	7	2	32	30	15	10	68	248
Riband Wave	banded	19	17	22	8	22	2	2	31	21	144
	typical	12	12	17	9	31	6	9	43	32	171

Dyfed		Douglas fir				Boundary vegetation	Open deciduous				Lost	Total
		Tk	Br	F	G		Tk	Br	F	Gf		
Mottled beauty	melanic	39	6	10	5	3	6	5	5	2	21	102
	banded	43	8	18	3	1	7	5	4	1	33	123
	typical	28	14	17	3	8	29	20	13	4	73	209
Tawny-barred angle	melanic	16	31	46	10	21	2	2	2	0	29	159
	typical	10	14	24	7	31	16	17	10	10	38	177

Tk, tree trunks; Br, tree branch; F, tree foliage; Tr, trees in chalk grassland; Sh, shrubs in chalk grassland; G, ground in conifer woodland; Gf, ground foliage.

Table A6 The background choices made by various morphs of four species of moth when offered a choice of equal areas of black or white card substrate.

Form and origin	Black	White	Across black/white boundary	Total
Mottled beauty (ex Surrey)				
melanic (f. *nigra*, f. *fuscata*)	68	19	23	110
banded (f. *conversaria*)	58	15	34	107
typical	71	20	27	118
Mottled beauty (ex New Forest)				
melanic (f. *nigra*, f. *fuscata*)	71	18	21	110
banded f. *conversaria*	59	25	28	112
typical	64	15	20	99
Mottled beauty (ex Dyfed)				
melanic (f. *nigericata*)	34	9	12	55
banded (f. *conversaria*)	71	20	37	128
typical	84	22	34	140
Willow beauty (ex Cambridge)				
melanic (f. *rebeli*, f. *perfumaria*)	79	30	18	127
typical	53	38	20	111
Willow beauty (ex Surrey)				
melanic (f. *rebeli*, f. *perfumaria*)	46	22	15	83
typical	30	25	13	68
Tawny-barred angle (ex Dyfed)				
melanic (f. *nigrofulvata*)	53	26	13	92
typical	86	27	18	131
Tawny-barred angle (ex Staffs)				
melanic (f. *nigrofulvata*)	38	17	12	67
typical	44	29	15	88
Riband wave (ex Surrey)				
banded (f. *aversata*)	29	52	28	109
typical (f. *remutata*)	28	69	16	113

Table A7 (a) Summary of recapture results following releases of different morphs of seven species of moth at fixed points along the boundary between contrasting habitat types. Equal numbers of each type of moth were released at each release point.

Species	Morph	Number released	Number recaptured				Total	
			YA	YB	CGA	CGB	Y	CG
Mottled beauty	melanic	381	49	17	8	11	66	19
	banded	171	16	10	7	7	26	14
	typical	843	37	29	42	30	66	72
Willow beauty	melanic	369	45	19	18	10	64	28
	intermediate	669	82	39	34	23	121	57
	typical	948	77	29	86	38	106	124
Riband wave	banded	222	31	12	20	6	43	26
	non-banded	846	72	30	75	18	102	93
Engrailed	melanic	111	14	6	10	2	20	12
	typical	297	15	9	23	8	24	31

Species	Morph	Number released	CCCA	CCCB	ODA	ODB	Total	
							CCC	OD
Tawny-barred angle	melanic	267	46	17	9	1	63	10
	typical	492	79	37	41	13	116	54
Dotted border	melanic	123	13	4	15	9	17	26
	typical	876	54	17	68	31	71	99
Common pug	melanic	165	23	10	17	12	33	29
	typical	102	14	7	14	9	21	23

Y, yew woodland; CG, chalk grassland; CCC, close canopy conifer woodland; OD, open deciduous woodland.
A, traps 10 m from boundary; B, traps 30 m from boundary.

(b) Summary of recapture results following releases of equal numbers of each of two or three morphs of seven species of Lepidoptera in each of two contrasting habitats.

Species	Total released	Number recaptured					
		In close canopy habitat			In open canopy habitat		
		melanic	banded	typical	melanic	banded	typical
Mottled beauty	900	38	32	17	8	11	27
Willow beauty	1200	melanic 49	intermediate 42	typical 28	melanic 13	intermediate 21	typical 33
Riband wave	800	banded 29	non-banded 20		banded 18	non-banded 27	
Engrailed	600	melanic 21	typical 18		melanic 16	typical 29	
Tawny-barred angle	800	57	36		11	20	
Dotted border	1200 (male)	52	48		47	45	
Common pug	800	21	30		26	24	

Table A8 Results of predation experiments in which two or more morphs of eight species of British Lepidoptera were placed out in 'natural' resting positions, for 48 hours.

	Melanic		Banded		Typical	
	Eaten	Not eaten	Eaten	Not eaten	Eaten	Not eaten
Mottled beauty						
Dyfed, close canopy	17	29	20	26	39	48
Dyfed, open canopy	24	27	18	22	30	40
Hants, close canopy	26	38	13	31	30	41
Hants, open canopy	31	33	24	27	36	42
Suffolk, close canopy	23	49	14	26	30	45
Suffolk, open canopy	31	40	20	22	39	47
Surrey, close canopy	68	138	27	58	82	130
Surrey, open canopy	95	106	46	39	116	107

	Melanic		Non-melanic	
	Eaten	Not eaten	Eaten	Not eaten
Willow beauty				
Dyfed, close canopy	21	36	19	38
Dyfed, open canopy	27	30	29	28
Hants, close canopy	16	34	24	26
Hants, open canopy	18	32	20	30
Surrey, close canopy	17	33	15	35
Surrey, open canopy	31	19	37	13
Tawny-barred angle				
Dyfed, close canopy	30	70	38	62
Dyfed, open canopy	48	52	43	57
Dotted border				
Hants, close canopy	30	70	46	54
Hants, open canopy	53	47	48	52
Suffolk, close canopy	21	29	23	27
Suffolk, open canopy	26	24	24	26
Common pug				
Suffolk, close canopy	13	87	18	82
Suffolk, open canopy	16	84	17	83
Engrailed				
Suffolk, close canopy	22	28	20	30
Suffolk, open canopy	31	19	24	26
Small engrailed				
Suffolk, close canopy	31	69	38	62
Suffolk, open canopy	37	63	31	69

	Banded		Non-Banded	
	Eaten	Not eaten	Eaten	Not eaten
Riband wave				
Surrey, close canopy	10	40	8	42
Surrey, open canopy	7	43	14	36

Table A9 Summary of results of multiple mark–release–recapture experiments, to obtain estimates of fitness and longevity, for four species of moth showing habitat-related differences in morph frequency, but not showing morph specific differences in dispersal rates across habitat boundaries.

Species	Form	Location	Habitat	Estimated longevity (days)	Fitness
Dotted border	melanic	Cambs	ODW	7.4	1
	non-melanic	Cambs	ODW	7.2	1.03
	melanic	Hants	CCCP	6.9	0.95
	non-melanic	Hants	CCCP	7.3	1
	melanic	Hants	ODW	7.1	0.97
	non-melanic	Hants	ODW	7.3	1
Small engrailed	melanic	Surrey	CCCP	5.3	1.29
	non-melanic	Surrey	CCCP	4.1	1
	melanic	Surrey	CG	4.7	0.87
	non-melanic	Surrey	CG	5.4	1
Riband wave	banded	Surrey	CCCP	6.3	0.94
	non-banded	Surrey	CCCP	6.7	1
	banded	Surrey	CG	6.4	0.98
	non-banded	Surrey	CG	6.5	1
Common pug	melanic	Suffolk	CCCP	4.2	0.98
	non-melanic	Suffolk	CCCP	4.3	1
	melanic	Suffold	ODW	4.6	1.05
	non-melanic	Suffolk	ODW	4.4	1

Glossary

Abdomen
The hindmost of the three main body divisions of an insect.

Adaptation
Any characteristic of an organism that enables it to cope better with conditions in its environment.

Adaptive landscape
A hypothetical three-dimensional representation of the interaction between genotype frequencies and fitness, in which the vertical axis is fitness. High points are called adaptive peaks and represent regions where all small changes in genotype result in lowered mean fitness of individuals, while valleys are regions of low fitness. Evolution is portrayed by the movement of a population uphill across an ever-changing landscape, searching for adaptive peaks with its progress being barred by the valleys. An adaptive landscape is very much a conceptual tool rather than an attempt to model reality.

Aestivation
Summer dormancy, when conditions become unfavourable.

Allele (allelomorph)
Form of a gene. Genes are considered alleles of one another when they occur in the same positions on the members of a chromosome pair (homologous chromosomes), have different effects in respect of a particular characteristic, and can mutate one to another.

Amino acids
The basic building blocks of proteins.

Arms race
A coevolutionary pattern involving two or more interacting species (e.g. predator and prey or parasite and host), where evolutionary change in one party leads to the evolution of a counter adaptation in the other.

Artificial selection
Human attempts to exaggerate natural traits by breeding selectively from those organisms which show these traits most strongly.

Autosome
A chromosome other than a sex chromosome.

Balanced polymorphism
A genetic polymorphism in which the various forms are maintained in the population at constant frequencies, or their frequencies cycle in a more or less regular manner, and alleles are maintained above frequencies that could be maintained by recurrent mutation.

Balanced view of the genome
View that most sexual species contain many polymorphic loci, that their genomes exhibit high levels of heterozygosity, and that much of this variation is maintained by a balance of selective advantage and disadvantage. This view was held by many selectionists during the first half of the twentieth century.

Basal
Concerning the base of a structure, that part nearest to the body.

Bilateral mosaic
An individual in which one-half displays one phenotype, the other half displaying a different phenotype.

Biparental inheritance
Genes and genetic elements in sexual organisms which are inherited from both parents (e.g. most nuclear genes) (cf. uniparental inheritance).

Centromere
The region on a chromosome that becomes attached to the spindle during cell division.

Chromosomal fission
The splitting into two parts of a single chromosome. If both parts are to be maintained, a new centromere must arise.

Chromosomal fusion
The unification of two chromosomes into a single chromosome with the loss of a centromere.

Chromosomes
Small elongated bodies, consisting largely of DNA and protein, in the nuclei of most cells, existing generally in a definite number of pairs for each species, and accepted to be the carriers of most hereditary qualities.

Classical view of the genome
View that the genomes of most sexual species are composed of monomorphic loci, with polymorphic loci being relatively rare, these polymorphic loci being the result of occasional mutations which are either being eliminated from the population, or are spreading through the population as a result of selection.

Cline
A gradual change, within an interbreeding population, in the frequencies of different genotypes, or phenotypes, of a species.

Clone
An organism that has been produced as a result of mitotic rather than meiotic cell division, that is to say, reproduction has been asexual, not sexual.

Codon
A sequence of three nucleotide bases along a DNA or RNA chain which represents the code of a single amino acid.

Crossing-over
An interchange of groups of genes between the members of a homologous pair of chromosomes.

Cryptic coloration
When the colours of an animal match those of the environment in which it lives.

Cytoplasm
The living substance of the cell, excluding the organelles within it.

Cytoplasmic genes
Genes located in the cytoplasm of cells rather than in the nucleus, usually applied to those associated with mitochondria and chloroplasts.

Deoxyribonucleic acid
The principal heritable material of all cells. Chemically it is a polymer of nucleotides, each nucleotide subunit consisting of the pentose sugar 2-deoxy-D-ribose, phosphoric acid and one of the five nitrogenous bases adenine, cytosine, guanine, thymine, or uracil.

Direct selection
In sexual selection when the mating choice made by an individual (usually the female) leads to an increase in the number of progeny it raises during its lifetime.

Directional selection
Selection in which individuals at one end of the distribution of a trait showing continuous variation are at an advantage over other individuals.

Disassortative mating
A preference to mate with individuals having a different genotype to ones own.

Disruptive selection
Selection in which individuals at both extremes of the distribution of a trait showing continuous variation are at an advantage over intermediate individuals.

DNA = Deoxyribonucleic acid.

DNA fingerprinting
The first genetic method capable of identifying individuals uniquely, based on minisatellite DNA sequences and invented by Professor Alec Jeffreys. Often used more loosely to refer to any genetic method which identifies individuals with high confidence.

Dominant
The opposite of recessive. The stronger of a pair of alleles, expressed as fully when present in a single dose (i.e. heterozygous) as it is when present in a double dose (i.e. homozygous).

Dorsal
Concerning the back or upperside of an animal.

Ecdysis
The moulting process, by which an insect changes its outer coat.

Eclosion
The metamorphosis of a pupa into an adult insect.

Effective population size
The effective size of the population is the size of an ideal population which would exhibit equivalent properties with respect to genetic drift.

Electrophoresis
A technique used to detect differences in proteins and polypeptide chains based on differences in electrical charge of the molecules.

Elytra
The hard and horny front wings of a beetle (singular elytron).

Endoparasite
A parasite which lives inside the body of its host.

Endosymbiotic
Organisms which live within the cell or body of another.

Enzyme
A protein that initiates or facilitates a chemical reaction between other substances.

Epidermis
The single layer of living cells which underlies and secretes the cuticle.

Epistasis
In terms of phenotype, epistasis is an interaction between two different genes such that the epistatic gene interferes with or blocks the expression of the other (which is referred to as the hypostatic gene). Also used in connection with fitness, where the relative fitnesses of the different genotypes at one locus depend upon the background at another.

Eukaryotes
Single or multicellular organisms in which the cell(s) contain a well defined nucleus containing multiple linear chromosomes and which is enclosed by a nuclear membrane.

Evolution
A cumulative, inheritable change in a population.

Exoskeleton
The external skeleton of an insect, made of cuticle.

F_1, F_2, etc.
First, or subsequent generation. Abbreviation for first filial generation, second filial generation, etc.

Fecundity
The number of eggs or offspring produce by an organism.

Female choice
One of Darwin's two mechanisms of sexual selection. The phenomenon whereby females exercise choice over which of two or more available males they mate with.

Fisherian mechanism of sexual selection
The spread of a female choice gene because of genetic correlation with the preferred male trait, the spread initially owing to a naturally selected advantage, and later being associated with a sexually selected advantage.

Fitness
The relative ability of an organism to transmit its genes to the next generation.

Fitness peaks
Genotype–environment combinations in which fitness is maximised such that any small genetic perturbation results in selection for a return to the original genotype (cf. adaptive landscape).

Founder effect
The effect on genetic variability of the small size of a colonising, or founding, population, which may comprise just one or a few individuals, and consequently contains only a small fraction of the genetic variation of the parental population.

Frequency dependent selection
Selection in which the fitness of a genetic form is dependent on the frequency of that form in the population relative to others.

Gamete
Reproductive cell.

Gametogenesis
Formation of male and female gametes or sex cells.

Gene
An hereditary factor or heritable unit which can transmit a characteristic from one generation to the next. Composed of DNA and usually situated in the thread-like chromosomes in the nucleus.

Gene conversion
A non-reciprocal interaction between homologous chromosomes which has the effect of replacing a sequence on one homologue with a similar sequence from the other.

Gene flow
Genetic exchange between populations resulting from the dispersal of gametes, zygotes, or individuals.

Genetics
The study of inheritance.

Genetic bottleneck
The reduction in genetic variability in a population resulting from a period of low population size.

Genetic drift
Random changes of the frequency of genes in populations.

Genetic linkage
The tendency for certain genes to be inherited together, instead of assorting independently, because they are situated close to one another on the same chromosome.

Genetic markers
Any allele (or sequence) that is used experimentally to identify a sequence, gene, chromosome, or individual.

Genetic polymorphism
The occurrence together in the same population, at the same time, of two or more discontinuous, heritable forms of a species, the rarest of which is too frequent to be maintained merely by recurrent mutation.

Genome
The entire complement of genetic material in a cell. In eukaryotes this word is sometimes used in referring to the material in just a single set of chromosomes.

Genotype
The genetic make-up of an individual in respect of one genetic locus, a group of loci or even its total genetic complement.

Germ cells
Those cells destined to become the reproductive cells.

Gradualists
Evolutionists who took the view that most evolutionary change resulted from the action of selection on heritable continuous variation exhibited by individuals in populations (cf. mutationists).

Habitat
The specific place and type of local environment that an organism lives in.

Habitat selection
Exhibited by an organism that preferentially resides in one type of habitat, rather than others.

Haemolymph
Insect blood.

Haploid
Of cells (particularly gametes) or individuals having a single set of chromosomes.

Hardy–Weinberg equilibrium
The state in which genotype frequencies in a population are in accord with expectations of the Hardy–Weinberg law.

Hardy–Weinberg law
A basic principle of population genetics which allows genotypic frequencies to be predicted from gene frequencies. The law states that in an effectively infinite population of sexually reproducing, randomly mating, diploid organisms containing a gene with two alleles having frequencies p and q, the frequencies of the homozygote genotypes will be p^2 and q^2 and that of the heterozygote will be 2pq within a single generation, and will not change thereafter in the absence of mutation, migration, and selection.

Hermaphrodite
An individual that bears some tissues that are identifiably male and others that are female, and that can produce mature male and female gametes.

Heritability
The proportion of the total variation in a trait that is due to genetic rather than environmental factors.

Heterogametic sex
The sex which carries sex chromosomes of different types and thus produces gametes of two types with respect to the sex chromosome they carry. In humans, males are the heterogametic sex carrying one X and one Y chromosome.

Heterozygote
An individual which bears different alleles of a particular gene, one from each parent.

Heterozygote advantage
The situation in which the fitness of heterozygote individuals exceeds that of either type of homozygote.

Hitch-hiking gene
Any gene that increases in frequency as a consequence of being associated by genetic linkage with a selectively advantageous gene, rather than as a consequence of its own fitness effect on its bearer.

Homologous
Genes or other sequences are said to be homologous if the similarities they share result from sharing a common ancestor.

Homozygote
An individual that bears two copies of the same allele at a given gene locus.

Homogametic sex
The sex which carries sex chromosomes that are same. In humans, females are the homogametic sex carrying two X chromosomes.

Hormone
Any type of chemical messenger responsible for the timing and regulation of metabolic, behavioural, or other processes.

Host
An organism which is being attacked by a parasite or parasitoid.

Host plant
A plant on which an animal feeds, or on which it resides.

Hybrid
Individual resulting from a mating between parents that are genetically unalike. Often used to describe the offspring of matings between the individuals of two different species.

Hypostasis
When an allele has its expression blocked by action of an allele at another locus. The opposite of epistasis.

Imago
The adult stage in the insect life.

Inbreeding
Reproduction between close relatives.

Inbreeding depression
The reduction in fitness of offspring from matings between close relatives, resulting from the increased likelihood of the same deleterious recessive genes being inherited from both parents and thus being expressed in their progeny.

Inclusive fitness
The fitness of an individual plus the effect that individual has on the fitness of other individuals, weighted by their relatedness to them.

Instar
The stage in an insect's life history between two moults. An insect which has recently hatched from an egg and which has not yet moulted is said to be the first instar nymph or larva. The adult (imago) is the final instar.

Inversion
Of a chromosome, in which a segment of the chromosome has been reversed.

Karyotype
The chromosome complement, in terms of number, size, and constitution, of a cell or an organism.

Keratin
Any of a group of nitrogenous substances that form the basis of hair, feathers, horns, nails, etc.

Kin selection
Selection acting on an individual in favour of survival not of that individual, but of its relatives which carry many of the same genes.

Lamarckian inheritance
The inheritance of characteristics of an organism, acquired as a result of environmental influence on the organism, during the organism's life-time.

Larva
The second stage in the life cycle of insects that undergo full metamorphosis.

Lek
A site where males of a species congregate in dense groups, and which are visited by females for the sole purpose of mating.

Lethal gene
A gene which, if expressed, is fatal to the individual bearing it before the individual reaches reproductive maturity

Locus
The position of a gene on a chromosome.

Male competition
One of Darwin's two mechanisms of sexual selection. Males compete among themselves (e.g. by aggressive displays or fighting) to gain access to females.

Mandibles
A pair of laterally moving horny jaws of an insect.

Meiosis
The process of two nuclear divisions and a single replication of the chromosome complement of a cell by which the number of chromosomes in resultant daughter cells is reduced to one-half during gamete production.

Meiotic spindle
A collection of microtubular fibres to which chromatids attach (at the centromere) during meiosis and are involved in the segregation of chromatids into daughter cells.

Melanism
Having dark or black pigmentation.

Melanocyte
A melanin bearing colour cell.

Messenger RNA
Ribonucleic acid molecules, produced during genetic transcription of DNA, and acting as a template for protein synthesis.

Metamorphosis
The changes that take place during an insect's life as it turns from an egg to an adult.

Metaphase plate
The equatorial region of a dividing cell where the chromosomes line up during metaphase.

Microsatellite
A block of short tandemly repeated motifs in which the basic repeat unit is five or fewer bases in length.

Mimicry
The resemblance shown by one organism to another for protective, or, more rarely, aggressive purposes.

Mimicry, Batesian
The resemblance shown by one species (the mimic) to another (the model) that is better protected by poisonous or distasteful qualities or active defence (sting, bite, etc.).

Mimicry, Müllerian
Resemblance of two or more species each of which has characteristics which are unpleasant to predators. The resemblance reduces the destruction of members of a Müllerian mimetic complex while inexperienced predators learn of their unpleasant properties.

Minisatellite
Any of a group of many dispersed arrays of short tandemly repeated motifs. The family is to an extent unified by the presence of a so-called core sequence, possibly related to the bacterial recombination signal chi, in every repeat unit. Minisatellites can exhibit extreme length variation due to the frequent gain and loss of repeat units.

Mitochondrial DNA
The circular, double-stranded genome of eukaryotic mitochondria.

Mitosis
The normal process of cell division in growth, involving the replication of chromosomes, and the division of the nucleus into two, each with the identical complement of chromosomes to the original cell.

Modifier genes
A series of genes, each with small effects, which influences the exact expression of a major gene.

Molecular clock
A theoretical clock based on the premise that the rates at which nucleotide (or amino acid) substitutions become fixed in evolutionary lineages is approximately constant for a given DNA sequence (or polypeptide chain) and reflects the time since taxa diverged.

Monogamy
Situation in which an individual has a single mating partner over a set time period.

Monomorphic
The opposite of polymorphic, a sequence, gene, or organism in which all individuals in a given sample are indistinguishable.

Morphology
The form and structure of an organism.

Mosaic
The presence in an individual of areas of two genetically different patterns.

mtDNA = Mitochondrial DNA.

Multiple alleles
A series of forms of a gene occurring at the same locus on a chromosome which have arisen by mutation.

Mutation
A sudden change in the genetic material controlling a particular character or characters of an organism. Such a change may be due to a change in the number of chromosomes, to an alteration in the structure of a chromosome, or to a chemical or physical change in an individual gene.

Mutationists
Evolutionists who took the view that most adaptations and species arose as the result of mutations (cf. selectionists, gradualists).

Natural selection
According to Charles Darwin, the main mechanism giving rise to evolution. The mechanism by which heritable traits which increase an organism's chances of sur-

vival and reproduction are more likely to be passed on to the next generation than less advantageous traits.

Neo-Darwinian theory/Neo-Darwinian synthesis
A development of Darwin's evolutionary theory refined by incorporating modern biological knowledge, particularly Mendelian genetics, during the mid-20th century.

Neutral mutation
A mutation that is selectively neutral, i.e. it has no adaptive significance.

Normal distribution
The spread of measurements of a trait that falls symmetrically in a bell-shape either side of the mean value.

Nucleotide
The structural unit of a nucleic acid.

Nucleotide base
The structural unit of a nucleic acid. The major nucleotide bases in DNA are the purines adenine and guanine and the pyrimidines cytosine and thymine, the last of which is replaced by uracil in RNA.

Nucleus
That portion of a cell which contains the chromosomes.

Ovipositor
The egg-laying apparatus of a female insect.

Parasite
An organism which lives in or on another organism, its host, obtaining resources at the host's expense.

Parasitoid
An organism (usually an insect) that lays its eggs inside another insect species, where the parasitoid develops, ultimately killing its host.

Parthenogenesis
Reproduction in which eggs develop without fertilisation.

Pathogen
Any disease causing organism, but most usually used for micro-organisms.

PCR = Polymerase chain reaction.

Phenotype
The observable properties of an organism, resulting from the interaction between the organism's genotype and the environment in which it develops.

Pheromone
Chemical substance which when released or secreted by an animal influences the behaviour or development of other individuals of the same species.

Phylogenetic
Relating to the pattern of evolutionary descent.

Phylogenetic tree
A diagrammatic representation of genetic distances between populations, species, or higher taxa, the branching of which is said to resemble a tree.

Phytophagous
Feeding on plants.

Pleiotropy
The situation in which a single gene effects two or more apparently unrelated phenotypic traits of an organism.

Polygamy
When organisms of either or both sexes have more than one mating partner within a single breeding season.

Polygenic
Of traits, whose phenotypic expression is controlled by many genes, none of which has an over-riding effect.

Polygenic trait
A character controlled polygenically is affected by a large number of genes each of which has a small effect.

Polymerase chain reaction
A method of amplifying specific DNA sequences by means of repeated rounds of primer-directed DNA synthesis.

Polymorphism
The occurrence of two or more distinctly different forms of a species in the same population.

Polyphagous
Feeding on a range of different types of foods.

Polyploid
Of individuals or cells, having more than two complete chromosome sets (e.g. triploid = three sets, tetraploid = four sets, hexaploid = six sets, etc.).

Population
A group of organisms of the same species living and breeding together.

Post-zygotic reproductive isolation mechanism
Any characteristic of an organism that prevents gene flow between it and members of other species by causing sterility, decreased viability, and/or decreased fitness in zygotes formed as a result of fertilisation between gametes of the organism and those of another species.

Pre-zygotic reproductive isolation mechanism
Any characteristic of an organism that prevents gene flow between it and members of other species by reducing the likelihood of zygote formation, usually

by decreasing the likelihood of copulation between the organism and a member of another species.

Primary sex organs
The gonads (testes and ovaries), their ducts, and associated glands.

Prokaryotes
Single-celled organisms which lack a membrane bound nucleus or any membrane bound organelles.

Pronotum
The dorsal surface of the first thoracic segment.

Proteins
Complex nitrogenous compounds whose molecules consist of numerous amino acid molecules linked together.

Pupa
The third stage in the life history of insects undergoing complete metamorphosis, during which the larval body is rebuilt into that of the adult.

Random genetic drift
Fluctuation in the frequencies of neutral genes and neutral alleles in a population due the fact that each generation is only a sample of the one it replaces.

Recessive
A recessive allele is not expressed phenotypically when present in a heterozygote, but only when in a homozygote. The opposite of dominant.

Recombination
The generation of new combinations of chromosomes and parts of chromosomes during meiosis through both crossing-over and the independent segregation of chromosomes.

Reflex bleeding
The defensive secretion produced by ladybirds (and some other insects), when threatened, from pores in the exoskeleton. The secretion consists of a bitter fluid, including haemolymph and noxious chemicals.

Replication
The duplication of genomic DNA, or RNA, as part of the processes of mitosis and meiosis.

Reproductive isolation
The situation in which individuals in a group of organisms breed with one another, but not with individuals of other groups. In sexually reproducing organisms true species are reproductively isolated from one another.

Ribonucleic acid
A polynucleotide consisting of a chain of sugar and phosphate units to which are attached various nitrogenous bases (adenine, cytosine, guanine, and uracil). This macromolecule has many diverse functions where DNA is the reference library,

RNA includes a range of individual books, summaries, pamphlets, and photo-copied sheets.

Ribosomes
Multicomponent structures, or organelles, found in all cells, that have a role in transcription, acting as a site of protein synthesis.

RNA = Ribonucleic acid.

Sampling error
The random statistical deviations that occur when drawing measurement infor-mation from a sample that is supposedly representative of a larger set.

Scutellary spot
The spot at the base of the elytra surrounding and covering the scutellum in some species of coccinellid.

Scutellum
The third of the major divisions of the dorsal surface of a thoracic segment; usually only obvious in the mesothorax. In coccinellids it is visible as a small tri-angular plate at the base of the elytra.

Secondary sexual characters
Characters which differ between the sexes, excluding the primary sex organs.

Segregation
The separation from one another of the pairs of genes constituting allelomorphs, and their passage, respectively, into different reproductive cells.

Selectionists
Evolutionists who took the view that most evolutionary change resulted from the action of natural selection on small heritable differences between the individuals in populations (cf. *mutationists*).

Sex chromosome
A chromosome which is present in a reproductive cell (or gamete) and which carries the factor for producing a male or female offspring. Such chromosomes are usually denoted by the letters X and Y. In humans females carry two X chromosomes, and males carry one X and one Y chromosome.

Sex-limited inheritance
That due to genes, situated in any chromosome, which are expressed in one sex only, although transmitted by both.

Sex linkage
The association of characters with sex, because the genes controlling them are situated in the sex chromosomes.

Sexual dimorphism
The existence, within a species, of morphological differences between the sexes.

Sexual selection
Selection which promotes traits that will increase an organism's success in mating and ensuring that its gametes are successful in fertilisation. This is distinct from natural selection which acts simply on traits which influence fecundity and survival.

'*Sexy sons*'
According to Fisher's theory of the evolution of female choice, females who choose to mate with males carrying a particular trait will produce 'sexy sons'; progeny who tend to carry the preferred trait and so be attractive to other females with the same mating preference.

Sibling
Brother or sister.

Speciation
The evolution of species.

Species
Groups of actually or potentially interbreeding natural populations which are reproductively isolated from other such groups.

Spermatheca
A small, sac-like branch of the female reproductive tract (of insects and other arthropods) in which sperm may be stored.

Sperm competition
The competition between sperm from two or more males, within a single female, for fertilisation of the ova.

Stabilising selection
Selection in which individuals that are intermediate in the distribution of a trait showing continuous variation are at an advantage over those towards the ends of the distribution.

Supergene
When a complex set of related traits are controlled by genes which lie so close to each other on the same chromosome that they behave as a single unit.

Switch gene
Any gene which controls the expression of one or more other genes in an on or off manner.

Taxonomy
The study of the classification of organisms.

Thorax
The body region of an insect behind the head and in front of the abdomen, comprising three segments and bearing the legs and wings.

Transfer RNA
RNA molecules involved in the genetic translation of messenger RNA into amino acid sequences of protein molecules.

Transient polymorphism
Genetic polymorphism in which the forms are changing in frequency in a directed manner.

Translocation
A chromosomal mutation characterised by the change in position of chromosome segments within the chromosome complement.

Transposition
A change in the position of a chromosomal segment without reciprocal exchange.

Ultra-selfish genes
Genes whose spread and maintenance occurs despite of and because they cause damage to the individual in which they occur.

Unequal crossing-over
The effect of a crossing-over between misaligned chromosomes, usually between tandemly repeated elements. The result is two chromosomes, one carrying a duplication and the other having this section deleted.

Uniparental inheritance
In eukaryotes, the transmission of genetic elements (particularly those of organelles such as mitochondria, ribosomes and chloroplasts) from only one parent (cf. biparental inheritance).

Ventral
Concerning the front or underside of an animal.

Viability
The ability of an organism to survive from zygote formation to reproductive maturity.

Wahlund effect
The deficiency of heterozygotes in subdivided populations relative to Hardy–Weinberg expectations based on a single large population.

X chromosome
The chromosome carrying genes concerned with sex determination. There are usually two X chromosomes in one sex (usually the female, but the male in birds and Lepidoptera) and a single X chromosome in the other.

Y chromosome
The partner of the X chromosome in one of the two sexes, usually the male, but the female in birds and Lepidoptera.

Zygote
The first cell of a new organism resulting from the fusion of two gametes.

References

Abbas, I., Nakamura, K., Katakura, H., and Sasaji, H. (1988). Geographical variation of elytral spot patterns in the phytophagous ladybird, *Epilachna vigintioctopunctata* (Coleoptera: Coccinellidae) in the province of Sumatra Barat, Indonesia. *Researches on Population Ecology*, **30**, 43–56.

Adams, R. (1972). *Watership Down*. Penguin, London.

Adkin, R. (1927*a*). Mongrel races of *Diacrisia mendica*. *Proceedings of the Entomological Society of London*, **2**, 15–16.

Adkin, R. (1927*b*). Mongrel races of *Diacrisia mendica*. *Proceedings of the Entomological Society of London*, **2**, 66.

Agodi, A., Stefani, S., Corsaro, C., Campanile, F., Gribaldo, S., and Sichel, G. (1996). Study of a melanic pigment of *Proteus mirabilis*. *Research in Microbiology*, **147**, 167–74.

Aldridge, D., Jones, C.W., Mahar, E., and Majerus, M.E.N. (1993). Differential habitat selection in polymorphic Lepidoptera in the Forest of Dean. *Entomologist's Record and Journal of Variation*, **105**, 203–14.

Allen, J.A. (1972). Evidence for stabilising and apostatic selection by wild blackbirds. *Nature*, **237**, 348–9.

Allen, J.A. (1974). Further evidence for apostatic selection by wild passerine birds: training experiments. *Heredity*, **33**, 361–72.

Allen, J.A. (1976). Further evidence for apostatic selection by wild passerine birds: 9:1 experiments. *Heredity*, **36**, 173–80.

Allen, J.A. and Clarke, B. (1968). Evidence for apostatic selection by wild passerines. *Nature*, **220**, 501–2.

Allison, A.C. (1954). Notes on sickle-cell polymorphism. *Annals of Human Genetics*, **19**, 39–57.

Allison, A.C. (1956). The sickle-cell and haemoglobin C genes in some African populations. *Annals of Human Genetics*, **21**, 67–89.

Andersson, M. (1982). Female choice selects for extreme tail length in a widowbird. *Nature*, **299**, 818–20.

Angus, R.A. (1989). Inheritance of melanistic pigmentation in the eastern mosquitofish. *Journal of Heredity*, **80**, 387–92.

Asami, T. and Grant, B. (1995). Melanism has not evolved in Japanese *Biston betularia* (Geometridae). *Journal of the Lepidopterists' Society*, **49**, 88–91.

Askew, R.R., Cook, L.M., and Bishop, J.A. (1971). Atmospheric pollution and melanic moths in Manchester and its environs. *Journal of Applied Ecology*, **8**, 247–56.

Averill, C.K. (1923). Black wing tips. *Condor*, **25**, 57–9.

Barkman, J.J. (1969). The influence of air pollution on bryophytes and lichens. *Air Pollution*, 197–209. PUDOC: Wageningen.

Barrett, C.G. (1902). *Lepidoptera of the British Isles*, Volume 8. Reeve & Co., London.

Bartle, J.A. and Sagar, P.M. (1987). Intraspecific variation in the New Zealand Bellbird *Anthornis melanura*. *Notornis*, **34**, 253–306.

Barwell, F.T. (1979). *Bearing Systems: Principles and Practice*. Oxford University Press, Oxford.

Bates, H.W. (1862). Contributions to an insect fauna of the Amazon Valley. Lepidoptera: Heliconidae. *Transactions of the Linnaean Society of London*, **23**, 495–566.

Bengston, S.A. and Hagen, R. (1975). Polymorphism in the two-spot ladybird *Adalia bipunctata* in western Norway. *Oikos*, **26**, 328–31.

Bengston, S.A. and Hagen, R. (1977). Melanism in the two-spot ladybird *Adalia bipunctata* in relation to climate in western Norway. *Oikos*, **28**, 16–19.

Benham, B.R., Lonsdale, D., and Muggleton, J. (1974). Is polymorphism in the two-spot ladybird an example of non-industrial melanism? *Nature,* **249**, 179–80.

Bennett, A.T.D. and Cuthill, I.C. (1994). Ultraviolet vision in birds: what is its function? *Vision Research*, **34**, 1471–8.

Berry, R.J. (1990). Industrial melanism and peppered moths (*Biston betularia* (L.)). *Biological Journal of the Linnaean Society*, **39**, 301–22.

Berry, R.J. and Willmer, P.G. (1986). Temperature and the colour polymorphism of *Philaenus spumarius* (Homoptera: Aphrophoridae). *Ecological Entomology*, **11**, 251–60.

Bishop, J.A. (1972). An experimental study of the cline of industrial melanism in *Biston betularia* (L.) (*Lepidoptera*) between urban Liverpool and rural North Wales. *Journal of Animal Ecology*, **41**, 209–43.

Bishop, J.A. (1980). Commentary on three papers on clines for melanism and polymorphism in moths. *Proceedings of the Royal Society of London B*, **210**, 273–5.

Bishop, J.A. and Cook, L.M. (1975). Moths, melanism and clean air. *Scientific American*, **232**, 90–9.

Bishop, J.A., Cook, L.M., Muggleton, J., and Seaward, M.R.D. (1975). Moths, lichens and air pollution along a transect from Manchester to north Wales. *Journal of Applied Ecology*, **12**, 83–98.

Bishop, J.A., Cook, L.M., and Muggleton, J. (1978a). The response of two species of moths to industrialization in northwest England. I. Polymorphism for melanism. *Philosophical Transactions of the Royal Society of London B*, **281**, 489–515.

Bishop, J.A., Cook, L.M., and Muggleton, J. (1978b). The response of two species of moths to industrialization in northwest England. II. Relative fitness of morphs and population size. *Philosophical Transactions of the Royal Society of London B*, **281**, 517–42.

Blair, W.F. (1941). Annotated list of mammals of the Tularosa Basin. *American Midland Naturalist*, 26, 218–29.

Blair, W.F. (1943). Ecological distribution of mammals of the Tularosa Basin, New Mexico. *Contributions from the Laboratory of Vertebrate Biology of the University of Michigan*, **20**, 20–4.

Blem, C.R. and Blem, L.B. (1995). The eastern cottonmouth (*Agkistrodon piscivorus*) at the northern edge of its range. *Journal of Herpetology*, **29**, 391–8.

Boardman, M., Askew, R.R., and Cook, L.M. (1974). Experiments on resting site selection by nocturnal moths. *Journal of Zoology (London)*, **172**, 343–55.

Bonser, R.H.C. (1995). Melanin and the abrasion resistance of feathers. *Condor*, **97**, 590–1.

Bonser, R.H.C. (1996). The mechanical properties of feather keratin. *Journal of Zoology (London)*, **239**, 477–84.

Bowater, W. (1914). Heredity of melanism in Lepidoptera. *Journal of Genetics*, **3**, 299–315.

Bowmaker, J.K. (1991). Photoreceptors, photopigments and oil droplets. In *Vision and Visual Dysfunction*. Vol. 6. *Perception of Colour* (ed. P. Gouras), pp. 108–27. Macmillan, London.

Brakefield, P.M. (1984). Ecological studies on the polymorphic ladybird *Adalia bipunctata* in the Netherlands. II Population dynamics, differential timing of reproduction and thermal melanism. *Journal of Animal Ecology*, **53**, 775–790.

Brakefield, P.M. (1985). Polymorphic Müllerian mimicry and interactions with thermal melanism in ladybirds and a soldier beetle: a hypothesis. *Biological Journal of the Linnaean Society*, **26**, 243–67.

Brakefield, P.M. (1987). Industrial melanism: do we have the answers? *Trends in Ecology and Evolution*, **2**, 117–22.

Brakefield, P.M. (1990). A decline of melanism in the peppered moth, *Biston betularia* in the Netherlands. *Biological Journal of the Linnaean Society*, **39**, 327–34.

Brakefield, P.M. and Lees, D.R. (1987). Melanism in *Adalia* ladybirds and declining air pollution in Birmingham. *Heredity*, **59**, 273–7.

Brakefield, P.M. and Liebert, T.G. (1990). The reliability of estimates of migration in the peppered moth *Biston betularia* and some implications for selection-migration models. *Biological Journal of the Linnaean Society*, **39**, 335–42.

Brakefield, P.M. and Willmer, P.G. (1985). The basis of thermal melanism in the ladybird *Adalia bipunctata*: differences in reflectance and thermal properties between morphs. *Heredity*, **54**, 9–14.

Brakefield, P.M., Gates, J., Keys, D., Kesbeke, F., Wijngaarden, P.J., Monteiro, A., French, V. and Carroll, S.B. (1996). Development, plasticity and evolution of butterfly eyespot patterns. *Nature*, **384**, 236–42.

Brower, J. van Z. (1958a). Experimental studies of mimicry in some North American butterflies. I. The Monarch *Danaus plexippus*, and the viceroy *Limenitis archippus archippus*. *Evolution*, **12**, 123–36.

Brower, J. van Z. (1958b). Experimental studies of mimicry in some North American butterflies. II. *Battus philenor*, *Papilio troilus*, *P. polyxenes*, and *P. glaucus*. *Evolution*, **12**, 123–36.

Bruge, H. (1992). A rare melanic form of *Dolichovespula media* Retzius 1783 (Hymenoptera: Vespidae). *Bulletin Annuel de Société Royale Belge Entomologique*, **128**, 16–18.

Burkhardt, D. and Maier, E. (1989). The spectral sensitivity of a passerine bird is highest in the UV. *Naturwissenschaften*, **76**, 82–3.

Burtt, E.H. (1979). Tips on wings and other things. In *The Behavioural Significance of Color* (ed. E.H., Burtt, Jr.), pp. 75–110. Garland STPM Press, New York.

Burtt, E.H. (1981). The adaptiveness of animal colors. *Bioscience*, **31**, 723–9.

Burtt, E.H. (1986). An analysis of physical, physiological and optical aspects of avian coloration with emphasis on wood-warblers. *Ornithological Monographs*, **38**, 1–126.

Cadbury, C.J. (1969). Melanism in moths with special reference to selective predation by birds. D.Phil. thesis, Oxford University.

Cain, A.J. and Currey, J.D. (1968). Climate and selection of banding morphs in *Cepaea* from the climatic optimum to the present day, *Philosophical Transactions of the Royal Society of London B*, **253**, 483–98.

Cain, A.J. and Sheppard, P.M. (1950). Selection in the polymorphic land snail *Cepaea nemoralis* (L.). *Heredity*, **4**, 275–94.

Cain, A.J. and Sheppard, P.M. (1954). Natural selection in *Cepaea. Genetics*, **39**, 89–116.

Chapman, T.A. (1888). On melanism in Lepidoptera. *Entomologist's Monthly Magazine*, **25**, 40.

Chardon, A., Cretois, I., and Hourseau, C. (1991). Skin colour typology and suntanning pathways. *International Journal of Cosmetic Science*, **13**, 191–208.

Chen, D. and Goldsmith, T.H. (1986). Four spectral classes of cone in the retinas of birds. *Journal of Comparative Physiology A*, **159**, 473–9.

Chen, D., Collins, J.S., and Goldsmith, T.H. (1984). The UV receptor of bird retinas. *Science*, **225**, 337–9.

Clarke, B. (1962). Natural selection in mixed populations of two polymorphic snails. *Heredity*, **17**, 319–45.

Clarke, C.A. (1979). *Biston betularia*, obligate f. *insularia* indistinguishable from f. *carbonaria* (Geometridae). *Journal of the Lepidoptera Society*, **33**, 60–4.

Clarke, C.A. and Sheppard, P.M. (1959). The genetics of some mimetic forms of *Papilio dardanus*, Brown, and *Papilio glaucus*, Linn. *Journal of Genetics*, **56**, 237–59.

Clarke, C.A. and Sheppard, P.M. (1960a). The evolution of mimicry in the butterfly *Papilio dardanus. Heredity*, **14**, 163–73.

Clarke, C.A. and Sheppard, P.M. (1960b). The evolution of dominance under disruptive selection. *Heredity*, **14**, 73–87.

Clarke, C.A. and Sheppard, P.M. (1963). Interactions between major genes and polygenes in the determination of the mimetic pattern of *Papilio dardanus. Evolution*, **17**, 404–13.

Clarke, C.A. and Sheppard, P.M. (1964). Genetic control of the melanic form *insularia* of the moth *Biston betularia* L. *Nature*, **202**, 215–16.

Clarke, C.A. and Sheppard, P.M. (1966). A local survey of the distribution of industrial melanic forms in the moth *Biston betularia* and estimates of the selective values of these in an industrial environment. *Proceedings of the Royal Society of London B*, **165**, 424–39.

Clarke, C.A. and Sheppard, P.M. (1971). Further studies on the genetics of the mimetic butterfly *Papilio memnon* L. *Philosophical Transactions of the Royal Society of London B*, **263**, 35–70.

Clarke, C.A. and Sheppard, P.M. (1972). The genetics of the mimetic butterfly *Papilio polytes* L. *Philosophical Transactions of the Royal Society of London B*, **263**, 431–58.

Clarke, C.A., Sheppard, P.M., and Thornton, I.W.B. (1968). The genetics of the mimetic butterfly *Papilio memnon* L. *Philosophical Transactions of the Royal Society of London B*, **254**, 37–89.

Clarke, C.A., Mani, G.S., and Wynne, G. (1985). Evolution in reverse: clean air and the peppered moth. *Biological Journal of the Linnaean Society*, **26**, 189–99.

Cloudsley-Thompson, J.L. (1964). Terrestrial animals in dry heat: arthropods. In *Handbook of Physiology*, Section 4 (ed. D.B. Dill), pp. 451–65. American Physiological Society, Washington.

Cockayne, E.A. (1926). Experimental melanic changes. *Entomologist's Record and Journal of Variation*, **38**, 44–5.

Cockayne, E.A. (1928a). Variation in *Callimorpha dominula* L. *Entomologist's Record and Journal of Variation*, **40**, 153–60.

Cockayne, E.A. (1928b). On larval variation. *Proceedings of the South London Entomological and Natural History Society*, (1927–28), 55–67.

Cockerell, T.D.A. (1887). On melanism. *Entomologist*, **20**, 58–9.

Conroy, B.A. and Bishop, J.A. (1980). Maintenance of the polymorphism for melanism in the moth *Phigalia pilosaria* in North Wales. *Proceedings of the Royal Society of London B*, **210**, 285–98.

Cook, L.M. and Jacobs, Th.M.G.M. (1983). Frequency and selection in the industrial melanic moth *Odontoptera bidentata*. *Heredity*, **51**, 487–94.

Cook, L.M. and Kettlewell, H.B.D. (1960). Radioative labelling of lepidopterous larvae: a method of estimating late larval and pupal mortality in the wild. *Nature*, **187**, 301–2.

Cook, L.M. and Mani, G.S. (1980). A migration-selection model for the morph frequency variation in the peppered moth over England and Wales. *Biological Journal of the Linnaean Society*, **13**, 179–98.

Cook, L.M., Askew, R.R., and Bishop, J.A. (1970). Increasing frequency of the typical form of the peppered moth in Manchester. *Nature*, **227**, 1155.

Cook, L.M., Mani, G.S., and Varley, M.E. (1986). Post-industrial melanism in the peppered moth. *Science*, **231**, 611–3.

Cook, L.M., Rigby, K.D., and Seaward, M.R.D. (1990). Melanic moths and changes in epiphytic vegetation in north-west England and north Wales. *Biological Journal of the Linnaean Society*, **39**, 343–54.

Cooke, F. and Cooch, F.G. (1968). The genetics of polymorphism in the goose *Anser caerulescens*. *Evolution*, **22**, 289–300.

Cooke, F. and McNally, C.M. (1975). Mate selection and colour preferences in lesser snow geese. *Behaviour*, **35**, 151–70.

Cooke, F., Finney, G.H., and Rockwell, R.F. (1976). Assortative mating in lesser snow geese (*Anser caerulescens*). *Behaviour Genetics*, **6**, 127–40.

Cooke, F., Finday, C.S., Rockwell, R.F., and Smith, J.A. (1985). Life history studies of the lesser snow goose (*Anser caerulescens caerulescens*). III. The selective value of plumage polymorphism: Net fecundity. *Evolution*, **39**, 165–77.

Cooke, N. (1877a). On melanism in Lepidoptera. *Entomologist*, **10**, 93–6.

Cooke, N. (1877b). Melanism in Lepidoptera. *Entomologist*, **10**, 151–3.

Cott, H.B. (1940). *Adaptive Coloration in Animals.* Methuen, London.

Cott, H.B. (1946). The edibility of birds. *Proceedings of the Zoological Society of London*, **116**, 371–524.

Cowie, R.H. and Jones, J.S. (1985). Climatic selection on body colour in *Cepaea*. *Heredity*, **55**, 261–7.

Creed, E.R. (1966). Geographical variation in the two-spot ladybird in England and Wales. *Heredity*, **21**, 57–72.

Creed, E.R. (1971a). Melanism in the two-spot ladybird, *Adalia bipunctata*, in Great Britain. In *Ecological Genetics and Evolution*, (ed. E.R. Creed), pp. 134–51. Blackwell Scientific, Oxford.

Creed, E.R. (1971b). Industrial melanism and smoke abatement. *Evolution*, **25**, 290–3.

Creed, E.R., Lees, D.R., and Bulmer, M.G. (1980). Pre-adult viability difference of melanic *Biston betularia* (L.) (*Lepidoptera*). *Biological Journal of the Linnaean Society*, **13**, 251–62.

Darwin, C.R. (1859). *On the Origin of Species by Means of Natural Selection, or the Preservation of Favoured Races in the Struggle for Life.* John Murray, London.

Darwin, C.R. (1871). *The Descent of Man, and Selection in Relation to Sex.* John Murray, London.

Darwin, C.R. (1887). *The Autobiography of Charles Darwin 1809–1882.* 1958 edn (ed. N. Barlow). Collins, London.

Dice, L.R. (1930). Mammal distribution in the Alanogordo region, New Mexico. *Occasional Papers of the Museum of Zoology, University of Michigan,* **213**, 1–32.

Dice, L.R. (1947). Effectiveness of selection by owls on Deermice (*Permycus maniculatus*) which contrast in colour with their background. *Contributions from the Laboratory of Vertebrate Biology of the University of Michigan,* **34**, 1–20.

Dieke, G.H. (1947). Ladybirds of the genus *Epilachna* (sens. lat.) in Asia, Europe and Australia. *Smithsonian Miscellaneous Collections,* **106**, 1–183.

Diver, C. (1929). Fossil records of Mendelian units. *Nature,* **124**, 183.

Dobrée, N.F. (1887). On melanism. *Entomologist,* **20**, 25–8.

Dobrzhanskii, F.G. (Dobzhansky, Th.G.) (1922*a*). Imaginal diapause in Coccinellidae. *Izv. Otd. prikl. Ent.,* **2**, 103–24.

Dobrzhanskii, F.G. (Dobzhansky, Th.G.) (1922*b*). Mass aggregations and migrations in Coccinellidae. *Izv. Otd. prikl. Ent.,* **2**, 229–34.

Dobzhansky, Th.G. (1933). Geographical variation in lady-beetles. *American Naturalist,* **67**, 97–126.

Dobzhansky, Th.G. (1951). *Genetics and the Origin of Species,* (3rd edn). Columbia University Press, New York.

Dobzhansky, Th.G. (1962). *Mankind Evolving.* Yale University Press, New Haven.

Dobzhansky, Th.G. and Sivertzev-Dobzhansky, N.P. (1927). Die geographische Variabilitat von *Coccinella septempunctata. Biologisches Zentralblatt,* **47**, 556–69.

Doncaster, L. (1906*a*). Mendel's law of heredity. *Entomologist's Record and Journal of Variation,* **18**, 19–20.

Doncaster L. (1906*b*). Collective inquiry as to progressive melanism in Lepidoptera. *Entomologist's Record and Journal of Variation,* **18**, 165–8, 206–8, 222–6, 248–54.

Dumont, H.J., Demirsoy, A., and Mertens, J. (1988). Odonata from south-east Anatolia Turkey collected in Spring 1988. *Notes on Odonatology,* **3**, 22–6.

Edleston, R.S. (1864). No title (first *carbonaria* melanic of moth *Biston betularia*). *Entomologist,* **2**, 150.

Edwards, J. (1914). On the variation in Britain of *Coccinella hieroglyphica,* L., with some collateral matter. *Entomologist's Monthly Magazine,* **50**, 139–43.

Emmerton, J. and Delius, J.D. (1980). Wavelength discrimination in the 'visible' and UV spectrum by pigeons. *Journal of Comparative Physiology A,* **141**, 47–52.

Faure, J.C. (1943*a*). The phases of the lesser army worm. *Farming in South Africa,* **18**, 69–78.

Faure, J.C. (1943*b*). Phase variation in the army worm *Lephygma exempta* Walk. *Scientific Bulletin of the Department of Agriculture of South Africa.* No. 234.

Findlay, C.S., Rockwell, R.F., Smith, J.A., and Cooke, F. (1985). Life history studies of the lesser snow goose (*Anser caerulescens caerulescens*). VI. Plumage polymorphism, assortative mating and fitness. *Evolution,* **39**, 904–14.

Fisher, J. (1952). *The Fulmar.* Collins, London.

Fisher, R.A. (1922). On the dominance ratio. *Proceedings of the Royal Society of Edinburgh,* **42**, 321–41.

Fisher R.A. (1927). On some objections to mimicry theory; Statistical and genetic. *Transactions of the Royal Entomological Society of London,* **75**, 269–78.

Fisher, R.A. (1930). *The Genetical Theory of Natural Selection.* Oxford University Press, Oxford.

Fisher, R.A. and Ford, E.B. (1947). The spread of a gene in natural conditions in a colony of the moth *Panaxia dominula. Heredity*, **1**, 143–74.

Fisher, R.A. and Ford, E.B. (1950). The 'Sewall Wright' effect. *Heredity*, **4**, 117–9.

Fishpool, L.D.C. and Popov, G.B. (1981). Grasshopper faunas of the savannas of Mali, Niger, Benin and Togo. *Bull. Inst. Fondam Afr. Noire Ser A Sci Nat.*, **43**, 275–410.

Ford, E.B. (1936). The genetics of *Papilio dardanus* Brown (Lep.). *Transactions of the Royal Entomological Society of London*, **85**, 435–66.

Ford, E.B. (1937). Problems of heredity in the Lepidoptera. *Biological Reviews*, **12**, 461–503.

Ford, E.B. (1940*a*). Polymorphism and taxonomy. In *The New Systematics* (ed. J.S. Huxley), pp. 493–513. Clarendon Press, Oxford.

Ford, E.B. (1940*b*). Genetic research in the Lepidoptera. *Annals of Eugenics*, **10**, 227–52.

Ford, E.B. (1945*a*). *Butterflies*, No. 1. New Naturalist Series. Collins, London.

Ford, E.B. (1945*b*). Polymorphism. *Biological Reviews*, **20**, 73–88.

Ford, E.B. (1955). *Moths*, No. 30. New Naturalist Series. Collins, London.

Ford, E.B. (1964). *Ecological Genetics*. Methuen, London.

Ford, E.B. (1975). *Ecological Genetics*, (4th edn). Chapman and Hall, London.

Ford, E.B. and Sheppard, P.M. (1969). The medionigra polymorphism of *Panaxia dominula. Heredity*, **24**, 112–34.

Forsman, A. (1995). Opposing fitness consequences of colour pattern in male and female snakes. *Journal of Evolutionary Biology*, **8**, 53–70.

Fraiers, T., Boyles, T., Jones, C.W., and Majerus, M.E.N. (1994). Short distance form frequency changes across habitat boundaries. *British Journal of Entomology and Natural History*, **7**, 47–52.

Frazer, J.F.D. and Rothschild, M. (1960). Defence mechanisms in warningly-coloured moths and other insects. *Proceedings of the International Congress of Entomology*, **11**, 249–56.

Fritz, U. (1995). On the intraspecific variation of *Emys orbicularis* (LINNAEUS, 1958) 5a. Taxonomy in Central and West Europe, on Corsica, Sardinia, the Italian Peninsula, and Sicily, and subspecies groups of *E. orbicularis* (Reptilia: Testudines: Emydidae). *Zoologische Abhandlungen (Dresden)*, **48**, 185–242.

Frost, S.K., Borchert, M., and Carson, M.K. (1989). Drug-induced and genetic hypermelanism effects on pigment cell differentiation. *Pigment Cell Research*, **2**, 182–90.

Garstka, W.R., Cooper, W.E. Jr, Wasmund, K.W. and Lovich, J.E. (1991). Male sex steroids and hormonal control of male courtship behavior in the yellow-bellied slideer turtle, *Trachemys scripta. Comparative Biochemistry Physiology A. Comparative Physiology*, **98**, 271–80.

Gershenzon, S.M. (1994). A melanistic form of the oak silkworm *Antheraea pernyi* (Lepidoptera, Attacidae). *Vestnik Zoologii*, **6**, 46–51.

Gilbert, L.E., Forrest, H.S., Schultz, T.D., and Harvey, D.J. (1988). Correlations of ultrastructure and pigmentation suggest how genes control development of wing scales of *Heliconius* butterflies. *Journal for Research on the Lepidoptera*, **26**, 141–60.

Goldschmidt, R.B. (1921). Erblichkeitsstudien an Schmetterlingen. III. Der Melanismus der Nonne, *Lymantria monacha* L. *Zeitschrift für induktive Abstammungs- und Vererbungslehre*, **25**, 89–163.

Goldsmith, T.H. (1991). Optimization, constraint and history in the evolution of eyes. *Quarterly Review of Biology*, **65**, 281–322.

Grant, B. and Howlett, R.J. (1988). Background selection by the peppered moth (*Biston betularia* Linn.): individual differences. *Biological Journal of the Linnaean Society*, **33**, 217–32.

Grant, B., Owen, D.F., and Clarke, C.A. (1995). Decline of melanic moths. *Nature*, **373**, 565.

Grant, B.S., Owen, D.F., and Clarke, C.A. (1996). Parallel rise and fall of melanic peppered moths in America and Britain. *Journal of Heredity*, **87**, 351–7.

Gunnarsson, B. (1987). Melanism in the spider *Pityohyphantes phrygianus* (C.L. Koch): the genetics and the occurrence of different colour phenotypes in a natural population. *Heredity*, **59**, 55–61.

Gunnarsson, B. (1993). Maintenance of melanism in the spider *Pityohyphantes phrygianus*: is bird predation a selective agent? *Heredity*, **70**, 520–6.

Guppy, C.S. (1986a). The adaptive significance of alpine melanism in the butterfly *Parnassius phoebus* (Lepidoptera: Papilionidae). *Oecologia*, **70**, 205–13.

Guppy, C.S. (1986b). Geographic variation in wing melanism of the butterfly *Parnassius phoebus* (Lepidoptera: Papilionidae). *Canadian Journal of Zoology*, **64**, 956–62.

Gustafson, E.J. and Vandruff, L.W. (1990). Behaviour of black and gray morphs of *Sciurus carolinensis* in an urban environment. *American Midland Naturalist*, **123**, 186–92.

Haldane, J.B.S. (1924). A mathematical theory of natural and artificial selection. *Transactions of the Cambridge Philosophical Society*, **23**, 19–41.

Haldane, J.B.S. (1956). The theory of selection for melanism in the Lepidoptera. *Proceedings of the Royal Society of London B*, **145**, 303–8.

Haase E., Ito, S., and Wakamatsu, K. (1995). Influences of sex, castration, and androgens on the eumelanin and pheomelanin contents of different feathers in wild mallards. *Pigment Cell Research*, **8**, 164–70.

Hamilton, W.J. III (1973). *Life's Color Code*. McGraw-Hill, New York.

Harris, A.C. (1988). Cryptic coloration and melanism in the sand-burrowing beetle, *Chaerodes trachyscelides* (Coleoptera: Tenebrionidae). *Journal of the Royal Society of New Zealand*, **18**, 333–40.

Harris, A.C. and Weatherall, I.L. (1991). Geographic variation for colour in the sand-burrowing beetle *Chaerodes trachyscelides* White (Coleoptera: Tenebrionidae) on New Zealand beaches analysed using CIELAB L values. *Biological Journal of the Linnaean Society*, **44**, 93–104.

Harrison, A. (1908). On *Aplecta nebulosa*. *Entomologist*, **41**, 314.

Harrison, A. (1909). On *Aplecta nebulosa*. *Proceedings of the Entomological Society of London*, (1908), lxi–xii.

Harrison, A. and Main, H. (1910). On *Aplecta nebulosa*. *Proceedings of the South London Entomological and Natural History Society*, (1909–10), 84.

Harrison, J.G. (1947). Colour change without moult as seen in the spring plumage of certain wading birds. *Bulletin of the British Ornithological Club*, **68**, 40–52.

Harrison, J.W.H. (1920). Genetic studies in the moths of the genus *Oporabia* (*Oporinia*) with special consideration of melanism in the Lepidoptera. *Journal of Genetics*, **9**, 195–280.

Harrison, J.W.H. (1926a). The inheritance of wing colour and pattern in the lepidopterous genus *Tephrosia* (*Ectropis*) II. Experiments involving melanic *Tephrosia bistortata* and typical *T. crepuscularia*. *Journal of Genetics*, **17**, 1–9.

Harrison, J.W.H. (1926b). Miscellaneous observations on the induction, incidence and inheritance of melanism in the Lepidoptera. *Entomologist*, **59**, 121–3.

Harrison, J.W.H. (1927a). The inheritance of melanism in hybrids between continental *Tephrosia crepuscularia* and British *T. bistortata*, with some remarks on the origin of parthenogenesis in interspecific crosses. *Genetica*, **9**, 467–80.

Harrison, J.W.H. (1927b). The induction of melanism in the Lepidoptera and its evolutionary signifigance. *Nature*, **119**, 127–9.

Harrison, J.W.H. (1927c). Melanism in the Lepidoptera and its evolutionary significance. *Nature*, **119**, 318.

Harrison, J.W.H. (1927d). Some thoughts on melanism and melanochroism in the Lepidoptera. *Vasculum*, **13**, 103–5.

Harrison, J.W.H. (1928a). A further induction of melanism in the Lepidopterist insect, *Selenia bilunaria* Esp., and its inheritance. *Proceedings of the Royal Society of London B*, **102**, 338–47.

Harrison, J.W.H. (1928b). Induced changes in the pigmentation of the pupae of the butterfly *Pieris napi* L., and their inheritance. *Proceedings of the Royal Society of London B*, **102**, 347–53.

Harrison, J.W.H. and Doncaster, L. (1914). On hybrids between moths of the geometrid sub-family Bistoninae, with an account of the behaviour of the chromosomes in gametogenesis in *Lycia (Biston) hirtaria*, *Ithysia (Nyssia) zonaria* and their hybrids. *Journal of Genetics*, **3**, 229–48.

Harrison, J.W.H. and Garrett, F.C. (1926). The induction of melanism in the Lepidoptera and its subsequent inheritance. *Proceedings of the Royal Society of London B*, **99**, 241–63.

Hasebroëk, K. (1925). Untersuchungen zum problem des neuzeitlichen Melanismus der Schmetterlinge. *Fermentforschung*, **8**, 199–226.

Hasebroëk, K. (1929). Über den Industrie und Grosstadt-melanismus der Schmetterlinge. *Zeitschrift für induktive Abstammungs- und Vererbungslehre*, **50**, 201–18.

Hawkes, O.A.M. (1920). Observations on the life-history, biology and genetics of the ladybird beetle *Adalia bipunctata* (Mulsant). *Proceedings of the Zoological Society of London*, **1920**, 475–90.

Heer, O. (1836). Einfluss des Alpenklimas auf die Farbe der Insekten. In *Metteilungen der Gebiet theoretische Erdkunne* (ed. F. Froebel and O. Heer), pp. 161–70. Zurich.

Heikertinger, F. (1932). Die Coccinellinden, ihr 'Ekelblut', ihre Warntracht und ihre Feinde. *Biologisches Zentralblatt*, **52**, 65–102, 385–412.

Herbert, P.D.N. and Emery, C.J. (1990). The adaptive significance of cuticular pigmentation in *Daphnia*. *Functional Ecology*, **4**, 703–10.

Heylaerts, F.J.M. (1870). Les macrolépidoptères des environs de Bréda. *Tijdschrift voor Entomologie*, **13**, 142–157.

Hodek, I. (1973). *Biology of Coccinellidae*. Junk, The Hague; Academia, Prague.

Hodek, I. and Honek, A. (1996). *Ecology of Coccinellidae*. Kluwer Academic Publishers, Dordrecht.

Hoffmeyer, S. (1948). *De Danske Spindera*. Universitetsforlaget I, Aarhus, Denmark.

Hoffmeyer, S. (1949). *De Danske Ugler*. Universitetsforlaget I, Aarhus, Denmark.

Holloway, G.J., de Jong, P.W., Brakefield, P.M., and de Vos, H. (1991). Chemical defense in ladybird beetles (Coccinellidae). I. Distribution of coccinelline and individual variation in defense in 7-spot ladybirds (*Coccinella septempunctata*). *Chemoecology*, **2**, 7–14.

Honek, A. (1975). Colour polymorphism in *Adalia bipunctata* in Bohemia (Coleoptera: Coccinellidae). *Entomologia Germanica*, **1**, 293–9.

Honek, A. (1984). Melanism in populations of *Philaenus spumarius* (Homoptera: Aphrophoridae) in Czechoslovakia. *Vestnik Ceske Spolecnosti Zoologicke*, **48**, 241–7.

Honek, A. (1993). Melanism in the land snail *Helicella candicans* (Gastropoda: Helicidae) and its possible adaptive significance. *Malacologia*, **35**, 79–87.

Honek, A. (1995). Dustribution and shell colour and banding polymorphism of the *Cepaea* species in Bohemia (Gastropoda: Helicidae). *Acta Societatis Zoologicae Bohemicae*, **59**, 63–77.

Hosino, Y. (1933). On variation in the pattern of *Harmonia axyridis*. *Zoological Magazine*, **45**, 255–67.

Hosino, Y. (1936). Genetical study of the lady-bird beetle, *Harmonia axyridis* Pallas Rep. II. *Japanese Journal of Genetics*, **12**, 307–20.

Houston, K.J. and Hales, D.F. (1980). Allelic frequencies and inheritance of colour pattern in *Coelophora inaequalis* (F.) (Coleoptera: Coccinellidae). *Australian Journal of Zoology*, **28**, 669–77.

Howlett, R.J. (1989). The genetics and evolution of rest site preference in the Lepidoptera. Ph.D. thesis, Cambridge University.

Howlett, R.J. and Majerus, M.E.N. (1987). The understanding of industrial melanism in the peppered moth (*Biston betularia*) (Lepidoptera: Geometridae). *Biological Journal of the Linnaean Society*, **30**, 31–44.

Hudson, G.V. (1928). *The Butterflies and Moths of New Zealand*. Wellington.

Hurst, G.D.D. and Majerus, M.E.N. (1993). Why do maternally inherited microorganisms kill males? *Heredity*, **71**, 81–95.

Hurst, G.D.D., Majerus, M.E.N., and Walker, L.E. (1992). Cytoplasmic male killing elements in *Adalia bipunctata* (Linnaeus) (Coleoptera: Coccinellidae). *Heredity*, **69**, 84–91.

Hurst, G.D.D., Majerus, M.E.N., and Walker, L.E. (1993). The importance of cytoplasmic male killing elements in natural populations of the two spot ladybird, *Adalia bipunctata* (Linnaeus) (Coleoptera: Coccinellidae). *Biological Journal of the Linnaean Society*, **49**, 195–202.

Hurst, G.D.D., Sharpe, R.G., Broomfield, A.H., Walker, L.E., Majerus, T.M.O., Zakharov, I.A., and Majerus, M.E.N. (1995). Sexually transmitted disease in a promiscuous insect. *Ecological Entomology*, **20**, 230–6.

Hurst, L.D. and Majerus, M.E.N. (in preparation). Added spice to the peppered moth.

Inagami, K. (1954). Mechanism of the formation of red melanin in the silkworm. *Nature*, **174**, 1105.

Ireland, H., Kearns, P.W.E., and Majerus, M.E.N. (1986). Interspecific hybridisation in the Coccinellidae: some observations on an old controversy. *Entomologist's Record and Journal of Variation*, **98**, 181–5.

Jacobs, G.H. (1992). UV vision in vertebrates. *American Zoologist*, **32**, 544–54.

James, D.G. (1986). Thermoregulation in *Danaus plexippus* L. (Lepidoptera: Nymphalidae): two cool climate adaptations. *General and Applied Entomology*, **18**, 43–7.

Jane, S.D. and Bowmaker, J.K. (1988). Tetrachromatic colour vision in the duck (*Anas playtrhynchos* L.): Microspectrophotometry of visual pigments and oil droplets. *Journal of Comparative Physiology A*, **162**, 225–35.

Johnson, R.H. (1910). *Determinate Evolution in the Colour-pattern of the Lady Beetles.* Carnegie Institute, Washington.

Jones, C.W. (1993). Habitat selection in polymorphic Lepidoptera. Ph.D. thesis, Cambridge University.

Jones, C.W., Majerus, M.E.N., and Timmins, R. (1993). Differential habitat selection in polymorphic Lepidoptera. *Entomologist*, **112**, 118–26.

Jones, D.A. (1989). 50 years of studying the scarlet tiger moth. *Trends in Ecology and Evolution*, **4**, 298–301.

Jones, J.S. (1973). The genetic structure of a southern peripheral population of the snail *Cepaea nemoralis* (L.). *Proceedings of the Royal Society of London B*, **183**, 371–84.

Jones, J.S. (1982). Genetic differences in individual behaviour associated with shell polymorphism in the snail *Cepaea nemoralis*. *Nature*, **298**, 749–50.

Jones, J.S. (1993). *The Language of the Genes*. HarperCollins, London.

de Jong, P.W., Holloway, G.J., Brakefield, P.M., and de Vos, H. (1991). Chemical defence in ladybird beetles (Coccinellidae). II. Amount of reflex fluid, the alkaloid adaline and individual variation in defence in 2-spot ladybirds (*Adalia bipunctata*). *Chemoecology*, **2**, 15–9.

Kane, W. de V. (1896). Observations on the development of melanism in *Camptogramma bilineata*. *Irish Naturalist*. **v**, 75–80.

Katakura, H., Abbas, I., Nakamura, K., and Sasaji, H. (1988). Records of epilachnine in crop pests (Coleoptera, Coccinellidae) in Sumatra Barat, Sumatra, Indonesia. *Kontyû*, **56**, 281–97.

Kayser, H. (1985). Pigments. In *Comprehensive Insect Physiology, Biochemistry, and Pharmacology*, Volume 10 (ed. G.A. Kerkut and L.I. Gilbert), pp. 367–415. Pergamon Press, New York.

Kearns, P.W.E. and Majerus, M.E.N. (1987). Differential habitat selection in the Lepidoptera: a note on deciduous versus coniferous woodland habitats. *Entomologist's Record and Journal of Variation*, **99**, 103–6.

Kearns, P.W.E., Tomlinson, I.P.M., Veltman, C.J., and O'Donald, P. (1992). Non-random mating in the two-spot ladybird (*Adalia bipunctata*). II. Further tests for female mating preference. *Heredity*, **68**, 385–9.

Kennicott, R. (1857). Quadrupeds of Illinois injurious and beneficial to the farmer. *Executive Document 65, 34th Congress, 3rd Session Report of the Commission on Patents*, 1856, 52–110.

Kettlewell, H.B.D. (1943–44). Temperature experiments on the pupae of *Heliothis peltigera* Schiff. and *Panaxia dominula* L. *Proceedings of the South London Entomological Society*, 1943–4, 69–81.

Kettlewell, H.B.D. (1955a). Selection experiments on industrial melanism in the Lepidoptera. *Heredity*, **9**, 323–42.

Kettlewell, H.B.D. (1955b). Recognition of appropriate backgrounds by the pale and black phases of the Lepidoptera. *Nature*, **175**, 943–4.

Kettlewell, H.B.D. (1956). Further selection experiments on industrial melanism in the Lepidoptera. *Heredity*, **10**, 287–301.

Kettlewell, H.B.D. (1958a). A survey of the frequencies of *Biston betularia* (L.) (Lepidoptera) and its melanic forms in Great Britain. *Heredity*, **12**, 51–72.

Kettlewell, H.B.D. (1958b). The importance of the micro-environment to evolutionary trends in the Lepidoptera. *Entomologist*, **91**, 214–24.

Kettlewell, H.B.D. (1958c). Industrial melanism in the Lepidoptera and its contribution to our knowledge of evolution. *Proceedings of the 10th International Congress of Entomology*, **2**, 831–41.

Kettlewell, H.B.D. (1965). Insect survival and selection for pattern. *Science*, **148**, 1290–5.

Kettlewell, H.B.D. (1973). *The Evolution of Melanism*. Clarendon Press, Oxford.

Kettlewell, H.B.D. and Berry, R.J. (1969). Gene flow in a cline. *Heredity*, **24**, 1–14.

Kettlewell, H.B.D. and Conn, D.T. (1977). Further background-choice experiments on cryptic Lepidoptera. *Journal of Zoology (London)*, **181**, 371–6.

Kettlewell, H.B.D. and Cook, L.M. (1960). Radioactive labelling of lepidopterous larvae: a method of estimating late larval and pupal mortality in the wild. *Nature*, **187**, 301–2.

Kettlewell, H.B.D., Cadbury, C.J., and Lees, D.R. (1971). Recessive melanism in the moth *Lasiocampa quercus* L. In *Ecological Genetics and Evolution* (ed. E.R. Creed), pp. 175–201. Blackwell Scientific, Oxford.

Kiltie, R.A. (1989). Wildfire and the evolution of dorsal melanism in fox squirrels, *Sciurus niger*. *Journal of Mammalogy*, **70**, 726–39.

Kiltie, R.A. (1992a). Camouflage comparisons among fox squirrels from the Mississippi River delta. *Journal of Mammalogy*, **73**, 906–13.

Kiltie, R.A. (1992b). Tests of hypotheses on predation as a factor maintaining polymorphic melanism in coastal-plain fox squirrels, *Sciurus niger* L. *Biological Journal of the Linnaean Society*, **45**, 17–38.

King, R.B. (1988). Polymorphic populations of the garter snake *Thamnophis sirtalis* near Lake Erie Canada/USA. *Herpetologica*, **44**, 451–8.

Kingsolver, J.G. and Wiernasz, D.C. (1987). Dissecting correlated characters adaptive aspects of phenotypic covariation in melanization patterns of *Pieris* butterflies. *Evolution*, **41**, 491–503.

Kingsolver, J.G. and Wiernasz, D.C. (1991). Seasonal polyphenism in wing-melanin pattern and thermoregulatory adaptation in *Pieris* butterflies. *The American Naturalist*, **137**, 816–30.

Kipling, R. (1902). *Just So Stories for Little Children*. Macmillan, London.

Kirby, W.F. (1882). *European Butterflies and Moths*. Cassell, Petter, Galpin & Co., London.

Koch, P.B. and Bueckmann, D. (1982–1983). Comparative study of the colour patterns and colour adaptation in nymphalid pupae (Lepiodoptera). *Zoologische Beitraege*, **28**, 369–402.

Komai, T. (1956). Genetics of ladybeetles. *Advances in Genetics*, **8**, 155–88.

Korschefsky, R. (1931). *Coleopterorum Catalogus*. Pars 118. Coccinellidae. I. Berlin.

Korschefsky, R. (1932). *Coleopterorum Catalogus*. Pars 120. Coccinellidae. II. Berlin.

Kristin, A. (1984). The diet and trophic ecology of the tree sparrow (*Passer montanus*) in the Bratislava area. *Folia Zoologika*, **33**, 143–57.

Kristin, A. (1986). Heteroptera, Coccinea, Coccinellidae and Syrphidae in the food of *Passer montanus* L. and *Pica pica* L. *Biologia (Bratislava)*, **41**, 143–50.

Kristin, A. (1991). Feeding of some polyphagous songbirds on Syrphidae, Coccinellidae and aphids in beech-oak forests. In *Behaviour and Impact of Aphidophaga* (ed. I. Połgár, R.J. Chambers, A.F.G. Dixon, and I. Kodek), pp. 183–6. SPB Academic Publishing, The Hague.

Kula, E. and Kralicek, M. (1995). The occurrence of melanism with butterflies in polluted area Decinsky Sneznik. *Lesnictvi (Prague)*. **41**, 257–64.

Kuranova, V.N. (1989). On melanism in the viviparous lizard and common adder. *Vestník Zoologické*, **2**, 59–61.

Lambert, D.M., Millar, C.D., and Hughes, T.J. (1986). On the classical case of natural selection. *Rivista di Biologia*, **79**, 11–49.

Lees, D.R. (1968). Genetic control of the melanic form *insularia* of the peppered moth *Biston betularia* L. *Nature*, **220**, 1249–50.

Lees, D.R. (1971). Distribution of melanism in the Pale Brindled Beauty moth, *Phigalia pedaria*, in Britain. In *Ecological Genetics and Evolution*, (ed. E.R. Creed), pp. 152–74. Blackwell Scientific, Oxford.

Lees, D.R. (1974). Genetic control of the melanic forms of the moth *Phigalia pilosaria* (*pedaria*). *Heredity*, **33**, 145–50.

Lees, D.R. (1975). Resting site selection in the geometrid moth *Phigalia pilosaria* (Lepidoptera: Geometridae). *Journal of Zoology (London)*, **176**, 341–52.

Lees, D.R. (1981). Industrial melanism: Genetic adaptation of animals to air pollution. In *Genetic Consequences of Man Made Change*, (ed. J.A. Bishop and L.M. Cook), pp. 129–76. Academic Press, London.

Lees, D.R. and Creed, E.R. (1975). Industrial melanism in *Biston betularia*: the role of selective predation. *Journal of Animal Ecology*, **44**, 67–83.

Lees, D.R. and Creed, E.R. (1977). The genetics of *insularia* forms of the peppered moth, *Biston betularia*. *Heredity*, **39**, 67–73.

Lees, D.R., Creed, E.R., and Duckett, J.G. (1973). Atmospheric pollution and industrial melanism. *Heredity*, **30**, 227–32.

Lehnert, M. and Fritz, K. (1993). The adder (*Vipera berus* L. 1758) in the northern Black Forest: Supplementary contribution. *Jahreshefte der Gesellschaft fuer Naturkunde in Wuerttemberg*, **148**, 135–57.

Levene, H. (1953). Genetic equilibrium when more than one ecological niche is available. *American Naturalist*, **87**, 311–3.

Liebert, T.G. and Brakefield, P.M. (1987). Behavioural studies on the peppered moth *Biston betularia* and a discussion of the role of pollution and epiphytes in industrial melanism. *Biological Journal of the Linnaean Society*, **31**, 129–50.

Lipson, C. (1967). *Wear Considerations in Design*. Prentice Hall, New Jersey.

Long, D.B. (1953). Effects of population density on larvae of Lepidoptera. *Transactions of the Royal Entomological Society of London*, **104**, 543–85.

Lus, Ya. Ya. (Lusis, J.J.) (1928). On the inheritance of colour and pattern in lady beetles *Adalia bipunctata* L. and *Adalia decempunctata* L. *Izvestie Byuro Genetiki, Leningrad*, **6**, 89–163.

Lus, Ya.Ya. (Lusis, J.J.) (1932). An analysis of the dominance phenomenon in the inheritance of the elytra and pronotum colour in *Adalia bipunctata*. *Trudy Laboratorii Genetika, Leningrad*, **9**, 135–62.

Lusis, J.J. (1947). Some rules of reproduction in populations of *Adalia bipunctata*: heterozygosity of lethal alleles in populations. *Doklady Akademii Nauk SSSR (Moskva)*, **57**, 825–8.

Lusis, J.J. (1961). On the biological meaning of colour polymorphism of ladybeetle *Adalia bipunctata* L. *Latvijas Ent.* **4**, 3–29.

Lusis, J.J. (1973). Taxonomic relations and geographical distribution of forms in beetles of the genus *Adalia* Mulsant. *The Problems of Genetics and Evolution*, **1**, 5–128.

Lythgoe, J.N. (1979). *The Ecology of Vision*. Oxford University Press, Oxford.

MacArthur, R.H. and Connell, J.H. (1966). *The Biology of Populations*. Wiley, New York.

Mader, L. (1926–37). *Evidenz der Palaarktischen Coccinelliden und ihrer Aberationen in Wort und Bild*. Vienna.

Madsen, T. and Stille, B. (1988). The effect of size dependent mortality on color morphs in male adders, *Vipera-berus*. *Oikos*, **52**, 73–8.

Magnus, D.B.E. (1958). Experimental analysis of the Fritillary butterfly *Argynnis paphia* L. (Nymphalidae). *Proceedings of the 10th international Congress of Entomology*, **2**, 405–8.

Maier, E.J. (1992). Spectral sensitivities including the UV of the passeriform bird *Leiothrix lutea*. *Journal of Comparative Physiology A*, **170**, 709–14.

Majerus, M.E.N. (1978). The control of larval colour variation in *Phlogophora meticulosa* Linn. (Lepidoptera: Noctuidae) and some of its consequences. Ph.D. thesis, University of London.

Majerus, M.E.N. (1980). The control of larval colour variation in the angleshades moth (*Phlogophora meticulosa* Linn.) Part II: the maintenance of the variation. *Bulletin of the Amateur Entomologists' Society*, **39**, 85–9.

Majerus, M.E.N. (1981). The inheritance and maintenance of the melanic form *nigrescens* of *Pachycnemia hippocastanaria* (Lepidoptera: Geometridae). *Ecological Entomology*, **6**, 417–22.

Majerus, M.E.N. (1982*a*). The inheritance of two melanic forms of the alder moth (*Acronicta alni*. L.). *Entomologist's Monthly Magazine*, **118**, 207–9.

Majerus, M.E.N. (1982*b*). Genetic control of two melanic forms of *Panolis flammea* (Lepidoptera: Noctuidae). *Heredity*, **49**, 171–7.

Majerus, M.E.N. (1983*a*). Larval colour variation in *Phlogophora meticulosa* Linn. (Lepidoptera: Noctuidae) I: the variation and its control in instars 1–3. *Proceedings of the British Entomological and Natural History Society*, **16**, 34–49.

Majerus, M.E.N. (1983*b*). Larval colour variation in *Phlogophora meticulosa* Linn. (Lepidoptera: Noctuidae) II: genetic control in instars 3–5. *Proceedings of the British Entomological and Natural History Society*, **16**, 63–76.

Majerus, M.E.N. (1983*c*). Some observations on lichen-marked larvae of the scalloped hazel (*Gonodontis bidentata* Clerck). *Entomologist's Record and Journal of Variation*, **95**, 21–3.

Majerus, M.E.N. (1986*a*). The genetics and evolution of female choice. *Trends in Ecology and Evolution*, **1**, 1–9.

Majerus, M.E.N. (1986*b*). Some notes on ladybirds from an acid heath. *Bulletin of the Amateur Entomologists' Society*, **45**, 31–7.

Majerus, M.E.N. (1989*a*). *The Cambridge Ladybird Survey—Results Supplement 1984–1989*. Department of Genetics, Cambridge.

Majerus, M.E.N. (1989*b*). Melanic polymorphism in the peppered moth *Biston betularia* and other Lepidoptera. *Journal of Biological Education*, **23**, 267–84.

Majerus, M.E.N. (1990). The importance of form frequency data to ecological genetics. *Bulletin of the Amateur Entomologists' Society*, **49**, 123–31.

Majerus, M.E.N. (1991). A rare melanic form of the 16 spot ladybird (*Micraspis 16-punctata* Linn.). *Entomologist's Monthly Magazine*, **127**, 176.

Majerus, M.E.N. (1993). Notes on the inheritance of a scarce form of the striped ladybird, *Myzia oblongoguttata* (Coleoptera: Coccinellidae). *Entomologist's Record and Journal of Variation*, **105**, 271–7.

Majerus, M.E.N. (1994*a*). *Ladybirds*, No. 81. New Naturalist Series. HarperCollins, London.

Majerus, M.E.N. (1994*b*). Female promiscuity maintains high fertility in ladybirds (Coleoptera:Coccinellidae). *Entomologist's Monthly Magazine*, **130**, 205–9.

Majerus, M.E.N. (1995). *The Current Status of Ladybirds in Britain: Final Report of the Cambridge Ladybird Survey—1984–1994.* 144 pp., Department of Genetics (University of Cambridge): Cambridge.

Majerus, M.E.N., (in preparation *a*). Evidence of differences in morph frequencies correlated to habitat type in fourteen species of British Lepidoptera.

Majerus, M.E.N. (in preparation *b*). The natural resting positions of some British Lepidoptera.

Majerus, M.E.N. (in preparation *c*). Evidence of morph specific dispersal differences across habitat boundaries in polymorphic Lepidoptera.

Majerus, M.E.N. (in preparation *d*). The evolution of morph specific habitat preferences in melanic Lepidoptera.

Majerus, M.E.N. (in preparation *e*). The origin and inheritance of melanism in *Alcis repandata* (Lepidoptera: Geometridae).

Majerus, M.E.N. (in preparation *f*). Declines in melanism in two species of moths in industrial areas.

Majerus, M.E.N. (1997). The private lives of ladybirds. *British Wildlife*, **8**, 233–42.

Majerus, M.E.N. and Kearns, P.W.E. (1989). *Ladybirds*, No. 10. Naturalists' Handbooks Series. Richmond Publishing, Slough.

Majerus, M.E.N. and Majerus, T.M.O. (1997). Predation of ladybirds by birds in the wild. *Entomologist's Monthly Magazine*, **133**, 55–61.

Majerus, M.E.N. and Zakharov, I.A. (in preparation). Does thermal melanism maintain melanic polymorphism in the 2 spot ladybird, *Adalia bipunctata* (Coleoptera : Coccinellidae)?

Majerus, M.E.N., O'Donald, P., and Weir, J. (1982*a*). Female mating preference is genetic. *Nature*, **300**, 521–3.

Majerus, M.E.N., O'Donald, P., and Weir, J. (1982*b*). Evidence for preferential mating in *Adalia bipunctata*. *Heredity*, **49**, 37–49.

Majerus, M.E.N., O'Donald, P., Kearns, P.W.E., and Ireland, H. (1986). The genetics and evolution of female choice. *Nature*, **321**, 164–7.

Majerus, M.E.N., Grigg, A., Jones, C.W., Salmon, F., Strathdee, A., and Dearnaley, N. (1994). Factors affecting habitat preferences in the Lepidoptera. *British Journal of Entomology and Natural History*, **7**, 129–37.

Majerus, M.E.N., Amos, W., and Hurst, G.D.D. (1996). *Evolution: the Four Billion Year War.* Longmans, Harlow.

Majerus, M.E.N., Majerus, T.M.O., and Cronin., A. (in preparation). The inheritance of a purple mutant of the seven-spot ladybird, *Coccinella 7-punctata* L.

Majerus, T.M.O., Harris, A., Walker, L.E., and Majerus, M.E.N. (in preparation *a*). Melanic polymorphism in a New Zealand population of the two-spot ladybird, *Adalia bipunctata* (Coleoptera: Coccinellidae).

Majerus, T.M.O., Majerus, M.E.N., Knowles, B., Wheeler, J., Betrand, D., Zakharov, I.A., and Hurst, G.D.D. (in preparation *b*). Variation in male killing behaviour in three populations of the ladybird *Harmonia axyridis* (Col: Coccinellidae).

Mani, G.S. (1980). A theoretical study of morph ratio clines with special reference to melanism in moths. *Proceedings of the Royal Society of London B*, **210**, 299–316.

Mani, G.S. (1982). A theoretical analysis of the morph frequency variation in the peppered moth over England and Wales. *Biological Journal of the Linnaean Society*, **17**, 259–67.

Mani, G.S. and Majerus, M.E.N. (1993). Peppered moth revisited: analysis of recent decreases in melanic frequency and predictions for the future. *Biological Journal of the Linnaean Society*, **48**, 157–165.

Mani, M.S. (1968). *Ecology and Biogeography of High Altitude Insects*. W. Junk, The Hague.

Mansbridge, W. (1909a). On *Aplecta nebulosa*. *Entomologist*. **42**, 127.

Mansbridge, W. (1909b). On *Aplecta nebulosa*. *Proceedings of the South London Entomological and Natural History Society*, (1909–10), 64–5.

Marples, N.M. (1990). The influence of predation on ladybird colour patterns. PhD. Thesis: University of Wales College of Cardiff.

Marples, N.M., Brakefield, P.M., and Cowie, R.J. (1989). Differences between the 7-spot and 2-spot ladybird beetles (Coccinellidae) in their toxic effects on a bird predator. *Ecological Entomology*, **14**, 79–84.

Maynard Smith, J. (1975). *The Theory of Evolution*, (3rd edn). Penguin Books, London.

Maynard Smith, J. and Hoekstra, R. (1980). Polymorphism in a varied environment: how robust are the models. *Genetic Research*, 35, 45–57.

Mayr, E. (1942). *Systematics and the Origin of Species*. Columbia University Press, New York.

Mayr, E. (1955). Is the Great White Heron a good species? *Auk*, 73, 71–7.

McDonald, C.G., Reimchen, T.E., and Hawryshyn, C.W. (1995). Nuptial colour loss and signal masking in *Gasterosteus*: An analysis using video imaging. *Behaviour*, 132, 963–77.

Mikkola, K. (1979). Resting site selection of *Oliga* and *Biston* moths (Lepidoptera: Noctuidae and Geometridae). *Acta Entomologica Fennici* 45, 81–7.

Mikkola, K. (1984). On the selective force acting in the industrial melanism of *Biston* and *Oliga* moths (Lepidoptera: Geometridae and Noctuidae). *Biological Journal of the Linnaean Society*, **21**, 409–421.

Mikkola, K. (1989). The first case of industrial melanism in the subArctic lepidopteran fauna, *Xestia gelida* f. *inferna*: new form of Noctuidae. *Notulae Entomologicae*, **69**, 1–3.

Mikkola, K. and Albrecht, A. (1988). The melanism of *Adalia bipunctata* around the Gulf of Finland as an industrial phenomenon (Coleoptera, Coccinellidae). *Annales Zoologici Fennici*, **25**, 177–85.

Mizer, A.V. (1970). On eating of beetles from Coccinellidae family by birds. *Vestnik Zoologii*, **6**, 21–4.

Monney, J-C., Luiselli, L., and Capula, M. (1996a). Body size and melanism in *Vipera aspis* in the Swiss Prealps and Central Italy and comparison with different alpine populations of *Vipera berus*. *Revue Suisse de Zoologie*, **103**, 81–100.

Monney, J-C., Luiselli, L., and Capula, M. (1996b). Correlates of melanism in a population of adders (*Vipera berus*) from the Swiss Alps and comparisons with other alpine populations. *Amphibia-Reptilia*, **16**, 323–30.

Morton Jones, F. (1932). Insect coloration and the relative acceptability of insects to birds. *Transactions of the Entomological Society of London*, **80**, 345–85.

Muggleton, J. (1978). Selection against the melanic morphs of *Adalia bipunctata* (two-spot ladybird): A review and some new data. *Heredity*, **40**, 269–80.

Muggleton, J. (1979). Non-random mating in wild populations of polymorphic *Adalia bipunctata*. *Heredity*, **42**, 57–65.

Muggleton, J., Lonsdale, D., and Benham, B.R. (1975). Melanism in *Adalia bipunctata* L.

(Col., Coccinellidae) and its relationship to atmospheric pollution. *Journal of Applied Ecology*, **12**, 451–64.

Müller, F. (1878). [Notes on Brazilian entomology]. *Transactions of the Entomological Society of London*, **Pt. 3**, 211–23.

The Munsell Book of Colour. (1966) Munsell Color Co. Inc., Baltimore.

Murray, K.F. (1928). *Entomologist's Record and Journal of Variation*, **40**, Pl. 3.

Murton, R.J. and Westwood, N.J. (1977). *Avian Breeding Cycles*. Clarendon Press, Oxford.

Murton, R.J., Westwood, N.J., and Thearle, R.J.P. (1973). Polymorphism and the evolution of a continuous breeding season in the pigeon, *Columba livia*. *Journal of Reproduction and Fertility, Supplement*, **119**, 563–77.

Needham, A.E. (1974). *The Significance of Zoochromes*. Springer-Verlag, New York.

Netshayev, V.A. and Kuznetsov, V.N. (1973). Avian predation on coccinellids in Primorie region. In *Entomofauna Sovetskogo Dalnego Vostoka. Vladivostok Tr. Biol.-pochv. Inst.* **9**, 97–9.

Newman, L.W. (1915). Proportions in mongrel families. *Proceedings of the Entomological Society of London*, (1915): lxxxi.

Newman, L.W. (1916). On *Aplecta nebulosa*. *Entomologist's Record and Journal of Variation*, **58**, 22.

Ng, S.M. (1986). Elytral colour patterns and their polymorphism in some aphidophagous ladybirds (Coleoptera: Coccinellidae) in Malaysia. *Malayan Nature Journal*, **39**, 149–56.

Nijhout, H.F. (1978). Wing pattern formation in Lepidoptera: a model. *Journal of Experimental Zoology*, **206**, 119–36.

Nijhout, H.F. (1981). The colour patterns of butterflies and moths. *Scientific American*, **245**, 104–15.

Nijhout, H.F. (1985). The developmental physiology of colour patterns in Lepidoptera. In *Advances in Insect Physiology 18* (ed. M.J. Berridge, J.E. Treherne, and V.B. Wigglesworth), pp. 181–247. Academic Press, Chicago.

Nijhout, H.F. (1991). *The Development and Evolution of Butterfly Wing Patterns*. Smithsonian Institution Press, Washington.

Nijhout, H.F. and Grunett, L.W. (1988). Colour pattern regulation after surgery on the wing disks of *Precis coenia* (Lepidoptera: Nymphalidae). *Development*, **102**, 377–85.

Novak, I. and Spitzer, K. (1986) Industrial melanism in *Biston betularia* (Lepidoptera: Geometridae) in Czechoslovakia. *Acta Entomologica Bohemoslov*, **83**, 185–91.

O'Donald, P. (1967). A general model of sexual and natural selection. *Heredity*, **31**, 145–56.

O'Donald, P. (1980). *Genetic Models of Sexual Selection*. Cambridge University Press, Cambridge.

O'Donald, P. (1983). *The Arctic Skua: a study of the ecology and evolution of a seabird*. Cambridge University Press, Cambridge.

O'Donald, P. and Majerus, M.E.N. (1985). Sexual selection and the evolution of preferential mating in ladybirds. I. Selection for high and low lines of female preference. *Heredity*, **55**, 401–12.

O'Donald, P. and Majerus, M.E.N. (1989). Sexual selection models and the evolution of melanism in ladybirds. In *Mathematical Evolutionary Theory* (ed. F.M.W. Feldman), pp. 247–69. Princeton University Press, Princeton, New Jersey.

O'Donald, P. and Majerus, M.E.N. (1992). Non-random mating in the two-spot ladybird, *Adalia bipunctata* III. New evidence of genetic preference. *Heredity*, **61**, 521–6.

O'Donald, P. and Muggleton, J. (1979). Melanic polymorphism in ladybirds maintained by sexual selection. *Heredity*, **43**, 143–8.

O'Donald, P., Derrick, M., Majerus, M.E.N., and Weir, J. (1984). Population genetic theory of the assortative mating, sexual selection and natural selection of the 2 spot ladybird, *Adalia bipunctata. Heredity*, **52**, 43–61.

Oertel, H. (1910). Merkwürdige Färbung eimer Rauoe von *Chaerocampa elpenor* L. *Internationale Entomologische Zeitschrift*, **4**, 48–9.

Onslow, H. (1921). Inheritance of wing colour in Lepidoptera, VI. *Diaphoria mendica* Clerck and var *rustica* Hübn. *Journal of Genetics*, **11**, 277–92.

Osawa, N. and Nishida, T. (1992). Seasonal variation in elytral colour polymorphism in *Harmonia axyridis* (the ladybird beetle): the role of non-random mating. *Heredity*, **69**, 297–307.

Owen, D.F. (1961). Industrial melanism in North American moths. *American Naturalist*, **95**, 227–33.

Owen, D.F. (1962). The evolution of melanism in six species of North American Geometrid Moths. *Annals of the Entomological Society of America*, **55**, 695–703.

Owen, D.F. (1980). *Camouflage and Mimicry*. Oxford University Press, Oxford.

Paim, S., Linhares, L.F., Mangrich, A.S., and Martin, J.P. (1990). Characterization of fungal melanins and soil humic acids by chemical analysis and IR spectroscopy. *Biol. Fertil. Soils*, **10**, 72–6.

Palmer, M.A. (1911). Some notes on heredity in the coccinellid genus *Adalia* Mulsant. *Annals of the Entomological Society of America*, **4**, 283–302.

Palmer, M.A. (1917). Additional notes on heredity and life history in the Coccinellid genus *Adalia* Mulsant. *Annals of the Entomological Society of America*, **10**, 289–302.

Parker, G.A. (1970). Sperm competition and its evolutionary consequences in the insects. *Biological Reviews*, **45**, 525–67.

Partridge, J.C. (1989). The visual ecology of avian oil droplets. *Journal of Comparative Physiology A*, **165**, 415–26.

Pekkarinen, A. and Teras, I. (1986). Melanism in *Bombus veteranus* and *Bombus soroeensis* (Hymenoptera: Apidae) in southern Finland. *Notulae Entomologicae*, **66**, 49–54.

Pocock, R.I. (1911). On the palatability of some British insects, with notes on the significance of mimetic resemblances. With notes on the experiments by E.B. Poulton. *Proceedings of the Zoological Society of London*, 1911, 809–68.

Popescu, C., Broadhead, E., and Shorrocks, B. (1978). Industrial melanism in *Mesopsocus unipunctatus* (Müll) (Psocoptera) in northern England. *Heredity*, **42**, 133–42.

Popescu, C. (1979). Natural selection in the industrial melanic psocid *Mesopsocus unipunctatus* (Müll) (Psocoptera) in northern England. *Heredity*, **42**, 133–42.

Porritt, G.T. (1907). Melanism in Yorkshire Lepidoptera. *Transactions of the British Association*, (1906), 316–25.

Porritt, G.T. (1926). The induction of melanism in the Lepidoptera and its subsequent inheritance. *Entomologist's Monthly Magazine*, **62**, 107–111.

Poulton, E.B. (1885). The essential nature of colouring of phytophagous larvae (and their pupae): with an account of some experiments upon the relation between the colour of such larvae and that of their food plant. *Proceedings of the Royal Society of London*, **38**, 269–315.

Poulton, E.B. (1886). A further enquiry into a special colour relation between the larvae of *Smerinthus ocellatus* and its food plants. *Proceedings of the Royal Society of London*, **40**, 135–73.

Poulton, E.B. (1887). An enquiry into the cause and extent of a special colour relation between certain exposed lepidopterous pupae and the surfaces which immediately surround them. *Philosophical Transactions of the Royal Society of London B*, **178**, 311–441.

Poulton, E.B. (1892). Further experiments upon the colour relation between certain lepidopterous larvae, pupae, cocoons and imagines, and their surroundings. *Transactions of the Entomological Society of London*, 1892, 293–487.

Poulton, E.B. (1893). The experimental proof that the colours of certain lepidopterous larvae are largely due to modified plant pigments derived from food. *Proceedings of the Royal Society of London*, **54**, 417–30.

Prater, A.J., Marchant, J.H., and Vuorinen, J. (1977). *Guide to the Identification and Aging of Holarctic Waders*. British Trust for Ornithology, Tring.

Prest, W. (1877). On melanism and variation in Lepidoptera. *Entomologist*, **10**, 129–31.

Prota, G. and Thomson, R.H. (1976). Melanin pimentation in mammals. *Endeavour*, **35**, 32–8.

Provine, W.B. (1985). The R.A. Fisher-Sewall Wright controversy. In *Oxford Surveys in Evolutionary Biology*, Vol. 2 (ed. R. Dawkins and M. Ridley), pp. 197–219. Oxford University Press, Oxford.

Provine, W.B. (1986). *Sewall Wright and Evolutionary Biology*. Chicago Press, Chicago.

Rapoport, E.H. (1969). Golger's rule and pigmentation of Collembola. *Evolution*, **23**, 622–6.

Regnart, H.C. (1933). Additions to our knowledge of chromosome numbers in the Lepidoptera. *Proceedings of the Univiversity of Durham Philosophical Society*, **9**, 79–83.

Rettenmeyer, C.W. (1970). Insect mimicry. *Annual Review of Entomology*, **15**, 43–74.

Revels, R. and Majerus, M.E.N. (1997). Grouping behaviour in overwintering 16 spot ladybirds (*Tytthaspis 16-punctata*) (Coleoptera: Coccinellidae). *Bulletin of the Amateur Entomologists' Society*, **413**, 143–4.

Rhamhalinghan, M. (1988). Seasonal variations in the color patterns of *Coccinella septempunctata* L. (Coleoptera: Coccinellidae) in Nilgiri Hills, India. *Journal of the Bombay Natural History Society*, **85**, 551–8.

Richards, A.M. (1983). The *Epilachna vigintioctopunctata* complex (Coleoptera: Coccinellidae). *International Journal of Entomology*, **25**, 11–41.

Riley, P.A. (1977). The mechanism of melanogenesis. *Symposia of the Zoological Society of London*, **39**, 77–95.

Robbins, L.S., Nadeau, J.H., Johnson, K.R., Kelly, M.A., Roselli-Rehfuss, L., Baack, E., Mountjoy, K.G., and Cone, R.D. (1993). Pigmentation phenotypes of variant extension locus alleles result from point mutations that alter MSH receptor function. *Cell*, **72**, 827–34.

Robinson, E.K. (1877). Causes of melanism in Lepidoptera. *Entomologist*, **10**, 131–2

Robinson, R. (1971). *Lepidoptera Genetics*. Pergamon Press, New York.

Rothschild, M. (1961). Defensive odours and Müllerian mimicry among insects. *Transactions of the Royal Entomological Society of London*, **113**, 101–22.

Rothschild, M. (1963). Is the Buff Ermine (*Spilosoma lutea* Hufn.) a mimic of the White

Ermine (*Spilosoma lubricipeda* L.)? *Proceedings ot the Royal Entomological Society of London A*, **38**, 159–64.

Rothschild, M. and Reichstein, T. (1976). Some problems associated with the storage of cardiac glycosides by insects. *Nova Acta Leopoldina*, Suppl. 7, 507–50.

Rothschild, M., Euw, J. von, and Reichstein, T. (1973). Cardiac glycosides in a scale insect (*Aspidiotus*), a ladybird (*Coccinella*) and a lacewing (*Chrysopa*). *Journal of Entomology A*, **48**, 89–90.

Sage, B.L. (1962). Albinism and melanism in birds. *British Birds*, **55**, 201–25.

Sage, B.L. (1963). The incidence of albinism and melanism in British Birds. *British Birds*, **56**, 409–16.

Sargent, T.D. (1966). Background selections of Geometrid and Noctuid moths. *Science*, **154**, 1674–5.

Sargent, T.D. (1968). Cryptic moths: effects on background selections of painting the circumocular scales. *Science*, **159**, 100–1.

Sargent, T.D. (1969). Background selections of the pale and melanic forms of the cryptic moth *Phigalia titea* (Cramer). *Nature*, **222**, 585–6.

Sargent, T.D. (1974). Melanism in moths of central Massachusetts (Noctuidae, Geometridae). *Journal of the Lepididopterists' Society*, **28**, 145–52.

Sargent, T.D. (1985). Melanism in *Phigalia titea* (Lepidoptera: Geometridae) in southern New England—a response to forest disturbance. *Journal of the New York Entomological Society*, **93**, 1113–20.

Sargent, T.D. and Keiper, R.R. (1969). Behavioural adaptations of cryptic moths. I. Preliminary studies of bark-like species. *Journal of the Lepidopterists' Society*, **23**, 1–9.

Sasaji, H. (1971). *Fauna Japonica Coccinellidae (Insecta: Coleoptera)*. Academic Press of Japan, Tokyo.

Scali, V. and Creed, E.R. (1975). The influence of climate on melanism in the two-spot ladybird, *Adalia bipunctata*, in central Italy. *Transactions of the Royal Entomological Society of London*, **127**, 163–9.

Schummer, R. (1983). Neue Ergebnisse uber die Structur von *Adalia bipunctata*—Gemeinschaften (Insecta: Coleoptera, Coccinellidae). *Tagungsbericht 2, Leipziger Sympos. urbane Okologie*, 59–61.

Scoble, M.J. (1992). *The Lepidoptera: Form, Function and Diversity*. British Museum (Natural History), London.

Scott, P.W. (1981). *Axolotls*. T.F.H. Publications, New Jersey.

Semler, D.E. (1971). Some aspects of adaptation in a polymorphism for breeding colours in the Threespine Stickleback (*Gasterosteus aculeatus*). *Journal of Zoology (London)*, **165**, 291–302.

Sergievsky, A.O. and Zakharov, I.A. (1989). Population response on the stress effects: the concept of two-stage response. (In *Ontogenesis, Evolution, Biosphere*), pp. 157–73. Nauka Publications, Moscow.

Shelford, R. (1916). *A Naturalist in Borneo*.

Sheppard, P.M. (1951). A quantitative study of two populations of the moth, *Panaxia dominula* (L.). *Heredity*, **5**, 349–78.

Sheppard, P.M. (1952). A note on non-random mating in the moth *Panaxia dominula* L. *Heredity*, **6**, 239–41.

Sheppard, P.M. and Cook, L.M. (1962). The manifold effects of the medionigra gene of the moth *Panaxia dominula* and the maintenance of polymorphism. *Heredity*, **17**, 415–26.

Shorten, M. (1954). *Squirrels*. New Naturalist Series. Collins, London.

Simmonds M.S.J. and Blaney, W.M. (1986). Effects of rearing density on development and feeding behaviour in larvae of *Spodoptera exempta*. *Jounal of Insect Physiology*, **32**, 1043–54.

Skinner, B. (1984). *Colour Identification Guide to Moths of the British Isles*. Viking, Harmondsworth.

Smith, M. (1964). *British Amphibians and Reptiles*. New Naturalist Series. Collins, London.

Sømme, L. (1989). Adaptations of terrestrial arthropods to the alpine environment. *Biological Reviews*, **64**, 367–407.

Sound, P. (1994). A melanistic individual of the common wall lizard (*Podarcis muralis*) from the Middle Rhine Valley. *Salamandra*, **30**, 221–2.

Stalker, J., Brunton C.F.A., and Majerus M.E.N. (in preparation). The characteristics of melanic moths and their backgrounds in the ultraviolet. *Biological Journal of the Linnaean Society*.

Standfuss, M. (1910). Die alternative oder discontinuerliche vererbung und ihre veranschaulichung an der Ergebnissen von Zuchexperimenten mit *Aglia tau* und deren Mutationen. *Pt. end. Natn.-Biblthk.* **1**, 5, 14, 21, 28.

Steward, R.C. (1976). Eperiments on resting site selection by the typical and melanic forms of the moth *Allophyes oxyacanthae* (Caradrinidae). *Journal of Zoology (London)*, **178**, 107–15.

Steward, R.C. (1977*a*). Industrial and non-industrial melanism in the peppered moth *Biston betularia* (L.). *Ecological Entomology*, **2**, 231–43.

Steward, R.C. (1977*b*). Multivariate analysis of variation within the *insularia* complex of the moth *Biston betularia*. *Heredity*, **39**, 97–109.

Steward, R.C. (1977*c*). Melanism and selective predation in three species of moths. *Journal of Animal Ecology*, **46**, 483–96.

Steward, R.C. (1977*d*). Industrial melanism in the moths *Diurnea fagella* (Oecophoridae) and *Allophyes oxyacanthae* (Caradrinidae). *Journal of Zoology (London)*, **183**, 47–62.

Steward, R.C. (1985). Evolution of resting behaviour in polymorphic industrial melanic moth species. *Biological Journal of the Linnaean Society*, **24**, 285–93.

Stewart, A.J.A. (1986). Nymphal colour pattern polymorphism in the leafhoppers *Eupteryx urticae* and *Eupteryx cyclops* (Hemiptera: Homoptera: Auchenorrhyncha): spatial and temporal variation in morph frequencies. *Biological Journal of the Linnaean Society*, **27**, 79–101.

Stewart, A.J.A. and Lees, D.R. (1988). Genetic control of color pattern polymorphism in British populations of the spittlebug Philaenus spumarius L. (Homoptera: Aphrophoridae). *Biological Journal of the Linnaean Society*, **34**, 57–80.

Stewart, A.J.A. and Lees, D.R. (1996). The colour pattern polymorphism in *Philaenus spumarius* (L.) (Homoptera: Cercopidae) in England and Wales. *Philosophical Transactions of the Royal Society of London B*, **351**, 69–89.

Stewart, L.A. and Dixon, A.F.G. (1989). Why big species of ladybird are not melanic. *Functional Ecology*, **3**, 165–77.

Sugumaran, M., Saul, S.J., and Ramesh, N. (1985). Endogenous protease inhibitors prevent undesired activation of prophenolase in insect hemolymph. *Biochemical and Biophysical Research Communications*, **132**, 1124–9.

Tan, C.-C. (1949). Seasonal variation of color patterns in *Harmonia axyridis*. *Proceedings of the 8th International Congress of Genetics*, 669–70.

Tan, C.-C. and Li, J.-C. (1934). Inheritance of the elytral color patterns in the lady-bird beetle, *Harmonia axyridis* Pallas. *American Naturalist*, **68**, 252–65.

Timofeeff-Ressovsky, N.W. (1940). Zur analyse des Polymorphismus bei *Adalia bipunctata* L. *Biologisches Zentralblatt*, **60**, 130–7.

Tinbergen, L. (1960). The natural control of insects in pine woods. I. Factors influencing of predation by songbirds. *Archives Neerlandaises de Zoologie*, **13**, 265–343.

Tolkien, J.R.R. (1954). *The Lord of the Rings*. George Allen and Unwin, London.

Tugwell, W.H. (1877). Melanism, &c., in insects. *Entomologist's Monthly Magazine*, **13**, 256–7.

Turner, J.R.G. (1984). Mimicry: The patability spectrum. In *The Biology of Butterflies* (ed. R.I. Vane-Wright and P.R. Ackery). pp. 141–61. Academic Press, London.

Tursch, B., Braekman, J.C., Daloze, D., Hootele, D., Losman, D., Karlsson, R., and Pasteels, J.M. (1973). Chemical ecology of arthropods—VI. Adaline, a novel alkaloid from *Adalia bipunctata* L. (Coleoptera: Coccinellidae). *Tetrahedron*, **29**, 201.

Tursch, B., Daloze, D., Braekman, J.C., Hootele, C., Cravador, A., Losman, D., and Karlsson, R. (1974). Chemical ecology of arthropods—IX. Structure and absolute configuration of hippodamine and convergine, two novel alkaloids from the American ladybug *Hippodamia convergens*. (Coleop, Cocc.). *Tetrahedron*, **30**, 409–12.

Tursch, B., Daloze, D., Braekman, J.C., Hootele, C., and Pasteels, J.M. (1975). Chemical ecology of arthropods—X. The structure of myrrhine and the biosynthesis of coccinelline. *Tetrahedron,* **31**, 1541–3.

Tursch, B., Daloze, D., Dupont, M., Hootele, C., Kaisin, M., Pasteels, J.M., and Zimmermann, D. (1971). Coccinelline, the defensive alkaloid of the beetle *Coccinella septempunctata*. *Chemia*, **25**, 307.

Tursch, B., Daloze, D., Pasteels, J.M., Cravador, A., Braekman, J.C., Hootele, C., and Zimmermann, D. (1972). Two novel alkaloids from the American ladybug *Hippodamia convergens* (Coleoptera: Coccinellidae). *Bulletin de la Societe Chimique Belgique*, 81, 649–50.

Tutt, J.W. (1891). *Melanism and Melanochroism in British Lepidoptera*. Swan Sonnenschein, London.

Tutt, J.W. (1896). *British Moths*. George Routledge, London.

Van Raay, T.J. and Crease T.J. (1995). Mitochondrial DNA diversity in an apomictic *Daphnia* complex from the Canadian High Arctic. *Molecular Ecology*, **4**, 149–61.

Verhoog, M.D., van Boven, A., and Brakefield, P.M. (1996). Melanic moths and the ability to encapsulate parasitoid eggs and larvae. *Proceedings of the Netherlands Entomological Society*, **7**, 127–33.

Walsingham, T. de G. (1885). On some probable causes of a tendency to melanic variation in the Lepidoptera of high latitudes. *Entomological Transactions of the Yorkshire Naturalists Union*, D, (1885), 113–40.

Watson, E.V. (1981). In *The Oxford Encyclopaedia of Trees of the World* (ed. B. Hora). Oxford University Press, Oxford.

Watt, W.B. (1968). Adaptive significance of pigment polymorphism in *Colias* butterflies, I: Variation of melanin in relation to thermoregulation. *Evolution*, **22**, 437–58.

Watt, W.B. (1969). Adaptive significance of pigment polymorphism in *Colias* butterflies, 2: Thermoregulation and photoperiodically controlled melanin variation in *Colias eurytheme*. *Proceedings of the National Academy of Sciences (USA)*, **63**, 767–74.

Watt, W.B. (1974). Adaptive significance of pigment polymorphism in *Colias* butterflies, 3: Progress in the study of the 'alba' variant. *Evolution*, **27**, 537–48.

West, B.K. (1977). Melanism in *Biston* (Lepidoptera: Geometridae) in the rural Appalachians. *Heredity*, **39**, 75–81.

West, B.K. (1988). *Biston betularia* L. (Lep.: Geometridae): melanism in decline. *Entomologist's Record and Journal of Variation*, **100**, 39–41.

West, B.K. (1992). *Hydriomena impluviata* D. & S. (Lep.: Geometridae): an extraordinarily rapid decline in melanism. *Entomologist's Record and Journal of Variation*, **104**, 329.

West, B.K. (1994). The continued decline of melanism in *Biston betularia* L. (Lep.: Geometridae) in N.W. Kent. *Entomologist's Record and Journal of Variation*, **106**, 229–32.

White, F.B. (1877). Melanochroism &c., in Lepidoptera. *Entomologist*, **10**, 126–30.

Whittle, P.D.J., Clarke, C.A., Sheppard, P.M., and Bishop, J.A. (1976). Further studies on the industrial melanic moth *Biston betularia* (L.) in the northwest of the British Isles. *Proceedings of the Royal Society of London B*, **194**, 467–80.

Wigglesworth, V.B. (1964). *The Life of Insects.* Weidenfeld & Nicholson, London.

Williams, H.B. (1931). Ten years in North East Survey. *Entomologist's Record and Journal of Variation*, **43**, 42–9.

Williams, H.B. (1933). Notes on *Boarmia repandata* and *B. rhomboidaria*. *Proceedings of the South London Entomological and Natural History Society*, (1932–1933), 1–10.

Williamson, M.H. (1959). Colour and genetics of the Black Slug. *Proceedings of the Royal Phylosophical Society of Edinburgh*, **27**, 87–93.

Wright, S. (1948). On the roles of directed and random changes in gene frequency in the genetics of populations. *Evolution*, **2**, 279–94.

Wright, S. (1978). *Evolution and the Genetics of Population, Volume 4. Variability Within and Among Natural Populations.* Chicago University Press, Chicago.

Wright, S. (1982). The shifting balance theory and macroevolution. *Annual Review of Genetics*, **16**, 1–19.

Yabe, T. (1994). Population structure and male melanism in Reeves' turtle, *Chinemys reevesii*. *Journal of Herpetology*, **15**, 131–7.

Zakharov, I.A. (1992). Investigation of genetic differentiation and population stability of *Adalia bipunctata* from Moscow. *Genetika*, **28**, 77–81.

Zakharov, I.A. and Sergievsky, S.O. (1980). Study of the genetic polymorphism of the two-spot ladybird, *Adalia bipunctata*, in the district of Leningrad. I. Season dynamics of polymorphism. *Genetika*, **16**, 270–5.

Zakharov, I.A. and Sergievsky, S.O. (1983). Study of the genetic polymorphism of two-spot ladybird *Adalia bipunctata* (L.) populations in Leningrad district. III. Polymorphism of Leningrad district populations. *Genetika*, **19**, 1144–51.

Zakharov, I.A., Hurst, G.D.D., Chernyshova, N.E., and Majerus, M.E.N. (1996). The maternally inherited male-killing bacterium in the Petersberg population of *Adalia bipunctata* does not belong to the genus *Rickettsia*. *Russian Journal of Genetics*, **32**, 1303–6.

Zellmer, I.D. (1995). UV-B tolerance of alpine and arctic *Daphnia*. *Hydrobiologia*, **307**, 153–9.

Index

Printed in the United Kingdom
by Lightning Source UK Ltd.
9526800001BA